Advances in
Molecular Biology and
Targeted Treatment
for AIDS

GWUMC Department of Biochemistry
Annual Spring Symposia
Series Editors:
Allan L. Goldstein, Ajit Kumar, and J. Martyn Bailey
The George Washington University Medical Center

ADVANCES IN MOLECULAR BIOLOGY AND
TARGETED TREATMENT FOR AIDS
Edited by Ajit Kumar

BIOLOGY OF CELLULAR TRANSDUCING SIGNALS
Edited by Jack Y. Vanderhoek

BIOMEDICAL ADVANCES IN AGING
Edited by Allan L. Goldstein

CARDIOVASCULAR DISEASE
Molecular and Cellular Mechanisms, Prevention, and Treatment
Edited by Linda L. Gallo

CELL CALCIUM METABOLISM
Physiology, Biochemistry, Pharmacology, and Clinical Implications
Edited by Gary Fiskum

DIETARY FIBER IN HEALTH AND DISEASE
Edited by George V. Vahouny and David Kritchevsky

EUKARYOTIC GENE EXPRESSION
Edited by Ajit Kumar

NEURAL AND ENDOCRINE PEPTIDES AND RECEPTORS
Edited by Terry W. Moody

PROSTAGLANDINS, LEUKOTRIENES, AND LIPOXINS
Biochemistry, Mechanism of Action, and Clinical Applications
Edited by J. Martyn Bailey

THYMIC HORMONES AND LYMPHOKINES
Basic Chemistry and Clinical Applications
Edited by Allan L. Goldstein

Advances in Molecular Biology and Targeted Treatment for AIDS

Edited by
Ajit Kumar
The George Washington University Medical Center
Washington, D.C.

Plenum Press   ·   New York and London

Library of Congress Cataloging-in-Publication Data

Advances in molecular biology and targeted treatment for AIDS / edited
by Ajit Kumar.
      p.   cm. -- (GWUMC Department of Biochemistry annual spring
symposia)
   Contains articles from the Xth International Spring Symposium in
the Health Sciences, held at the George Washington University, May
15-18, 1990, organized by the George Washington University Dept. of
Biochemistry and Molecular Biology and the National Cooperative Drug
Discovery Groups of the National Institute of Allergy and Infectious
Diseases (NIAID).
   Includes bibliographical references and index.
   ISBN-13: 978-1-4684-5930-2   e-ISBN-13: 978-1-4684-5928-9
   DOI: 10.1007/978-1-4684-5928-9

   1. AIDS (Disease)--Molecular aspects--Congresses. 2. AIDS
(Disease)--Genetic aspects--Congresses. 3. Antiviral agents-
-Congresses. 4. HIV (Viruses)--Congresses. I. Kumar, Ajit.
II. George Washington University. Dept. of Biochemistry and
Molecular Biology. III. National Institute of Allergy and
Infectious Diseases (U.S.). National Cooperative Drug Discovery
Groups. IV. International Spring Symposium in the Health Sciences
(10th : 1990 : George Washington University). V. Series.
   [DNLM: 1. Acquired Immunodeficiency Syndrome--genetics-
-congresses. 2. Acquired Immunodeficiency Syndrome--therapy-
-congresses. 3. Proteins--congresses. 4. Virus Replication-
-genetics--congresses. WD 308 A2445 1990]
RC607.A26A3434 1991
616.97'92--dc20
DNLM/DLC
for Library of Congress   91-3048
                                                                 CIP

Proceedings of the Tenth International Washington Spring Symposium
at The George Washington University, held May 15-18, 1990, in Washington, D.C.

© 1991 Plenum Press, New York
Softcover reprint of the hardcover 1st edition 1991
A Division of Plenum Publishing Corporation
233 Spring Street, New York, N.Y. 10013

Since the discovery of HIV-1 as the etiologic agent of acquired immunodeficiency syndrome (AIDS) in the early 1980s, remarkable progress has been made in both the basic understanding of the biological processes leading to AIDS and an accelerated effort in finding new treatments. As is often the case in rapidly advancing fields, most of the scientific discussions are best handled in specialized groups. The effort to organize a meeting on advances in molecular biology and targeted treatment for AIDS was an experiment of sorts to gather experts in selected areas of overlapping interests where advances in basic biology and its application in the development of new drugs could be discussed. Of necessity, the scope of the meeting had to be limited to maintain a certain focus. Important areas of rapid development in AIDS research, such as the vaccine development, epidemiology, animal models, etc., had to be left out for more specialized meetings. The result, from all accounts, appeared to be quite a successful gathering, which provided a forum for informal discussions among scientists from industry and academic institutions.

A remarkable feature of the AIDS virus is its genetic complexity and how some of its seemingly "extra genes" manage to regulate the normal functions of the host and most importantly its immune system. The first one of these auxiliary genes to be identified, the <u>tat</u> gene, was discovered about the same time as AZT, which proved to be an efficient and specific inhibitor of viral replication. Not surprisingly, considerable effort and hope is being put on our basic understanding of the viral molecular genetic mechanisms to help design the next generation of antiviral therapy. Representative articles in this volume roughly follow the sequence of events in viral infection, each of which is being intensely studied as potential targets for development of antivirals: The first six articles focus on viral gene replication and integration; the second group of thirteen articles deals principally with the functions of the viral regulatory genes, tat, rev, nef, vif, vpu; the third group of seven articles deals with viral envelope and nucleocapsid organization; and the final six articles deal with antivirals and prospects of gene therapy. This volume therefore should be useful to a wider group of investigators and physicians actively pursuing new approaches to treatment.

This volume is based on the contributed articles from participants of the Xth International Spring Symposium in the Health Sciences entitled "Advances in Molecular Biology and Targeted Treatments for AIDS" (held at The George Washington University campus in Washington, D.C., May 15-18), which was jointly organized by The George Washington University Department of Biochemistry and Molecular Biology and the National Cooperative Drug Discovery Groups of The National Institute of Allergy and Infectious Diseases (NIAID). Clearly, the highlights of the meeting were a scintillating keynote address by Dr. Harold Varmus, who urged the group to join in their efforts and share their reagents. Coinciding with the twentieth anniversary of the discovery of reverse transcriptase, Dr. Howard Temin's (Distinguished Scientist Award) lecture characteristically focused on the challenges facing the immune system evading rapid diversity of the virus. A special lecture by Dr. Anthony Fauci provided the foundation for satisfaction in the progress made toward our understanding of the problem and its future challenges.

I am particularly grateful to Dr. Margaret I. Johnston of NIAID and members of the program committee, Drs. Karen Beemon (Johns Hopkins), John McGowan (NIAID), Arnold Rabson (Institute of Biomedical Research, N.J.), Peter Shank (Brown University), and Ron Swanstrom (University of North Carolina, Chapel Hill), for their efforts in making this an exciting meeting.

A. Kumar

CONTENTS

MECHANISTIC ANALYSIS OF HIV-1 REVERSE TRANSCRIPTASE

John Abbotts and Samuel H. Wilson

Laboratory of Biochemistry
National Cancer Institute, National Institutes of Health
Bethesda, MD 20892

INTRODUCTION

Two biological issues have formed the basis for this laboratory's investigation of the reverse transcriptase (RT) of human immunodeficiency virus, type 1 (HIV-1). The first issue lies with the fact that the reverse transcriptase is an obvious target for drug therapy against HIV (Mitsuya and Broder, 1987). The approach in this regard has been to identify fundamental characteristics of the enzyme, with the idea that such information may facilitate rational drug design. We describe here work to identify the kinetic features of the HIV reverse transcriptase, including studies employing as substrate the triphosphates of the inhibitor AZT and one of its analogues.

The second biological issue lies with the observation that HIV shows considerable genetic diversity (for review, see Coffin, 1986). One wishes to determine if the reverse transcriptase could provide a source of this diversity through error-prone replication. We will describe the work in this laboratory and in collaboration which suggests a mechanism by which the HIV reverse transcriptase could produce errors. Also described are efforts to express and purify a homodimer form of the enzyme, which may facilitate investigation of subunit function and other studies that would benefit from the availability of a single polypeptide species. Because of space limitations, this article focuses on work in this laboratory, and is not meant as a comprehensive review.

KINETIC CHARACTERISTICS OF HIV-1 REVERSE TRANSCRIPTASE

The availability of highly purified HIV-1 reverse transcriptase (di Marzo Veronese et al., 1986) allowed detailed mechanistic studies of this enzyme. The first set of experiments in this laboratory attempted to elucidate a steady-state kinetic mechanism for polymerization by the HIV reverse transcriptase. A system with poly (rA) as template, oligo (dT) as primer, and dTTP as substrate was employed. If steady-state kinetics could be readily applied to this enzyme, we expected a set of internally consistent kinetic constants and linear double-reciprocal primary plots. We found these expectations were fully met with purified HIV reverse transcriptase. Substrate initial velocity studies together with product inhibition by pyrophosphate suggested an ordered reaction mechanism with enzyme binding first to template-primer. (Majumdar et al., 1988)

Products of synthesis in this system were separated by electrophoresis and visualized by autoradiography. We examined chain lengths of labeled DNA products formed during a short incubation period (200 s) at a high ratio of template-primer to enzyme. These conditions were chosen so that each product molecule would represent just one cycle of enzyme binding, synthesis, and termination. Under these conditions, one determined a chain elongation rate of approximately 2-4 nucleotide/s/enzyme, which was consistent with data obtained for the steady-state rate from initial rate measurements. We found that HIV reverse transcriptase conducts processive DNA synthesis, but exhibits some probability of terminating synthesis after each dTMP addition to the nascent chain. After the third and subsequent dTMP additions, the amount of termination was essentially constant and was equal to a termination probability of ~0.01, or one termination for every 100 nascent product molecules at any one chain length. Termination after incorporation of the first dTMP residue was ~20-fold higher. This observation of higher termination after the first incorporation event supports the idea that the initiation of synthesis is kinetically distinct from subsequent elongation. The experimental results are consistent with the kinetic scheme displayed below (Majumdar et al., 1988):

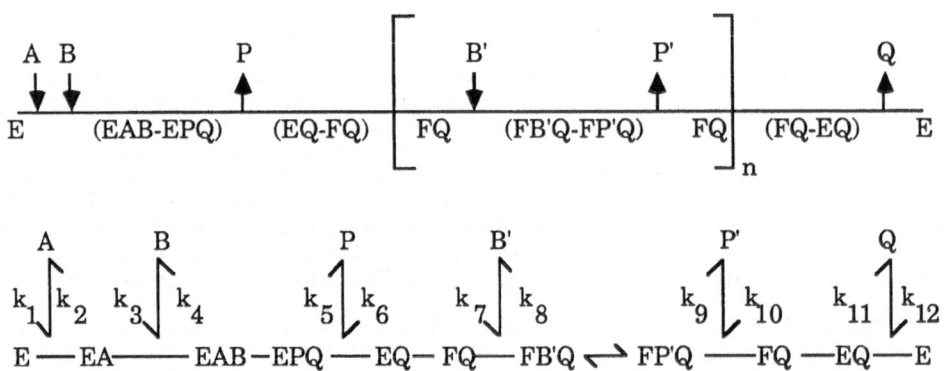

The steps inside brackets represent processive synthesis. E and F are stable forms of the enzyme. A is the template-primer complex. B and B' are, respectively, the first and all subsequent dNTP added to the enzyme during one cycle of free enzyme binding to A, processive synthesis and dissociation from Q (i.e., termination). P and P' are the first and all subsequent pyrophosphate molecules released, respectively. Q is the polynucleotide product (extended primer). This reaction scheme has been applied for analysis of two other processive DNA polymerases, *E. coli* Pol I large fragment, and mouse polymerase α, where the initiation of synthesis is kinetically different from subsequent processive synthesis (Detera et al., 1981; Detera and Wilson, 1982).

With the additional assumption that free enzyme does not form an enzyme:product complex with the template-primer, the steady-state rate equation for this mechanism can be simplified to the general equation for a terreactant-terproduct reaction:

$$v = \frac{[A][B]V_{max}}{K_mB\,K_mA + K_mA[B] + K_mB'[A] + [A][B]} \tag{1}$$

where the constants are

$$K_mB = \frac{k_2(k_5+k_4)}{k_3k_5} \tag{2}$$

$$K_mA = \frac{k_5k_9k_{11}}{k_1(k_9k_{11}+k_5k_{11}+k_5k_9)} \tag{3}$$

$$K_mB' = \frac{k_{11}(k_5k_7k_9+k_7k_9k_4+k_3k_5k_9+k_3k_5k_8)}{k_3k_7(k_9k_{11}+k_5k_{11}+k_5k_9)} \tag{4}$$

and

$$V_{max} = \frac{k_5k_9k_{11}E_t}{k_9k_{11}+k_5k_{11}+k_5k_9} \tag{5}$$

K_mB represents the Michaelis constant for addition of the first dNMP residue to the primer. K_mB' represents a consensus Michaelis constant for subsequent dNMP additions during each cycle of enzyme binding to A, processive polymerization, and termination.

Further studies revealed that $d(C)_{28}$ was a linear competitive inhibitor of DNA synthesis by the HIV reverse transcriptase, against poly $r(A)\cdot$oligo $d(T)$ as template-primer, indicating that $d(C)_{28}$ and the template-primer combine with the same form of the enzyme in the reaction scheme, the free enzyme. The phosphorothioate oligodeoxynucleotide $Sd(C)_{28}$ also is a linear competitive inhibitor against template-primer, but the K_i for inhibition (~2.8 nM) was ~200-fold lower than the K_i for inhibition by $d(C)_{28}$. Substrate kinetic studies of DNA synthesis by the HIV reverse transcriptase using $Sd(C)_{28}$ as primer and poly $r(I)$ as template revealed that K_m for the phosphorothioate primer was 24 nM. These results enabled calculation of rate values for the HIV reverse transcriptase for enzyme-primer association ($k_{on} = 5.7 \times 10^8$ M^{-1} s^{-1}) and dissociation ($k_{off} = 1.6$ s^{-1}). (Majumdar et al., 1989)

Annealing the phosphorothioate primer to poly $r(I)$ template inhibited the HIV reverse transcriptase to a similar degree as did the primer alone, and poly $r(I)$ alone caused no inhibition of DNA synthesis on a poly $r(A)\cdot$oligo $d(T)$ template-primer. Thus, enzyme binds the model primer as tightly as it binds a complex of the same primer annealed to template. The interpretation is that the initial step in the template-primer recognition process is primer binding to free enzyme. The template clearly directs association of the nucleotide substrate, but free enzyme recognizes the primer rather than the template. (Majumdar et al., 1989)

Various phosphorothioate oligonucleotides are being evaluated as potential antiviral agents for acquired immunodeficiency syndrome (Yarchoan and Broder, 1987; Matsukura et al., 1987). Antiviral effects of phosphorothioate oligodeoxynucleotides have been found and can be placed in two general categories: oligonucleotide sequence dependent and sequence independent (Matsukura et al., 1987). A proposed mechanism for sequence-dependent antiviral activity involves "antisense" hybridization to viral mRNA. Based on the findings described with the $Sd(C)_{28}$ primer, a plausible mechanism for the sequence independent activity could be reverse transcriptase-phosphorothioate oligonucleotide interaction. This primer also inhibited the activity of cellular DNA polymerases α and γ, with K_i values similar to that observed against the HIV reverse transcriptase. (Majumdar et al., 1989). Therefore, phosphorothioate oligonucleotides are potentially toxic for the host cell, although such toxicity does not appear to be strong. (Matsukura et al., 1987)

Subsequent studies of the HIV reverse transcriptase were carried out with dNTP analogues, 3'-azido-dTTP (AZTTP) and 3'-amino-dTTP (NH$_2$TTP). The nucleoside structures of these compounds are indicated in Figure 1. In the model DNA synthesis system with poly r(A)·oligo d(T) as template-primer, AZTTP was a strong inhibitor of the HIV-1 reverse transcriptase. In Figure 2, double-reciprocal plots are linear at each fixed concentration of AZTTP in the range of 10-60 nM, and the patterns converge to the same point on the ordinate. The replot of slope with each fixed concentration of AZTTP also is linear, as shown in the right hand panel of Figure 2. The results indicate that AZTTP blocks enzymatic activity through competitive inhibition with the normal substrate (dTTP) for binding to the primer-enzyme complex. (Kedar et al., 1990) The K_i value for AZTTP inhibition, 20 nM, was considerably lower than the K_m for dTTP incorporation, 2-3 µM (Majumdar et al., 1988).

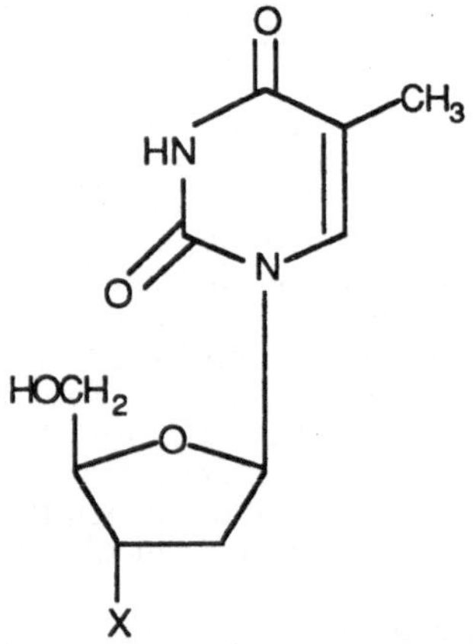

Figure 1.   Structure of AZT and NH$_2$T. These thymidine analogues differ only at the position indicated by X. For 3'-azido-3'-deoxythymidine (AZT), X = N$_3$; for 3'-amino-3'-deoxythymidine (NH$_2$T), X = NH$_2$.

Kinetic studies with AZTTP as substrate revealed a K_m for incorporation of 2.9 µM, and a k_{cat} of 0.27 s^{-1}. That the K_i for inhibition (20 nM) is 2 orders of magnitude less than K_m may indicate that binding of AZTTP to enzyme-template complex is a more important factor in interfering with dTTP incorporation than actual AZTTP incorporation. These measured values for AZTTP incorporation and mathematical treatment enable the calculation of kinetic constants for enzyme binding to template-primer. When data for dTTP incorporation are then included, values for steps in a simplified representation of the first two complexes in the reaction pathway can be assigned as follows (Kedar et al., 1990):

$$E + T\text{-}P \underset{\substack{k_{off} \\ 39\,sec^{-1}}}{\overset{\substack{k_{on} \\ 2.6 \times 10^{8}\,M^{-1}sec^{-1}}}{\rightleftharpoons}} E{\overset{T\text{-}P}{}} + dTTP \underset{\substack{K_D \\ 185\,nM}}{\rightleftharpoons} E{\overset{T\text{-}P}{\underset{dTTP}{}}} \overset{\substack{k_{cat} \\ 3\,sec^{-1}}}{\longrightarrow}$$

The k_{on} value for the normal primer oligo dT is little different from the value for the phosphorothioate primer $Sd(C)_{28}$; both values are similar to the theoretical diffusion-controlled rate of collision for macromolecules in this size range, $\sim10^{9}\,M^{-1}\,s^{-1}$. The association rates of the HIV reverse transcriptase for primer are similar to values reported for several tRNA/synthetase interactions (Pingoud et al., 1975) and similar to values reported for the interaction of *E. coli* RNA polymerase and TAC promoters (Mulligan et al., 1985) and λ promoter (Hawley et al., 1985).

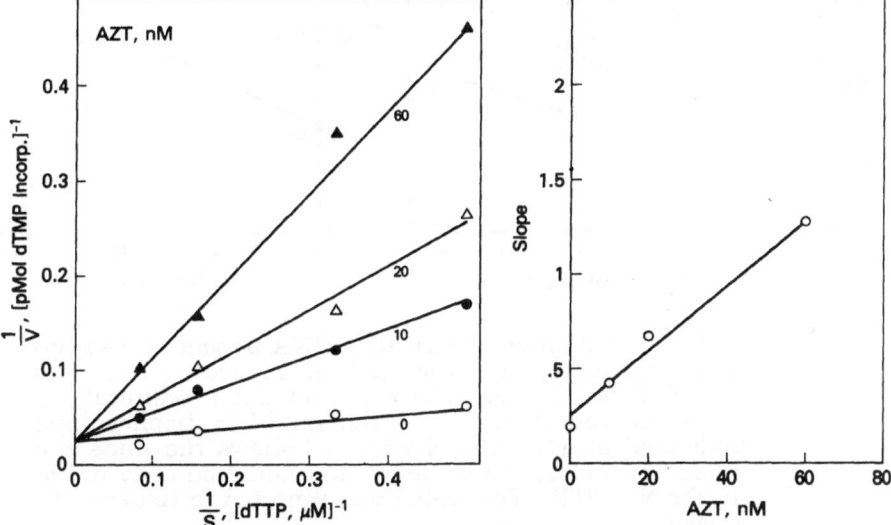

Figure 2. AZTTP inhibition of HIV RT. DNA polymerase assays were conducted with poly r(A) as template, oligo d(T) as primer, dTTP-Mg as nucleotide substrate, and 56 ng (480 fmol) of HIV-1 reverse transcriptase. Incorporation of labeled dTTP was measured by acid-precipitable counts collected on nitrocellulose filters (Kedar et al., 1990). The left panel is the reciprocal plot with dTTP as variable substrate and the concentration of the inhibitor AZT-triphosphate is indicated in nM. The right panel is the slope replot, which allows determination of 20 nM as K_i for AZTTP. This value was reported in Kedar et al. (1990).

A kinetic study of the inhibition by the analogue NH_2TTP is shown in Figure 3. In this case, a primary double-reciprocal plot reveals that the pattern of inhibition is noncompetitive, since the lines at different fixed concentrations of inhibitor extrapolate to different points on the ordinate. The replots of slopes and intercepts in the right-hand panel of Figure 3 are linear also and indicate linear noncompetitive inhibition with K_i value of 42 nM. Taken together, these results with AZTTP and NH_2TTP indicate that the two analogues block DNA synthesis by HIV reverse transcriptase by interacting with different forms of the enzyme. This result was rather unexpected since the two analogues are closely related in chemical composition. To confirm this observation, a "mixed inhibition" experiment was conducted. The presence of 10 nM NH_2TTP increased the slope for AZTTP inhibition, indicating again that the two substrates act on different enzyme forms. (Kedar et al., 1990)

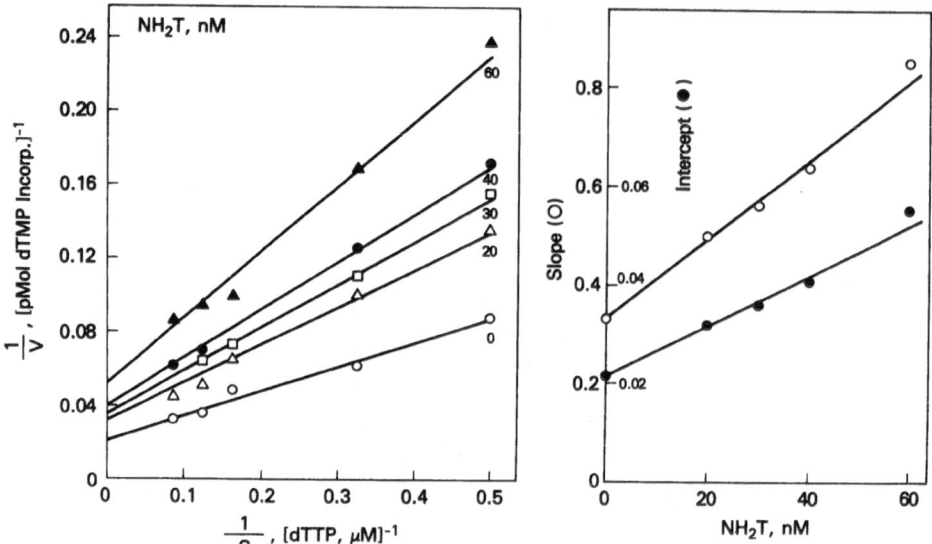

Figure 3. NH_2TTP inhibition of HIV RT. DNA polymerase assays were conducted as described in Figure 2, but with NH_2TTP as inhibitor. The left panel is the reciprocal plot with the concentration of the inhibitor triphosphate indicated in nM. The right panel shows the slope and intercept replots, which allow determination of 42 nM as K_i for NH_2TTP. This value was reported in Kedar et al. (1990).

Examination of products of DNA synthesis in the presence of each inhibitor also showed differential effects. As AZTTP concentration increased, the size distribution of products shifted toward smaller lengths, such that the average length of product molecules decreases. With increasing concentrations of NH_2TTP, the overall size distribution of products did not change, but the number of product molecules at each length decreased. These results are consistent with different modes of inhibition by the two analogues. Studies of NH_2TTP as substrate showed a K_m of 2.3 µM, similar to AZTTP, but the steady-state rate of incorporation of AZTTP was about 4-fold higher than the rate with NH_2TTP. (Kedar et al., 1990).

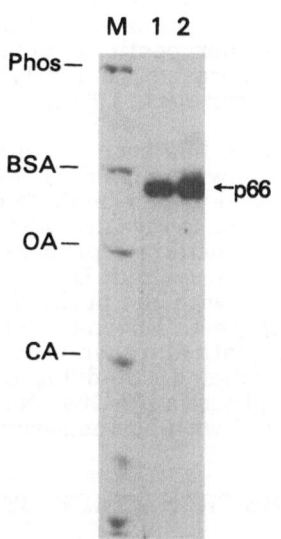

Figure 4. UV cross-linking of HIV-1 RT and the primer analogue
d(T)$_8$. An autoradiogram after SDS-PAGE is shown. p66
(pool IV) was incubated with [^{32}P]d(T)$_8$ at 25 °C for 20
min and the mixture (15 μl) was irradiated at 3.75 x 10^4
erg/mm^2 in a Stratagene UV Stratalinker 1800. The
final p66 homodimer concentration was 0.5 μM and the
final [^{32}P]d(T)$_8$ concentration was either 0.5 μM (lane 1)
or 1 μM (lane 2). The positions of labeled marker
proteins are indicated in the left-hand lane.

In the absence of data on the metabolic fate of AZT, including conversion to NH_2T, the therapeutic implications of our findings are not clear. Larder et al. (1989) recently reported HIV clinical isolates resistant to AZT but still sensitive to other reverse transcriptase inhibitors, including presumed chain terminators. Although in vitro mutagenesis studies have produced reverse transcriptases that were more resistant to AZTTP than wild type (Larder et al., 1987), purified RT from these AZT-resistant isolates showed no significant difference in AZTTP sensitivity (Larder et al., 1989; Larder and Kemp, 1989). Such results would be consistent with a situation where an AZT metabolite, acting by a different mechanism from AZTTP, was important for the inhibition of viral replication in cells.

These clinical results, and this laboratory's findings with the dNTP analogues, provide at least a suggestion that combined therapy with agents that act upon different forms of the HIV reverse transcriptase occurring during its reaction pathway may be useful, and AZTTP and NH_2TTP may be considered prototypes for such an approach. Unfortunately, NH_2TTP may not seem suitable as a practical therapeutic candidate since the triphosphate also inhibits cellular DNA polymerase α, and NH_2T is strongly toxic to cells in culture (for discussion, see Kedar et al., 1990).

Enzyme-primer binding also has been examined using the ultraviolet (UV) light cross-linking approach with oligo $d(T)_8$ as a model primer, labeled on the 5' end with ^{32}P. This model primer binds to the kinetically significant primer binding site on the enzyme, as revealed with competition studies with several authentic, previously characterized primers (data not shown), and is abundantly crosslinked to the RT protein at UV fluences that do not degrade the enzyme. An example of this with the purified p66 form of the RT to be described below is shown in Figure 4. Enzyme-$d(T)_8$ complexes were allowed to form in the absence of a template and then were covalently crosslinked by exposure to UV light. The labeled p66-$d(T)_8$ adduct was displayed by SDS-polyacrylamide gel electrophoresis (PAGE). No label was seen in the p66 region when the RT was replaced with bovine serum albumin (not shown).

FIDELITY OF HIV-1 REVERSE TRANSCRIPTASE

In initial experiments to examine the misinsertion propensity of HIV reverse transcriptase, we used a DNA template and a method termed DNA "bypass" synthesis applied by Beattie and colleagues (Hillebrand et al., 1984; Hillebrand and Beattie, 1985; Lai and Beattie, 1988). With this method, template-directed DNA synthesis is carried out with individual dNTPs omitted from the reaction mixture. Displaying reaction products by gel electrophoresis reveals a strong termination site before the first position at which the omitted nucleotide is normally inserted. Depending on the ability of a given polymerase to misinsert at such sites, product molecules of longer size are also seen.

Typical results with this method are shown in Figure 5, comparing the HIV enzyme with the reverse transcriptase (RT) of avian myeloblastosis virus (AMV) and *E. coli* Pol I large fragment (Pol I lf). The template is illustrated at the top of the figure: A ^{32}P-labeled primer is hybridized to $\phi X174$ single stranded DNA upstream of a region where the template is lacking in secondary structure, based on previous enzymatic studies and computer projections (Weaver and DePamphilis, 1982). In normal reactions with all four dNTPs, processive DNA synthesis is observed with both reverse transcriptases (Figure 5, lanes 13-16). These enzymes also show preferential sites for termination of processive synthesis; their termination patterns are similar, but not identical (for a general discussion of termination, see Abbotts et al., 1988).

When an individual dNTP is omitted from the reaction mixture, Pol I lf shows a strong termination band at the first position where the omitted substrate would have been inserted. Weak bands also are seen at downstream positions, suggesting that there is some bypass synthesis due to misinsertion. The incoming substrate for the first two template residues is dGTP, and very little synthesis can be seen when dGTP is omitted; however, longer exposure of the gel reveals faint bands at positions 649 and 638, corresponding to sites where dGTP is the incoming substrate (data not shown). The reverse transcriptases, in contrast, show no detectable synthesis when dGTP is omitted. For the reactions where dATP, dCTP, and dTTP are omitted, the reverse transcriptases show much more bypass than Pol I lf. Thus, the bypass patterns for the reverse transcriptases are similar to each other and distinctly different from Pol I lf.

The tendency of a given polymerase to bypass each "minus dNTP" site can be quantified from band intensities (Figure 5) by comparing the relative proportion of product molecules at the first "minus substrate" site to all product molecules of greater length. One can then calculate a bypass frequency, reflecting the probability that an enzyme will synthesize beyond the first "minus dNTP" site rather than terminate synthesis. Table 1 displays the bypass frequencies for the three enzymes examined. One sees for Pol I lf low levels of bypass synthesis for each minus nucleotide reaction. For the -dATP, -dCTP, and -dTTP reactions, the reverse transcriptases exhibit bypass frequencies that are at least an order of magnitude larger. In several cases, the bypass frequency of HIV reverse transcriptase approaches or exceeds 50 percent, indicating that the enzyme readily continues synthesis in the absence of the correct nucleotide. Only in the minus dGTP situation is bypass not seen with the reverse transcriptase; Pol I lf shows a small but detectable amount of bypass synthesis in this case.

These experiments indicate that the purified HIV reverse transcriptase conducts error-prone replication on a DNA template. Based on observations for bypass at a single position on the φX DNA template, the fidelity of HIV reverse transcriptase appears similar to that of AMV reverse transcriptase, and considerably worse than Pol I lf (Figure 5). One exception is the minus dGTP reaction, where both viral enzymes show bypass frequencies that are less than that of Pol I lf. This is consistent with an earlier observation by Hillebrand and Beattie (1985) that there was no striking difference between AMV reverse transcriptase and Pol I in bypass synthesis on an M13 template in the absence of dGTP, although these workers cautioned that relative fidelity of polymerases may change with template position.

Another group has also reported extensive bypass synthesis by HIV-1 reverse transcriptase; when these workers examined base substitution fidelity by measuring reversion of a φX174 *amber* phenotype, they determined an error rate for HIV reverse transcriptase of 1 per 4000 bases incorporated, compared to 1/9000 for AMV reverse transcriptase (Preston et al., 1988). Other workers, examining base substitutions by reversion of a nonsense codon on an M13 template, found an error rate of 1/18,000 for HIV RT, compared to 1/24,000 for AMV RT (Roberts et al., 1988). These same groups have also found that the HIV reverse transcriptase can efficiently elongate primers with mispaired 3' ends (Roberts et al., 1988; Perrino et al., 1989). Thus, the HIV reverse transcriptase could provide a source of genetic diversity through error-prone replication.

The observation that the HIV reverse transcriptase terminates DNA synthesis in a non-random manner (Figure 5; Abbotts et al., 1988), and the availability of a system which can measure the mutation spectrum of any DNA-polymerizing enzyme, allowed a collaboration to examine the influence of termination on mutations (Bebenek et al., 1989). That the tendency of a

Table 1. Bypass frequency of DNA synthesizing enzymes[a]

Omitted substrate	Bypass frequency in percent for:		
	Pol I lf	HIV RT	AMV RT
dATP	2.4	58	68
dCTP	1.6	33	75
dGTP	<1.0	0	0
dTTP	5.6	68	48

[a] Bands in Figure 5 were traced by densitometry, and areas under peaks were determined by cutting and weighing. Bypass frequency was determined by the relative amount of DNA synthesis products extended beyond the first "minus dNTP" site for a given substrate, compared to products representing termination at the first "minus" site. The bypass frequencies for the minus dGTP condition reflect detectable but small bypass for Pol I lf, and no detectable bypass synthesis by the reverse transcriptases.

---→

Figure 5. Bypass synthesis on a φX174 DNA template (facing page). The template-primer system, with extended primer indicated, is displayed at the top of the figure. DNA synthesis reactions (20 μl) were carried out with ^{32}P-labeled primer hybridized to φX plus strand DNA (200 fmol template-primer); 20 mM Tris-HCl, pH 8.0; 6 mM MgCl$_2$, 10 mM dithiothreitol; 50 μM each dATP, dCTP, dGTP, and dTTP; and 10 fmol of HIV reverse transcriptase. Reaction mixtures were incubated at 35 °C. Products of reaction were separated by electrophoresis on a 12% polyacrylamide, 7 M urea gel, and were visualized by autoradiography. Incubation times were 20 min, except for lanes 13 and 16. Unextended primer and some positions on the φX template are indicated. For lanes 1-4, Pol I lf was present at 50 fmol, with substrate conditions as follows: lane 1, dATP omitted; lane 2, dCTP omitted; lane 3, dGTP omitted; lane 4, dTTP omitted. For lanes 5-8, HIV reverse transcriptase was present at 10 fmol, with substrate conditions: lane 5, dATP omitted; lane 6, dCTP omitted, lane 7, dGTP omitted; lane 8, dTTP omitted. For lanes 9-12, AMV RT was present at 10 fmol, with substrate conditions: lane 9, dATP omitted; lane 10, dCTP omitted; lane 11, dGTP omitted; lane 12, dTTP omitted. Lane 13, 10 fmol HIV RT, all four dNTPs present, 3 min incubation; lane 14, same conditions as lane 13 but 20 min incubation; lane 15, 10 fmol AMV RT, all four dNTPs present, 20 min incubation; lane 16, same conditions as lane 15 but 3 min incubation. Lanes A, C, G, and T are sequencing reactions by the Sanger (1977) method, on the φX template using the labeled primer. These lanes and lanes 12-16 are shown at darker exposure than lanes 1-11, which were displayed on the same autoradiogram. For lane 4, the dot above the first termination site is a contaminant, but a light band can also be seen in that vicinity.

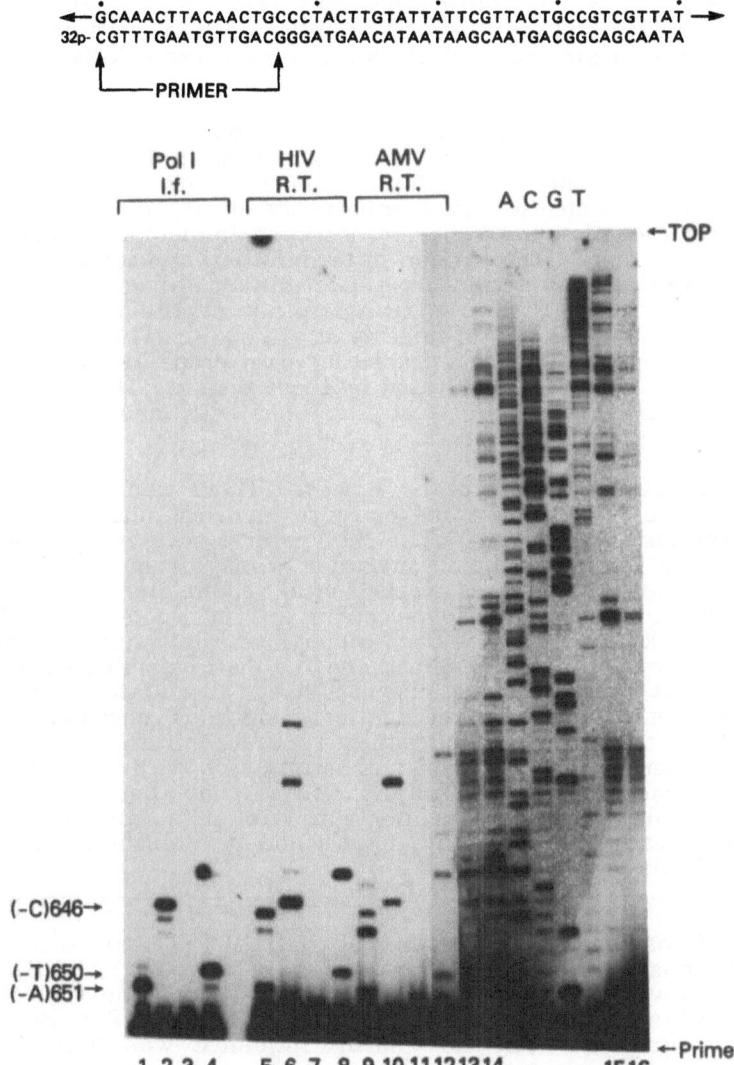

polymerase to dissociate from the template-primer during synthesis might affect fidelity was suggested by several observations: Kinetic studies with DNA polymerase α (Detera et al., 1981), the Pol I large fragment (Detera and Wilson, 1982), and HIV-1 reverse transcriptase (Majumdar et al., 1988) indicate that initiation and processive synthesis have distinctly different enzymatic properties; it would be of interest to determine if the kinetic differences are reflected in differences in error rate. The model for energy relay proofreading (Hopfield, 1980) predicts that the first nucleotide incorporated during DNA synthesis should be more error-prone that subsequent incorporations. In addition, previous work with eucaryotic DNA polymerases in the M13mp2 mutation assay indicated that more processive enzymes are more accurate, particularly for one-base frameshifts (Kunkel, 1985).

With the M13mp2 mutation system, it was possible to examine both the mutation spectrum of HIV-1 reverse transcriptase, as well as the termination pattern characteristic of the enzyme (Bebenek et al., 1989). If insertion of the first nucleotide were inherently error-prone, one would expect to observe mutation hot spots at positions immediately following strong termination sites. A comparison of termination and mutation sites suggested that reinitiation is not a strong determinant of base substitution errors for either the HIV-1 reverse transcriptase or the AMV reverse transcriptase (Bebenek et al., 1989; Roberts et al., 1989). Thus, Hopfield's model (1980) or other kinetic models predicting error-prone incorporation of the first nucleotide, were not corroborated by these results. This conclusion is consistent with earlier observations at single loci with mammalian DNA polymerases (Grosse et al., 1983; Abbotts and Loeb, 1984).

With HIV-1 reverse transcriptase, however, there was a correlation between termination in runs of three or more identical nucleotides and frameshift mutations in such runs. The HIV reverse transcriptase shows an overall mutation frequency that is an order of magnitude greater than those of other reverse transcriptases (Roberts et al., 1988), and the dominant features of the HIV enzyme's mutation spectrum are "hot spots" for one-base frameshifts (Bebenek et al., 1989). Each major frameshift site is within a homopolymer sequence of 3 to 5 residues, and the distinguishing feature of the major frameshift sites is the presence of a strong termination site. The observation of frameshifts with runs of identical residues is consistent with a model proposed by Streisinger and colleagues (1966) for mutations caused by slippage between template and primer, and termination within a run could facilitate such a mechanism. The implication is that the HIV reverse transcriptase is able to tolerate primer-template misalignment during synthesis, or may cause such misalignment when it reinitiates synthesis (Bebenek et al., 1989).

Figure 6. Construction and features of the HIV-1 RT expression construct pRC-RT (facing page). The HIV HXB2 proviral DNA is depicted at the top of the figure illustrating the relative position of the RT coding sequence. The RT coding sequence was flanked precisely by start and stop codons (Becerra et al., unpublished), and the resulting construct was cloned into the pRC23 vector. Numbers refer to nucleotide sequence of the proviral DNA. Restriction endonuclease sites noted are: BamHI (B); EcoRI (E); DraI (D); ScaI (S). In pRC-RT, the RT coding sequence (diagonal lines), the λ P_L promoter, the Shine-Delgarno sequence (S.D.), restriction sites and antibiotic markers are illustrated.

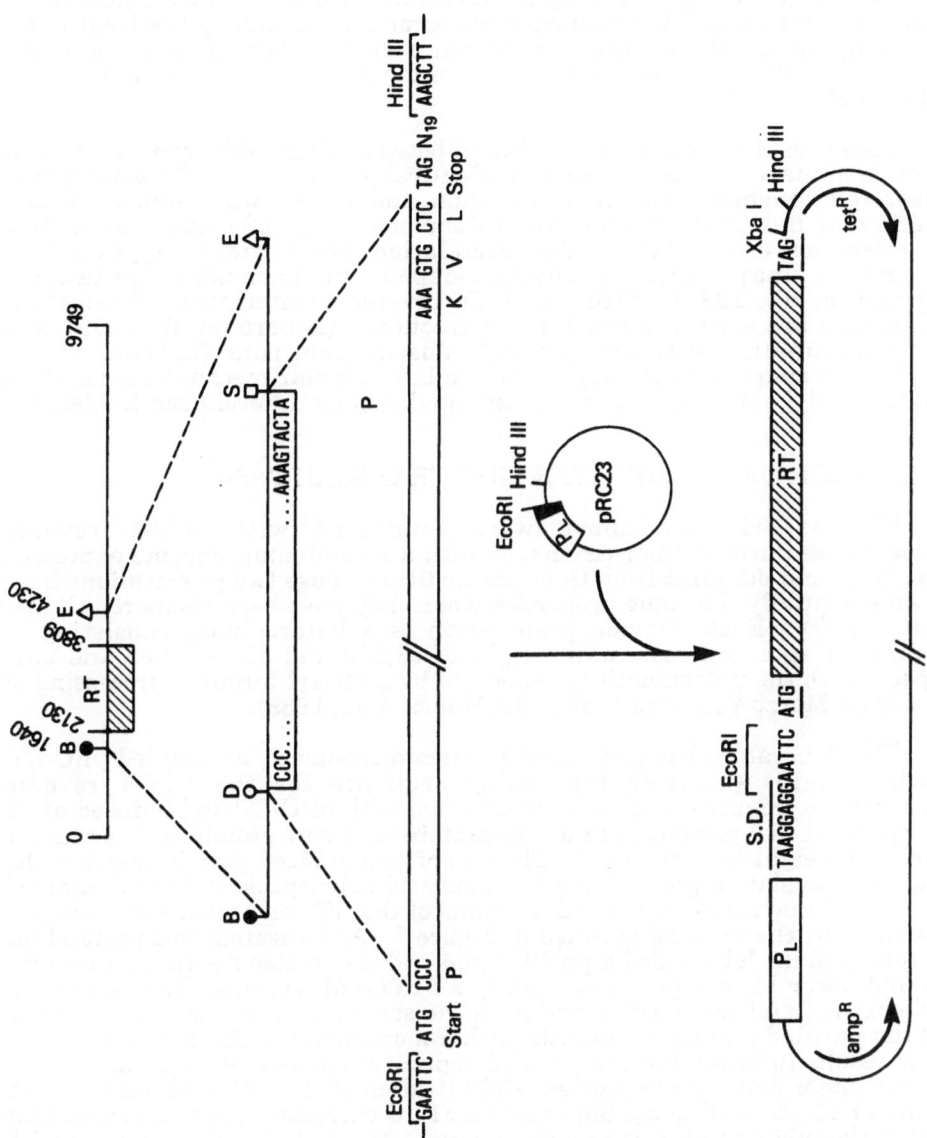

On the M13 template, mutation was still a low-probability event, compared to termination. For example, at a position where the enzyme terminated synthesis about half the time, the error rate for plus-one errors was much less than one percent. In addition, the correlation between termination probability and one-base frameshifts was not quantitative. The highest error rates did not necessarily occur at sites exhibiting the highest termination probability. Such quantitative differences are not inconsistent with the concept that processivity is important for fidelity, since a number of parameters are expected to influence the error rate, including the length and base composition of the run, the influence of neighboring bases, and the structure of the polymerase and its contacts with the template-primer (Bebenek et al., 1989).

These results indicate a correlation between frameshift error rates and major termination sites in runs of identical nucleotides. To establish a causal relationship between termination and errors will require further experiments to modulate termination frequency at specific sites and observe the effect on error rates. The correlation nonetheless suggests that termination may facilitate slippage-dependent frameshift mutations. (Bebenek et al., 1989) With the AMV reverse transcriptase, frameshifts within runs are seen at a much lower frequency (Roberts et al., 1989), and strong termination sites are generally absent from runs (Bebenek et al., 1989). This suggests that enzymes which have similar mechanisms for DNA synthesis might produce errors during synthesis by different mechanisms.

EXPRESSION OF THE HIV-1 REVERSE TRANSCRIPTASE

The experiments above were conducted with HIV-1 reverse transcriptase purified from virions, or with a recombinant enzyme expressed in bacteria and obtained from Genetics Institute; these two preparations have shown essentially the same properties when they have been compared (Kedar et al., 1990). Each enzyme preparation is a heterodimer, consisting of subunits of apparent molecular weight of 66,000 and 51,000 (p66 and p51, respectively); the p51 subunit is generated by carboxyl-terminal processing of the p66 (di Marzo Veronese et al.,1986; Huber et al., 1989).

This laboratory has generated a plasmid construct, designated pRC-RT, which contains precisely the coding sequence for the HIV-1 reverse transcriptase (Figure 6). E. coli transformed with pRC-RT and induced at 42 °C expressed ~10 percent of total cell protein as a new protein with apparent molecular weight (M_r) of 66,000. The size of this induced protein matches the predicted translation product (a 560-amino acid polypeptide of 64,484) encoded by the 1680 nucleotide open reading frame of the RT gene construct. The p66 is purified by the protocol outlined in Figure 7. As indicated, this protocol for the recombinant RT yielded a purified p66 and also a side fraction containing p66 and lower M_r polypeptides, mainly a 51,000 M_r species. The p66 in the crude extract and the purified p66 have the same apparent molecular weight, and the purified p66 immunoreacts with a monoclonal antibody to HIV-1 RT. These results indicate that the purified peptide of ~66,000 M_r is RT and is free of other major polypeptide species. Gel filtration of the p66 preparation at ~1 mg/ml (or 15 µM p66) using Superose-12 FPLC chromatography revealed that most of the p66 protein corresponds to native M_r of ~120,000 in solution with 75 mM NaCl at pH 7.0; a small amount (~10 percent) of p66 monomer was present also. This same result was found even when 120,000 M_r fractions were re-chromatographed. Therefore, most of the native p66 polypeptide appears to be in dimer structure in our solution, and this enzyme will be referred to here as a homodimer. (Becerra, Kumar, Widen, Karawya, Abbotts, Hughes, Shiloach, and Wilson, unpublished results)

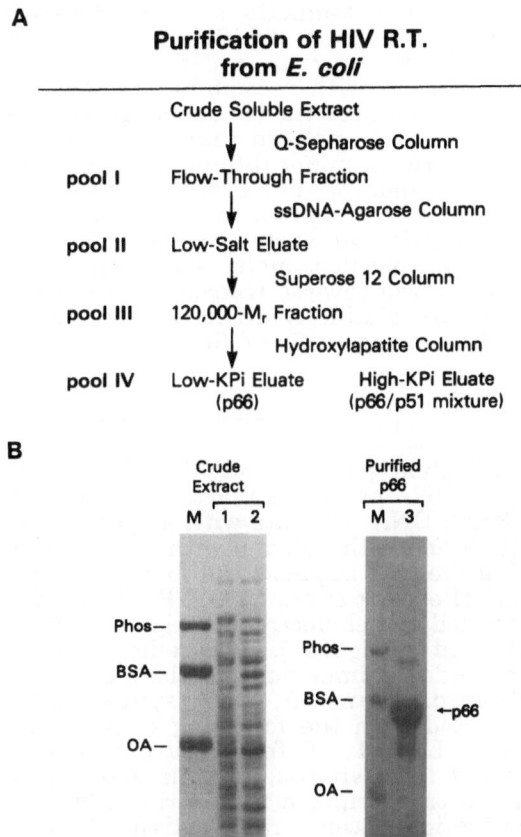

A

**Purification of HIV R.T.
from *E. coli***

Crude Soluble Extract
↓ Q-Sepharose Column
pool I   Flow-Through Fraction
↓ ssDNA-Agarose Column
pool II   Low-Salt Eluate
↓ Superose 12 Column
pool III   120,000-M_r Fraction
↓ Hydroxylapatite Column
pool IV   Low-KPi Eluate   High-KPi Eluate
(p66)   (p66/p51 mixture)

B

Crude Extract   Purified p66

M   1   2   M   3

Phos—
BSA—
OA—
CA—

Phos—
BSA—   ←p66
OA—

Figure 7.   Purification scheme and SDS-PAGE analysis of RT expression and final enzyme termed p66. Panel A, the steps in the purification scheme are summarized. Panel B, the photograph shows Coomassie blue stained SDS-PAGE analysis of *E. coli* crude extacts (on the left) and of an intentional excess of purified pool IV p66 (on the right). Lane 1, protein from uninduced cells; lane 2, protein from the same amount of cells induced at 42 °C for 2 h; lane 3, 50 µg p66 pool IV; lanes with marker proteins are designated by M.

The final fraction of purified p66 is free of Pol I contamination, as determined from activity gel analysis and Western blotting with anti-Pol I antibody (data not shown). To confirm the absence of Pol I activity, we conducted processivity analysis for DNA synthesis. We have found that several different DNA-polymerizing enzymes show patterns of termination of synthesis that are characteristic of the individual enzyme (Abbotts et al., 1988; Bebenek et al., 1989; Kumar et al., 1990). Here we conducted experiments on an M13 DNA template, and the p66 enzyme was compared with HIV-1 reverse transcriptase purified from virions (Figure 8). The termination pattern for the enzyme purified from virions is essentially similar to the pattern seen with recombinant heterodimer HIV-1 reverse transcriptase expressed in bacteria (Bebenek et al., 1989). Purified p66 enzyme showed a termination pattern that was very similar to that of the enzyme purified from virions; minor differences can be seen, however. The p66 preparation may also be moderately more processive than the heterodimer from virions; for approximately the same amount of activity, p66 shows less termination at early positions, and a slightly increased concentration of longer product molecules. The basis for the modest differences between the two reverse transcriptases is not known, but their termination patterns are distinctly different from that of Pol I lf on this template (Figure 8, Panel A, lanes 5 and 6).

Figure 8. Termination patterns on M13 DNA template (facing page). DNA synthesis reactions were carried out on an M13mp2 DNA template with a synthetic primer labeled at its 5' end with ^{32}P, and with its 3' hydroxyl at position 115 of the *lacZ* sequence; all four dNTPs were present at 1 mM (Bebenek et al., 1989). Products of synthesis were examined by gel electrophoresis and autoradiography as described in Figure 5. Autoradiograms are shown with unextended primer and positions on the M13 template indicated. Panel A, DNA synthesis reactions were conducted with the following enzymes and incubation times: Lane 1, 10 fmol HIV-1 reverse transcriptase, purified from virions, 10 min; lane 2, 10 fmol HIV-1 reverse transcriptase from virions, 30 min; lane 3, p66 preparation of comparable activity, 10 min; lane 4, p66, 30 min; lane 5, 5 fmol Pol I lf, 10 min; lane 6, 5 fmol Pol I lf, 30 min. Panel B, DNA synthesis reactions were conducted with the following enzymes and incubation times: Lane 1, 10 fmol HIV-1 reverse transcriptase from virions, 10 min; lane 2, 10 fmol HIV-1 reverse transcriptase from virions, 30 min; lane 3, p66 preparation, 10 min; lane 4, 1 ng protein of ssDNA-agarose column fraction number 35, 1 min; lane 5, 1 ng protein of fraction 35, 3 min; lane 6, 1.7 ng protein of ssDNA-agarose column fraction number 40, 1 min; lane 7, 1.7 ng protein of fraction 40, 3 min; lane 8, 5 ng protein of partially purified fraction (see text for details), 1 min; lane 9, 5 ng protein of partially purified fraction, 3 min; lane 10, 50 fmol Pol I lf, 10 min; lane 11, 50 fmol Pol I lf, 30 min. All lanes in Panel B were displayed on the same gel; lanes 10 and 11 were located in a different region of the gel from the other lanes, and ran with a slightly different mobility; in addition, lanes 10 and 11 represent shorter autoradiography exposures compared to the other lanes.

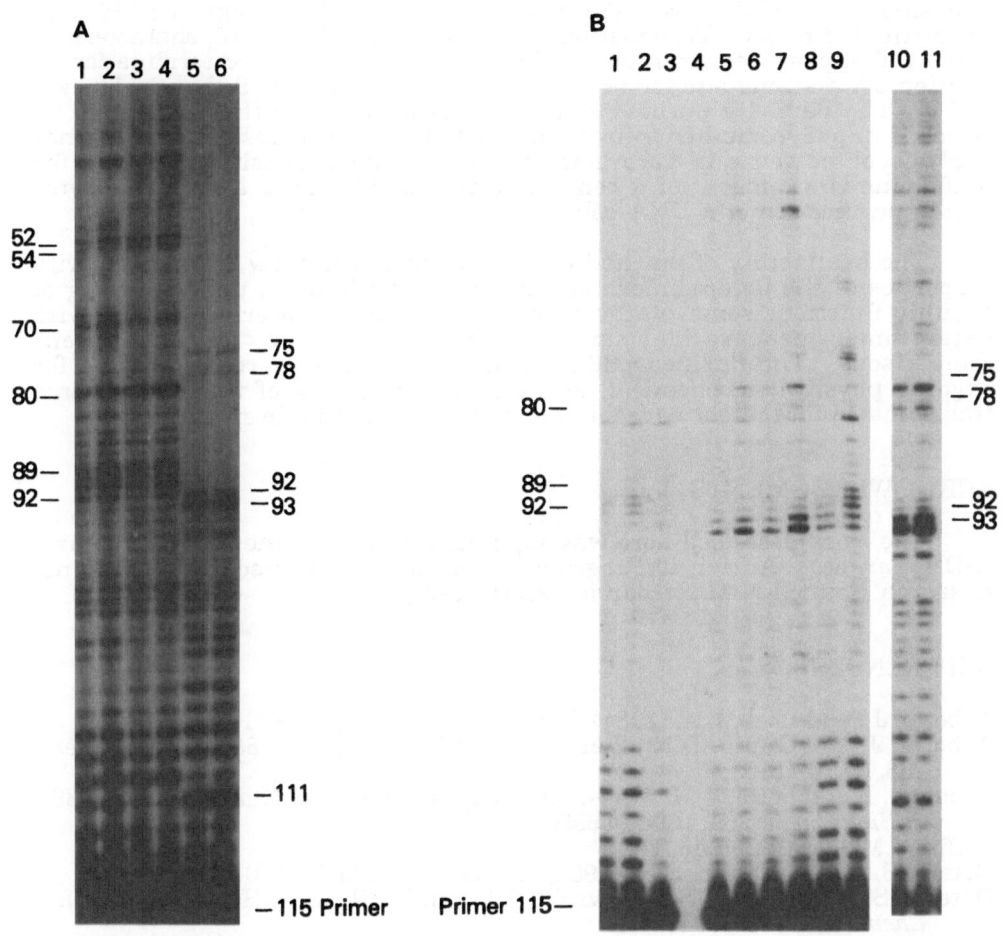

Figure 8, Panel B shows the patterns obtained for several different fractions during the purification. We examined ssDNA-agarose fractions 35 and 40, which were enriched for Pol I (data not shown). We also examined a partially purified fraction, obtained by stepwise elution rather than gradient elution, of a ssDNA-agarose column. Activity gel analysis of this fraction indicated the presence of both Pol I and reverse transcriptase (data not shown). The patterns generated by ssDNA-agarose fractions 35 and 40 (lanes 4 through 7) are similar to the pattern of Pol I lf (lanes 10 and 11), confirming that Pol I is a dominant DNA polymerase activity in these fractions. The pattern seen with the partially purified fraction (lanes 8 and 9) appears to represent a hybrid of the patterns seen with p66 and Pol I lf individually. For example, the termination pattern among positions 89-93 shows strong stops at positions 92 and 93, associated with Pol I lf, as well as stops at 89-91 that are attributed to p66. Termination sites at positions 75 and 78, characteristic of Pol I lf, appear weaker in lane 9 than in lanes 5 and 7; and termination at position 80, a strong site for p66, appears stronger in lane 9 than in lanes 5 and 7. The partially purified samples in lanes 8 and 9 thus appear to be enriched in p66, compared to fractions 35 and 40. These results suggest that analysis of patterns of termination can provide a sensitive method for confirming the removal of a contaminating activity from a known enzyme during purification of a DNA polymerase.

The availability of purified p66 homodimer will allow comparison with properties of the heterodimer, and may thus facilitate an understanding of subunit function; some of the properties we have described above with heterodimer preparations may provide a baseline for subsequent comparisons. The purified p66 homodimer is also a source of enzyme for ongoing physical, biochemical, and enzymatic studies of the HIV reverse transcriptase that might be facilitated by a sole polypeptide species.

ACKNOWLEDGMENTS

The work described here was supported in part by the NIH Intramural AIDS Targeted Antiviral Program. J.A. was supported by a National Research Council-NIH Research Associateship.

REFERENCES

Abbotts, J., and Loeb, L.A. (1984) *J. Biol. Chem.* 259, 6712-6714.
Abbotts, J., SenGupta, D.N., Zon, G., and Wilson, S.H. (1988) *J. Biol. Chem.* 263, 15094-15103.
Bebenek, K., Abbotts, J., Roberts, J.D., Wilson, S.H., and Kunkel, T.A. (1989) *J. Biol. Chem.* 264, 16948-16956.
Coffin, J.M. (1986) *Cell* 46, 1-4.
Detera, S.D., and Wilson, S.H. (1982) *J. Biol. Chem.* 257, 9770-9780.
Detera, S.D., Becerra, S.P., Swack, J.A., and Wilson, S.H. (1981) *J. Biol. Chem.* 256, 6933-6943.
di Marzo Veronese, F., Copeland, T.D., DeVico, A.L., Rahman, R., Oroszlan, S., Gallo, R.C., and Sarngadharan, M.G. (1986) *Science* 231, 1289-1291.
Grosse, F., Krauss, G., Knill-Jones, J.W., and Fersht, A.R. (1983) *EMBO J.* 2, 1515-1519.
Hawley, D.K., Johnson, A.D., and McClure, W.R. (1985) *J. Biol. Chem.* 260, 8618-8626.
Hillebrand, G.G., McCluskey, A.H., Abbott, K.A., Revich, G.G., and Beattie, K.L. (1984) *Nucleic Acids Res.* 12, 3155-3171.
Hillebrand, G.G., and Beattie, K.L. (1985) *J. Biol. Chem.* 260, 3116-3125.
Hopfield, J.J. (1980) *Proc. Natl. Acad. Sci. U.S.A.* 77, 5248-5252.

Huber, H.E., McCoy, J.M., Seehra, J.S., and Richardson, C.C. (1989) *J. Biol. Chem.* 264, 4669-4678.

Kedar, P.S., Abbotts, J., Kovács, T., Lesiak, K., Torrence, P., and Wilson, S.H. (1990) *Biochemistry* 29, 3603-3611.

Kumar, A., Abbotts, J., Karawya, E.M., and Wilson, S.H. (1990) *Biochemistry* 29, 7156-7159.

Kunkel, T.A. (1985) *J. Biol. Chem.* 260, 12866-12874.

Lai, M.-D., and Beattie, K.L. (1988) *Biochemistry* 27, 1722-1728.

Larder, B.A., and Kemp, S.D. (1989) *Science* 246, 1155-1158.

Larder, B.A., Darby, G., and Richman, D.D. (1989) *Science* 243, 1731-1734.

Larder, B.A., Purifoy, D.J.M., Powell, K.L., and Darby, G. (1987) *Nature* 327, 716-717.

Majumdar, C., Abbotts, J., Broder, S., and Wilson, S.H. (1988) *J. Biol Chem* 263, 15657-15665.

Majumdar, C., Stein, C.A., Cohen, J.S., Broder, S. and Wilson, S.H. (1989) *Biochemistry* 28, 1340-1346.

Matsukura, M., Shinozuka, K., Zon, G., Mitsuya, H., Reitz, M., Cohen, J.S., and Broder, S. (1987) *Proc. Natl. Acad. Sci. U.S.A.* 84, 7706-7710.

Mitsuya, M., and Broder, S. (1987) *Nature* 325, 773-778.

Mulligan, M.E., Brosius, J., and McClure, W.R. (1985) *J. Biol. Chem.* 260, 3529-3538.

Perrino, F.W., Preston, B.D., Sandell, L.L, and Loeb, L.A. (1989) *Proc. Natl. Acad. Sci. U.S.A.* 86, 8343-8347.

Pingoud, A., Boehme, D., Riesner, D., Kownatzki, R., and Maass, G. (1975) *Eur. J. Biochem.* 56, 617-622.

Preston, B.D., Poiesz, B.J., and Loeb, L.A. (1988) *Science* 242, 1168-1171.

Roberts, J.D., Bebenek, K., and Kunkel, T.A. (1988) *Science* 242, 1171-1173.

Roberts, J.D., Preston, B.D., Johnston, L.A., Soni, A., Loeb, L.A., and Kunkel, T.A. (1989) *Mol. Cell. Biol.* 9, 469-476.

Sanger, F., Nicklen, S., and Coulson, A.R. (1977) *Proc. Natl. Acad. Sci. U.S.A.* 74, 5463-5467.

Streisinger, G. Okada, Y., Emrich, J., Newton, J., Tsugita, A., Terzaghi, E., and Inouye, M. (1966) *Cold Spring Harbor Symp. Quant. Biol.* 31, 77-84.

Weaver, D.T., and DePamphilis, M.L. (1982) *J. Biol. Chem.* 257, 2075-2086.

Yarchoan, R., and Broder, S. (1987) *New Eng. Jour. Med.* 316, 557-564.

ANALYSES OF HIV INTEGRATION COMPONENTS

Kathryn S. Jones, Joseph Kulkosky and Anna Marie Skalka

Fox Chase Cancer Center, Institute for Cancer Research
7701 Burholme Avenue
Philadelphia, PA 19111

INTRODUCTION

The single-stranded genomic RNA of retroviruses is reverse-transcribed to produce double-stranded linear DNA shortly after entry of the virion into the host cell. For HIV, this reaction has provided the first effective target for anti-viral therapy. Following reverse transcription, viral DNA is stably integrated into the host cell genome. The study of integration is important since this event appears to be obligatory for efficient replication of all retroviruses including HIV and, as with reverse transcription, this reaction has no obvious cellular counterpart. Thus, the integration reaction represents another attractive target for the development of specific inhibitors. Although the requirements for efficient integration of HIV DNA have not yet been thoroughly defined, general features can be inferred from studies with more simple retroviruses, such as the avian sarcoma/leukosis (ASLV) and murine leukemia (MLV) viruses. For these viruses, integration is dependent upon 1) cis-acting sequences at the ends of the linear viral DNA and 2) activities of the integration protein (IN) encoded in the 3' portion of the retroviral pol gene (Skalka, 1988; Varmus and Brown, 1989; Kulkosky and Skalka, 1990). Recent studies from our laboratory have demonstrated that the ASLV IN is, in fact, both necessary and sufficient to direct integrative recombination in vitro (Katz et al., 1990).

In this report we summarize our studies which characterize the components believed to participate in the integrative recombination reaction of HIV. These include 1) analysis of sequences within the recognition sites at the termini of HIV DNA and 2) expression and preliminary functional analyses of HIV IN in a eukaryotic expression system.

RESULTS

Terminal Nucleotides Deduced for the Preintegrated Linear Form of HIV DNA. Studies of integration in the avian and murine systems suggest specificity in the DNA sequences recognized by the integration machinery (Skalka, 1988; Varmus and Brown, 1989). The termini of integrated retroviral DNAs, including HIV, invariably are the dinucleotides 5' TG...||...CA 3'. For avian and murine retroviruses, these termini arise by loss of two nucleotides from the ends of both LTRs prior to or during the integration of linear viral DNA into host sequences. Inspection of the plus and minus strand DNA initiation sites for HIV (see Fig. 1) led to a prediction that only a single nucleotide might be lost from the ends of this viral DNA (Varmus and Brown, 1989). However, the actual number of the terminal bases which extend beyond the conserved TG/CA dinucleotides in unintegrated forms had not been determined.

Advances in Molecular Biology and Targeted Treatment for AIDS
Edited by A. Kumar, Plenum Press, New York, 1991

We utilized the double LTR circular form of unintegrated HIV DNA to molecularly clone and identify these terminal sequences in unintegrated viral DNA extracted from infected human T cells. Studies of simple retroviruses have shown that most of the double LTR circular DNA species formed during normal infection result from the simple ligation of the full-length, blunt-ended linear species. Therefore, the junction sequences at the site of joining in such molecules would include the terminal nucleotides that have not yet been subjected to removal by IN.

To clone individual LTR-LTR junctions, a population of DNA molecules containing this site was prepared by the polymerase chain reaction (PCR) using oligodeoxynucleotide primers complementary to sequences near each side of the joined LTRs of the HIV DNA templates (see Fig. 1). Each primer was designed to incorporate an EcoRI site at the ends of the amplified products to facilitate insertion into the pUC19 cloning vector. Fourteen individual joined LTR junction fragments were cloned in this manner and their sequences determined. A number of individual clones were sequenced since studies with murine and avian retroviruses had previously shown that variation exists at the site of LTR joining.

As shown in Table 1, a majority of the cloned segments contained a common set of four nucleotides, [U5] GT-AC [U3], between the highly conserved CA/TG dinucleotides (shown in bold) that delineate the termini of integrated DNA. These data suggest that unintegrated linear HIV DNA, synthesized in infected T-cells, contains two terminal nucleotides beyond the conserved CA/TG. Models which may account for the incorporation of two base pairs at both the U5 and U3 ends of HIV DNA during reverse transcription, as well as the origins of the sequences and deletions in the other LTR-LTR junction clones, are discussed in detail elsewhere (Kulkosky et al., 1990).

Expression and Functional Analysis of HIV IN. By analogy with other retroviruses, HIV IN is believed to be sequestered in the infecting subviral particle during the early stages of infection. We wanted to devise a system that would allow us to study HIV IN which is expressed as an independent protein in eukaryotic cells. In previous studies in our laboratory, a eukaryotic expression vector containing an SV40 origin of replication was used to transiently express high levels of RSV IN in COS cells in the absence of other viral proteins (Morris-Vasios et al., 1988). Our initial attempts to express the IN protein of HIV involved the construction of an analogous plasmid (pBC12BI/HIN), illustrated in Fig. 2A. When COS cells were transfected with this plasmid, we were unable to detect expression of HIV IN. This was surprising, since the HIV IN coding region is in precisely the same position as that of RSV IN in the plasmid which expressed readily detectable levels of the RSV protein. Since the only difference between these two plasmids is the IN coding region itself, we reasoned that the HIV IN coding region may contain sequences that interfere with expression. Although such sequences have not yet been reported in this region, a number of recent studies have shown that both the gag and the env genes of HIV do contain such sequences, referred to as cis-acting repressive sequences, or CRS (Dayton et al., 1988; Emerman et al., 1989; Hadzopoulou-Cladaras et al., 1989; Hammarskjöld et al., 1989; Rosen et al., 1988). When CRS are present within a transcript, eukaryotic expression is dramatically reduced. These studies also demonstrated that it is possible to override the repressive effect of these sequences by providing both the HIV protein rev in trans, and the rev-responsive element (RRE), in cis (Sodroski et al., 1988; Malim et al., 1989; Rosen et al., Hammarskjöld et al., 1989; Dayton et al., 1988).

To determine whether we could express HIV IN using RRE sequences, we placed a portion of the HIV-1 genome that contains this element into pBC12BI/HIN. A schematic diagram of the resultant plasmid, pBC12BI/HIN+RRE, is shown in Fig. 2A. The RRE sequences are 3' to the coding region of IN, but within the RNA transcript in the sense orientation. HIV IN was expressed in cells co-transfected with pBC12BI/HIN+RRE and a rev-expressing vector, pSVrev (Cochrane et al., 1990). An immunoprecipitation of metabolically labelled COS cells lysed 48 hours after transfection is shown in Fig. 2B. Expression of detectable levels of HIV IN by the pBC12BI/HIN+RRE plasmid requires the presence of rev (Fig. 2B; compare lanes 1 and 2). No expression of HIV IN was seen in cells transfected with either the plasmid lacking the RRE sequences (lane 3) or the rev-expressing vector alone (lane 4). These requirements strongly suggest that CRS, which have previously been reported to be present in the HIV in the gag and env genes, are present in the IN portion of pol as well.

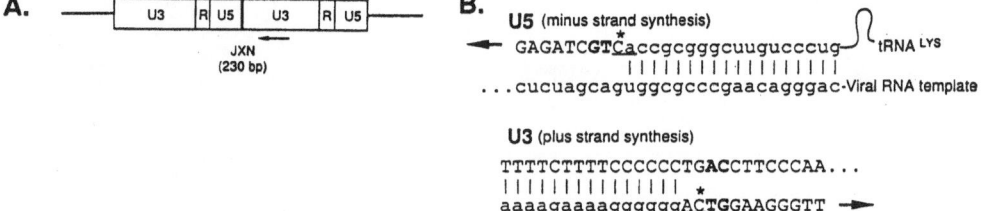

A.

| U3 | R | U5 | U3 | R | U5 |

JXN
(230 bp)

B.

U5 (minus strand synthesis)

GAGATCG**TC**accgcgggcuuguccug⌇ tRNA LYS
||||||||||||||||||
...cucuagcaguggcgcccgaacagggac-Viral RNA template

U3 (plus strand synthesis)

TTTTCTTTTCCCCCCTG**ACCTTCCCAA**...
|||||||||||||||||
aaaagaaaaggggggA̲C̲TGGAAGGGTT ➡

Fig. 1. PCR amplification strategy and sites for HIV DNA synthesis

A. Schematic representation of HIV joined LTR DNA template. Arrows designate the approximate location of the PCR primers used to amplify 230 bp HIV LTR/LTR junction fragments. Heavy line indicates site of LTR joining. B. Sequence of initiation sites at the internal boundaries of the HIV LTRs for minus (U5) and plus (U3) strand viral DNA synthesis. The single deoxynucleotides with an asterisk indicate the sequence predicted to occur between the boundaries of the integrated provirus (GT or AC, bold) and the end of the sequence complementary to the tRNA[lys] primer (Top) or the end of the polypurine tract (Bottom). Data in Table 1 suggest that the underlined pairs of nucleotides are incorporated at each terminus of HIV DNA during reverse transcription. DNA sequence is represented by capital letters, and the RNA sequence by lower case letters.

Table 1. Compilation of HIV-1 LTR-LTR junction sequences.

# Independent Clones	U5	Junction Sequences	U3
5	CTCTAG**CA**	GT AC	TGGAAGGG**TT** *
4	CTCTAG**CA**	AC	TGGAAGGG**TT**
1	CTCTAG^A**CA**	AC	TGGAAGGG**TT**
1	CTCTAG**CA**	G AC	TGGAAGGG**TT**
1	CTCTAG**CA**	G	(T)
1	CTCTAG		TT
1	CTCTAG**CA**	GT AC TTGG	TGGAAGGG**TT**

The sequence for the plus strand (5′ to 3′) of the joined LTR junctions is listed. From left to right, the sequences represent nucleotides derived from U5 to the highly conserved CA dinucleotide (bold); nucleotides at the junction; nucleotides derived from U3 beginning with the conserved TG. The asterisk denotes variance from the published sequence.

23

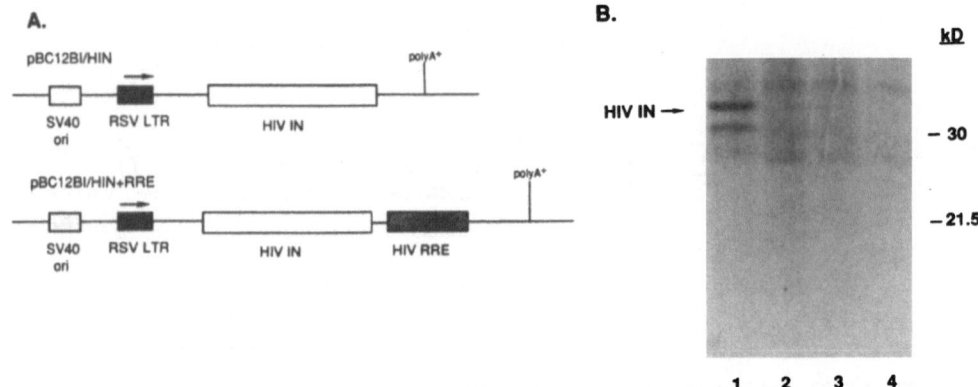

Fig. 2. Expression of HIV IN. **A.** Schematic diagrams of HIV IN-encoding plasmids. For details on the derivation and construction of these plasmids, see Jones and Skalka. Abbreviations and symbols: SV40 ori, SV-40 origin of replication; RSV LTR, long terminal repeat of RSV (used here as the promoter); HIV IN, coding region of that protein; HIV RRE, region which includes the rev-responsive element of HIV-1; poly A[+], polyadenylation site. **B.** Immunoprecipitation of HIV IN from transfected cells. COS cell cultures were transfected as described previously (Morris-Vasios et al., 1988) with 3 μg of the expression plasmid. Forty-eight hours after transfection, the cells were metabolically labelled with ^{35}S-methionine, lysed, and clarified by centrifugation. Samples of each cleared lysate were incubated with 10λ of rabbit polyclonal antisera directed against bacterially produced HIV IN at 4°C for 2 hours, and immune complexes were formed by incubation with protein A-agarose overnight at 4°C. After washing and dissociating the complex, the precipitated proteins were size fractionated by electrophoresis on a 15% SDS-polyacrylamide gel, fixed, treated with a fluorographic agent (Amplify), dried, and visualized by autoradiography. Lanes: 1, pBC12BI/HIN+RRE cotransfected with 5 μg/plate of pSVrev; 2, pBC12BI/HIN+RRE transfected without pSVrev; 3, pBC12BI/HIN; 4, no expression plasmid, transfected with 5 μg/plate of pSVrev.

Subcellular Localization of HIV IN. We have previously observed that the IN protein of RSV IN, when independently expressed, localizes to the nucleus. We wanted to use our HIV IN expression system to determine whether this is a common property of IN proteins. Immunocytochemical analysis was performed on COS cells co-transfected with pBC12BI/HIN+RRE and pSVrev. The results show that HIV IN protein localizes to the nucleus of these cells (Fig. 3).

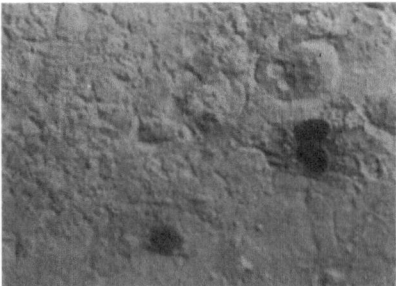

Fig. 3. Subcellular localization of HIV IN protein by immunocytochemistry. COS cells were plated onto glass slides coated with poly D-lysine at a concentration of 0.1 mg/ml and transfected with pBC12BI/HIN+RRE and pSVrev to express HIV IN. Sixty hours later, the cells were fixed with paraformaldehyde and visualized using the Vectastain Elite kit (Vector) according to the manufacturer's specifications. The primary antiserum used at a dilution of 1:1500 was produced in rabbits injected with purified, denatured, bacterially produced HIV-1 IN (gift of S. LeGrice).

DISCUSSION

Our results provide insight into the nature of the components required for HIV integration. We have deduced that sequences at the termini of linear HIV DNA, like those of avian and murine retroviruses, extend 2 base pairs beyond the conserved TG/CA dinucleotides. Such an extension in all retroviral DNAs studied thus far raises the possibility that this structure may be optimal for integration. These data establish the probable end of the plus strand oligoribonucleotide primer as shown in Figure 1. This prediction has recently been verified independently by Huber and Richardson (1990). In addition, we predict that the mechanism which leads to a 2 nucleotide extension during reverse transcription of the HIV minus strand is not equivalent to that of either the avian or murine systems. For HIV we proposed either use of a truncated tRNA^{lys} primer or retention of the terminal riboA of tRNA^{lys} at this site, perhaps due to failure of removal by HIV RT RNAse H (Kulkosky et al., 1990).

While constructing a vector system to study eukaryotic expression of HIV IN, we discovered that the IN coding region contains cis-acting repression sequences. It seems likely that the requirement of rev for expression in this system reflects a similar requirement in vivo: in rev⁻ HIV mutants, IN, along with the other structural proteins, are not expressed at detectable levels (Terwilliger et al., 1988). Since the IN coding region is located on the same RNA transcript and the same polypeptide as the gag proteins, the CRS located in both these regions presumably affect the expression of this polypeptide.

We also observed that, like RSV IN, independently expressed HIV IN localizes to the nucleus. The presence of these IN proteins in the nucleus could either be a manifestation of the DNA binding capability of this protein or may be due to the presence of karyophillic signals. These observations have led us to speculate that, in addition to functioning as an integrase and endonuclease, IN may play a direct role in the transport of retroviral DNA to a nuclear location for integration into the host DNA. Experiments are currently underway to determine whether this is indeed the case. If so, this would be another step in the life cycle of HIV which could be a future target of anti-viral therapeutics.

ACKNOWLEDGEMENTS

This work was supported by National Institutes of Health grants CA-49042, CA-06927, RR-05539, the W. W. Smith Charitable Trust, a grant from the Pew Charitable Trust and an appropriation from the Commonwealth of Pennsylvania.

REFERENCES

Cochrane, A. W., Perkins, A., and Rosen, C., 1990, Identification of sequences important in the nucleolar localization of human immunodeficiency virus rev: Relevance of nucleolar localization to function, J. Virol., 64:881.

Dayton, A. I., Terwillinger, E. F., Potz, J., Kowalski, M., Sodroski, J. G., and Haseltine, W. A., 1988, Cis-acting sequences responsive to the rev gene product of the human immunodeficiency virus, JAIDS 1:441.

Emerman, M., Vazeux, R., and Peden, K., 1989, The rev gene product of human immunodeficiency virus affects envelope-specific RNA localization, Cell 57:1155.

Hadzopoulou-Cladaras, M., Felber, B. K., Cladaras, C., Athanassopoulos, A., Tse, A., and Pavlakis, G. N., 1989, The rev (trs/art) protein of human immunodeficiency virus type I affects viral mRNA and protein expression via a cis-acting sequence in the env regions, J. Virol. 63:1265.

Huber, H. E., and Richardson, C. C., Processing of the primer for plus strand DNA synthesis by human immunodeficiency virus 1 reverse transcriptase, J. Biol. Chem. 265:10565.

Jones, K.S., and Skalka, A.M., HIV integration protein: Independent expression and subcellular location in eukaryotic cells. Submitted for publication.

Hammarskjöld, M.-L., Heimer, J., Hammarskjöld, B., Sangwan, I., Albert, L., and Rekosh, D., 1989, Regulation of human immunodeficiency virus env expression by the rev gene product, J. Virol. 63:1959.

Katz, R. A., Merkel, G., Kulkosky, J., Leis, J. and Skalka, A. M., 1990, The avian retroviral IN protein is both necessary and sufficient for integrative recombination in vitro, Cell, in press.

Kulkosky, J., Katz, R., and Skalka, A. M., 1990, Terminal nucleotides of the preintegrative linear form of HIV-1 DNA deduced from the sequence of circular DNA junctions, JAIDS, 3:852.

Kulkosky, J., and Skalka, A. M., 1990, HIV DNA integration: Observations and inferences, JAIDS, 3:839.

Malim, M. H., Hauber, J., Le, S.-Y., Maizel, J. V., and Cullen, B. R., 1989, The HIV-1 rev trans-activator acts through a structured target sequence to activate nuclear export of unspliced viral mRNA, Nature 338:254.

Morris-Vasios, C., Kochan, J. P., and Skalka, A. M., 1988, Avian sarcoma-leukosis virus pol-endo proteins expressed independently in mammalian cells accumulate in the nucleus but can be directed to other cellular components, J. Virol. 62:349.

Rosen, C. A., Terwilliger, E., Dayton, A., Sodroski, J. G., and Haseltine, W. A., 1988, Intragenic cis-acting art gene-responsive sequences of the human immunodeficiency virus, Proc. Natl. Acad. Sci. USA 85:2071.

Skalka, A. M., 1988, Integrative recombination of retroviral DNA, in: "Genetic Recombination," R. Kuperlapati, and G. R. Smith, eds., American Society for Microbiology, Washington, DC.

Sodroski, J., Goh, W. C., Rosen, C., Dayton, A., Terwilliger, E., and Haseltine, W., 1986, A second post-transcriptional transactivator gene required for HTLV-III replication, Nature (London) 321:412.

Terwilliger, E., Burghoff, R., Sia, R., Sodroski, J., Haseltine, W., and Rosen, C., 1988. The rev gene product of human immunodeficiency virus is required for replication, J. Virol. 62:655.

Varmus, H., and Brown, P., 1989, Retroviruses, in: "Mobile DNA," D. Berg, and M. Howe, eds., American Society for Microbiology, Washington, DC.

DNA CLEAVING ACTIVITY OF PURIFIED HUMAN IMMUNODEFICIENCY VIRUS INTEGRATION PROTEIN

Paula A. Sherman and James A. Fyfe

Wellcome Research Laboratories
Research Triangle Park, N.C. 27709

INTRODUCTION

An essential step in the life cycle of retroviruses is insertion of a double-stranded DNA copy of the viral RNA genome into the host cell DNA, to form the provirus. The insertion event depends on at least one viral protein, the integration protein (IN), which is a product of the viral *pol* gene. Mutations in the IN coding region of *pol* result in integration-negative retroviruses that will no longer replicate.[1-3] The proviral DNA is identical to the precursor viral DNA except for the loss of two base pairs at each end, at the points of attachment to cellular DNA. One proposed function for IN is the removal of these two bases from the 3'-termini of both strands of the viral DNA, in preparation for integration.

We have undertaken an investigation of the IN from the human immunodeficiency virus (HIV). HIV IN was expressed at a high level in *Escherichia coli*, and purified to near homogeneity using selective extraction followed by butyl-sepharose and heparin-sepharose chromatography.[4] The purified protein was free of detectable contaminating endonucleases of host bacterial origin. Here we describe the DNA cleaving activity of purified HIV IN.

PURIFIED HIV IN CLEAVES HETEROLOGOUS PLASMID DNA

The IN from avian sarcoma-leukosis viruses (ASLV) has been investigated in detail. This protein has been observed to convert supercoiled (form I) heterologous plasmid DNA to nicked (form II) and linear (form III) DNA[5,6]; to date, it is the only retroviral IN for which such a DNA cleaving activity has been demonstrated. We assessed the ability of our purified HIV IN to cleave supercoiled plasmid pBR322, and compared its DNA cleaving activity with that of avian IN. As shown in Fig. 1, the plasmid-cleaving activities of the two proteins were very similar. In the presence of Mn^{2+}, both proteins converted supercoiled plasmid to nicked and linear forms. Under these conditions the two proteins exhibited similar specific activities. The preference for Mn^{2+} over Mg^{2+} as the metal cofactor was a distinguishing feature of both proteins; this property was not observed for endonucleases of host bacterial origin.

Advances in Molecular Biology and Targeted Treatment for AIDS
Edited by A. Kumar, Plenum Press, New York, 1991

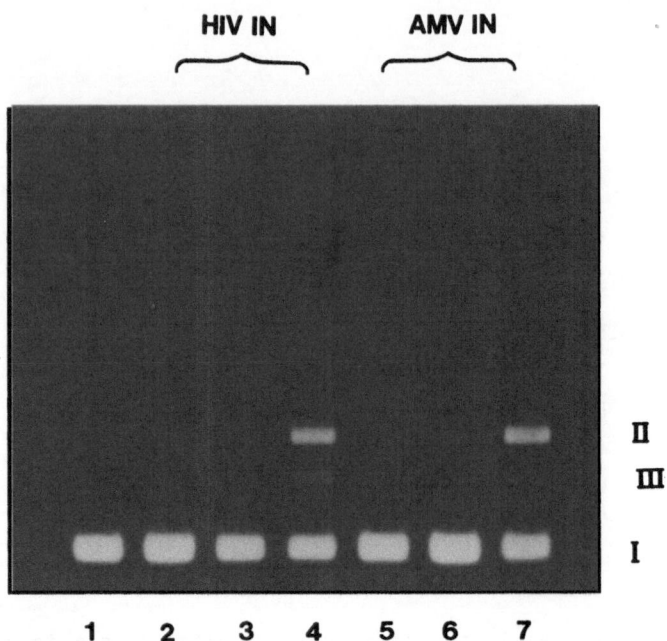

Fig. 1. Comparison of the endonucleolytic activities of HIV and avian myeloblastosis virus (AMV) IN's with a heterologous plasmid DNA substrate. Reaction mixtures (10 μl) contained 0.2 μg of supercoiled pBR322, 20 mM Tris-HCl (pH 8), 5 mM 2-mercaptoethanol, and 0.2 μg of protein. Incubations were for 30 min at 37°C. Reaction products were electrophoresed on a 1% agarose gel. Lane 1, no protein; lanes 2-4, HIV IN; lanes 5-7, AMV IN; lanes 2 and 5, no added metal; lanes 3 and 6, 2 mM MgCl2; lanes 4 and 7, 2 mM MnCl2. I, supercoiled; II, nicked; III, linear.

HIV IN SELECTIVELY CLEAVES SYNTHETIC OLIGONUCLEOTIDES THAT MIMIC VIRAL DNA

Retroviral IN's may be responsible for the removal of two bases from each of the 3'-termini of the linear viral DNA, in preparation for insertion into cellular DNA. We examined the ability of purified HIV IN to cleave synthetic double-stranded oligonucleotides which mimic the U3 and U5 termini of HIV DNA. As shown in Fig. 2, HIV IN selectively cleaved both of these substrates. Two nucleotides were removed from the 3'-ends of the U5 plus strand and the U3 minus strand; in both cases, cleavage was adjacent to a conserved CA. Reaction selectivity was further demonstrated by the lack of cleavage of an HIV U5 substrate on the complementary (minus) strand, an analogous substrate which mimics the U3 terminus of an avian retrovirus, and an HIV U5 substrate in which the conserved CA was replaced by a TA.

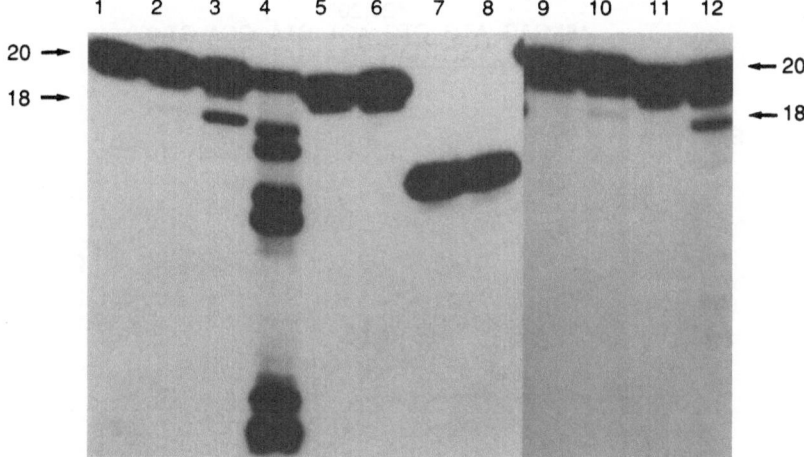

A

HIV U5

(T) ↓
*5' - TGT GGA AAA TCT CTA GCA GT - 3'
3' - ACA CCT TTT AGA GAT CGT CA - 5'

ASLV U3

5' - AAT GTA GTC TTA TGC - 3'
3' - TTA CAT CAG AAT ACG - 5'*
↑

HIV U3

5' - ACT GGA AGG GCT AAT TCA CT - 3'
3' - TGA CCT TCC CGA TTA AGT GA - 5'*
↑

B

Fig. 2. Selective oligonucleotide cleaving activity of HIV IN. Reaction
mixtures (10 μl) contained approximately 1 pmol of DNA (5' end-
labeled on the appropriate strand), 20 mM Tris-HCl (pH 8), 5 mM
2-mercaptoethanol, 1 mM $MgCl_2$ (contributed from the end-labeling
reaction, 1 mM $MnCl_2$ as indicated, and 0.2 μg of protein.
Incubations were for 1 hr at 37°C. (*A*) Sequences of double-
stranded synthetic oligonucleotide substrates. Expected
cleavage sites (adjacent to the conserved CA dinucleotide) are
indicated by arrows; labeled strands are indicated by asterisks.
The C to T change in the conserved CA is indicated above the
sequence for the HIV U5 substrate. (*B*) Analysis of reaction
products on 20% denaturing polyacrylamide gels. Lanes 1-4,
HIV U5 with plus strand end-labeled; lanes 5 and 6, same except
minus strand labeled; lanes 7 and 8, ASLV U3 with minus strand
labeled; lanes 11 and 12, HIV U3 with minus strand labeled; lane
4, G+A Maxam-Gilbert sequencing reaction; lanes 9 and 10, same
as 1-4 except with conserved C changed to T, lanes 1, 5, 7, 9,
and 11, no protein; lane 2, 1 mM $MgCl_2$; lanes 3, 6, 8, 10, and
12, 1 mM $MgCl_2$ plus 1 mM $MnCl_2$.

A

20:20
```
*5'-TGT GGA AAA TCT  CTA GCA GT-3'
 3'-ACA CCT TTT AGA  GAT CGT CA-5'
```

15:15
```
*5'-AAA ATC  TCT AGC AGT-3'
 3'-TTT TAG  AGA TCG TCA-5'
```

12:12
```
*5'-ATC TCT  AGC AGT-3'
 3'-TAG AGA  TCG TCA-5'
```

10:10
```
*5'-CTC TAG CAG T-3'
 3'-GAG ATC GTC A-5'
```

"20:20"
```
*5'-GAG ATG CTC ACT CTA GCA GT-3'
 3'-CTC TAC GAG TGA GAT CGT CA-5'
```

B

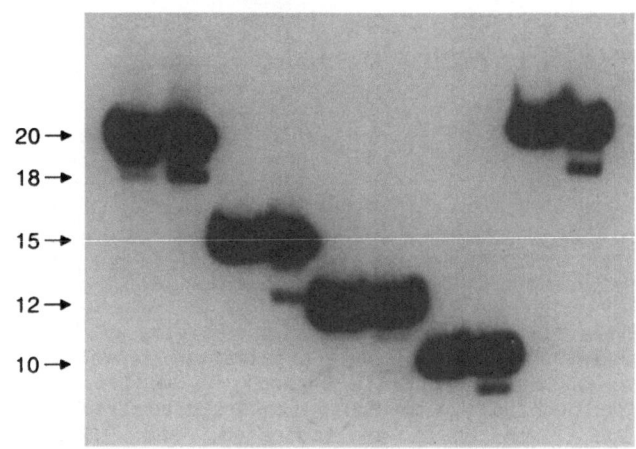

Fig. 3.   Effect of substrate length on HIV U5 cleavage.   Reactions were
as described in the legend to Fig. 2, except with incubations
at 25°C.   All reaction mixtures contained 1 mM MgCl$_2$ plus 1 mM
MnCl$_2$.   (A)   Sequences of double-stranded synthetic oligonucleotide substrates.   Labeled strands are indicated by
asterisks.   "20:20" is the substrate consisting of 10 base
pairs of random sequence (underlined) linked to 10 base pairs
of wild-type HIV sequence.   (B)   Analysis of reaction products
on 20% denaturing polyacrylamide gels.   The first lane in each
set of two contained a reaction mixture with no protein; the
second lane contained HIV IN.   Lanes 1 and 2, 20:20; lanes 3
and 4, 15:15; lanes 5 and 6, 12:12; lanes 7 and 8, 10:10; lanes
9 and 10, "20:20."

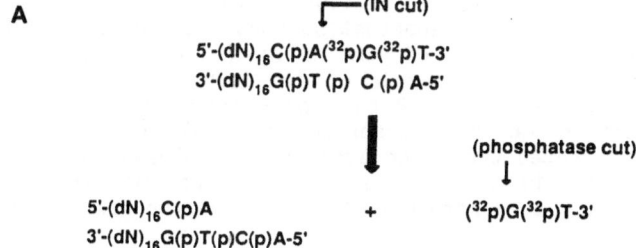

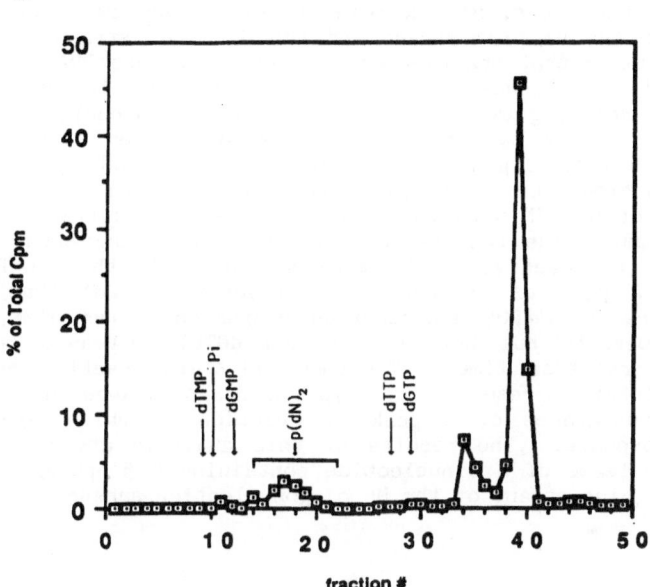

Fig. 4. Analysis of product formation with anion exchange chromatography. Substrate for the reaction was prepared by filling in a synthetic duplex 18:20-mer (3'-terminal GT missing in the plus strand) with Klenow enzyme and α-phosphate labeled dNTPs. The reaction with IN was as described in Fig. 2. (A) The labeled oligonucleotide substrate and expected cleavages with IN and phosphatase. (B) Radiochromatogram of the reaction mixture. The column (Mono Q, Pharmacia) was equilibrated with 20 mM Tris-HCl (pH 7.5) and eluted with a gradient of equilibration buffer and 50 mM Tris-HCl (pH 7.5)/1 M NaCl/6 M urea (0-12% from fraction 3-26 and 12-60% from fraction 26-45). Recovery of total counts chromatographed was 90%. Elution positions of standards (Pharmacia) are indicated with arrows.

31

LENGTH REQUIREMENT FOR U5 OLIGONUCLEOTIDE CLEAVAGE

The DNA sequence requirements for the IN-mediated terminal cleavage reaction were further investigated with a series of synthetic oligonucleotides of decreasing size which mimic the U5 terminus of HIV DNA. Reactions were performed at 25°C, which is below the calculated "melting temperature" of the shortest substrate tested ($T_m\sim30°C$). As shown in Fig. 3, the efficiency of the oligonucleotide cleaving reaction decreased with decreasing length of the substrate. A low level of cleavage was observed with a 12 base pair substrate. Cleavage of a 10 base pair substrate was also observed, but only one base was removed, instead of the expected two. The addition of 10 base pairs of apparently random sequence to 10 base pairs of wild-type HIV sequence partially restored activity, with a return to the original two base cleavage.

PRODUCTS OF THE HIV IN-MEDIATED OLIGONUCLEOTIDE CLEAVING REACTION

As described above, HIV IN selectively removes two nucleotides from the 3'-end of the U5 plus strand and the U3 minus strand. It has also been demonstrated[4] that the cleavage reaction produces DNA fragments with 3'-OH termini. The question arises as to whether IN functions as an endonuclease, removing two bases in the form of a dinucleotide, or as an exonuclease, removing two bases in a stepwise fashion. To address this question, an HIV U5 oligonucleotide substrate with radiolabel at the two 3'-terminal nucleotides of the plus strand was prepared. The products of the reaction of HIV IN with this substrate were fractionated on an anion exchange column. The results of this experiment are shown in Fig. 4. Very little radioactivity chromatographed with the dGMP or dTMP standards. A peak of radioactivity (fractions 16-20) that comigrated with a mixture of 5'-PO4 dinucleotide standards (after dTMP, dGMP, and inorganic phosphate but before dTTP and dGTP) increased in a linear fashion with reaction time. Treatment of this reaction product with alkaline phosphatase resulted in complete disappearance of the peak and concomitant appearance of a peak at fraction 10 which co-eluted with inorganic phosphate. The results of this analysis are consistent with IN-mediated release of a dinucleotide containing a 5'-phosphate from the 3'-end of the plus strand of the U5 oligonucleotide substrate.

SUMMARY

Purified HIV IN converts heterologous supercoiled plasmid DNA to nicked and linear forms. The protein also carries out a reaction analogous to one of the expected partial reactions in the integrative recombination event, that is, the removal of specific sequences at the termini of the linear viral DNA molecule in preparation for joining to the host cell DNA. When a 20 base pair synthetic oligonucleotide which mimics the U5 terminus of HIV DNA is the substrate, the products of the reaction are an 18:20-mer with a 3'-OH on the plus strand, and a dinucleotide containing a 5'-PO4. Ten base pairs of HIV U5 viral DNA sequence ("20:20" substrate) appear to be sufficient for selective cleaving by HIV IN.

REFERENCES

1. P. Schwartzberg, J. Colicelli, and S. P. Goff, Construction and analysis of deletion mutations in the *pol* gene of Maloney murine leukemia virus: a new viral function required for productive infection, Cell 37:1043 (1984).

2. L. Donehower and H. E. Varmus, A mutant murine leukemia virus with a single missense codon in *pol* is defective in a function affecting integration, Proc. Natl. Acad. Sci. USA 81:6461 (1984).

3. A. T. Panganiban and H. M. Temin, The retrovirus *pol* gene encodes a product required for DNA integration: identification of a retrovirus *int* locus, Proc. Natl. Acad. Sci. USA 81:7885 (1984).

4. P. A. Sherman and J. A. Fyfe, Human immunodeficiency virus integration protein expressed in *Escherichia coli* possesses selective DNA cleaving activity, Proc. Natl. Acad. Sci. USA 87:5119 (1990).

5. D. P. Grandgenett, A. C. Vora, and R. D. Schiff, A 32,00-dalton nucleic acid-binding protein from avian retrovirus cores possesses DNA endonuclease activity, Virology 89:119 (1978).

6. J. Leis, G. Duyk, S. Johnson, M. Longiaru, and A. Skalka, Mechanism of action of the endonuclease associated with the αβ and ββ forms of avian RNA tumor virus reverse transcriptase, J. Virol. 45:727 (1983).

A THERMODYNAMIC ANALYSIS OF THE BINDING OF NUCLEIC ACID TO HIV-1 REVERSE TRANSCRIPTASE

George R. Painter†, Lois L. Wright†, C. Webster Andrews†, Nancy Cheng*, Sam Hopkins* and Phillip A. Furman*

From the Divisions of Organic Chemistry† and Virology*
Burroughs Wellcome Co.
Research Triangle Park, NC   27709

INTRODUCTION

Human immunodeficiency virus type 1 (HIV-1) encodes a Mg^{+2}-dependent reverse transcriptase (E.C.2.7.7.49) that synthesizes a double-stranded DNA copy of genomic RNA. The enzyme purified from virions has been shown to consist of two polypeptides of molecular weights 66,000 and 51,000 (Hoffman et al., 1985; Di Marzo Veronese et al., 1986). The 66 kD subunit has both polymerase and RNase H activities. Like other retroviral reverse transcriptases, the associated RNase H activity is located on the carboxy terminal portion of the polypeptide (Johnson et al., 1986; Hansen et al., 1987; Tisdale et al., 1988). The 51 kD subunit is derived from the 66 kD polypeptide by cleavage at a protease sensitive site on the linker between the polymerase and RNase H domains (Lowe et al., 1988). Catalytically active HIV-1 reverse transcriptase has been cloned and expressed in E. coli (Larder et al., 1987). The recombinant, heterodimeric enzyme is kinetically indistinguishable from the native enzyme purified from virus.

The bireactant-biproduct mechanism proposed for the synthesis of DNA by HIV-1 reverse transcriptase is similar to the mechanism proposed for several well-characterized DNA polymerases (Majumdar et al., 1988). In this mechanism, binding of the reactants is ordered with the template-primer binding first, followed by the binding of the 2'-deoxynucleoside-5'-triphosphate (dNTP) that has Watson-Crick complementarity to the next appropriate template residue. It is proposed that the resulting ternary complex undergoes a conformational change to form the catalytically competent complex. Incorporation of the bound nucleotide into the nascent chain then occurs via an S_N2 displacement of pyrophosphate by the 3' hydroxyl at the primer terminus (Hopkins et al., 1989).

Very little information presently exists on the structural requirements for reactant binding, and no information exists on the topography of the binding sites on the heterodimeric enzyme. As a first step in establishing structure-function relationships for HIV-1 reverse transcriptase, we have initiated studies to determine the site specificity of and molecular forces responsible for initial binding of homopolymeric template-primers. We report the use of endogenous tryptophan fluorescence to quantify the interaction of the enzyme with various nucleic acids and to determine the role of hydrophobic and electrostatic forces in the stabilization of the binary complex. Efforts to create a three-dimensional model of the nucleic acid binding cleft based on the crystal structure analysis of the Klenow fragment of *E. coli* DNA polymerase I (Steitz *et al.*, 1983) are also outlined. Any insight into the forces and structural moieties involved in the binding of genomic RNA will conceivably aid in the design of chemotherapeutic agents whose mode of action is to intercede in or modify the binding or utilization of the natural template-primer by the enzyme.

USE OF ENDOGENOUS TRYPTOPHAN FLUORESCENCE TO ANALYZE THE INTERACTION OF NUCLEIC ACID WITH HIV-1 REVERSE TRANSCRIPTASE

Within the heterodimeric enzyme there is a total of 36 tryptophan residues, 17 in each of the two polymerase domains and two in the RNase H domain. Tryptophan has been used extensively as a fluorescent probe to monitor protein conformation in solution, with the assumption that the fluorescence quantum yield of the residue in a protein depends on its local environment (Lakowicz, 1983). In addition, the fluorescence of aromatic amino acids has been widely used to probe protein-DNA interactions (Lakowicz, 1983; Helene, 1971).

The fluorescence emission spectrum of a 5 nM solution of the enzyme in 50 mM TRIS, 5 mM MgCl$_2$, pH 7.6 at 25° is shown in Figure 1. The emission maximum at 338 nm is blue shifted relative to that of free L-tryptophan under the same conditions (Figure 1, curve b), suggesting that the average microenvironment of the indole sidechains is shielded from the aqueous medium and somewhat hydrophobic (Eftink and Ghiron, 1976). Denaturation of the protein by exposure to 8 M urea results in the λ_{max} red shifting to 350 nm (Figure 1, curve c), presumably due to loss of tertiary structure and a concomitant increase in the solvent accessible surface of the indole fluorophores. The shift in the λ_{max} of tryptophan to higher wavelengths upon exposure to polar solvents has been attributed to specific complex formation between the polar solvent and the indole fluorophore (Walker *et al.*, 1966; Hershberger *et al.*, 1981).

The association of either single- or double-stranded nucleic acid with the enzyme results in a decrease in the intensity of the fluorescence emission from tryptophan. Consequently, the extent to which the nucleic acid binds to the protein can be determined by monitoring the fluorescence emission of a fixed concentration of the HIV-1 reverse transcriptase and titrating with nucleic acid. The binding reaction can be described by equation 1.

$$K_d = \frac{[RT]\,[N]}{[RT\text{-}N]} \qquad \{1\}$$

K_d is the apparent dissociation constant, [RT] is the concentration of free protein, [RT-N] is the concentration of bound protein, and [N] is the concentration of unbound nucleic acid. The proportion of nucleic acid bound protein as described by equation 1 is related to measured fluorescence intensity by equation 2

$$\Delta F/\Delta F_{max} = [RT\text{-}N]/[RT]_{tot} \qquad \{2\}$$

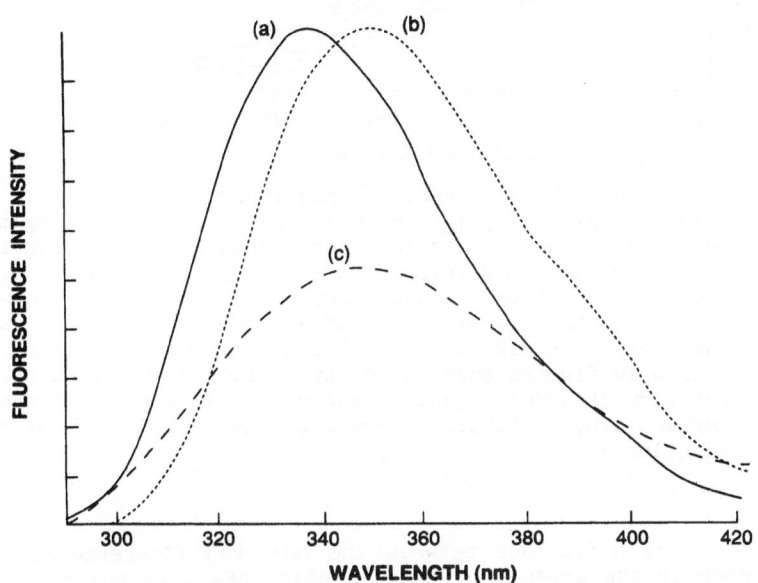

Fig. 1. The fluorescence emission spectra of (a) 5 nM HIV-1 reverse
transcriptase in 50 mM TRIS, 5 mM MgCl₂, pH 7.6 at 25°, (b)
an equivalent concentration of L-tryptophan under identical
conditions, and (c) the protein after exposure to an 8 M
solution of urea. Excitation was carried out at 280 nm. The
protein used in this study was derived from the recombinant
plasmid kindly provided by B. Larder, Wellcome Research
Laboratories, Beckenham, England. Induction with IPTG and
purification of the enzyme from *E.coli* was performed as
previously described (Tisdale *et al.*, 1988; Larder *et al.*, 1987).
The protein was greater than 98% pure, as judged by SDS-PAGE,
and consisted of equal molar proportions of p66 and p51.

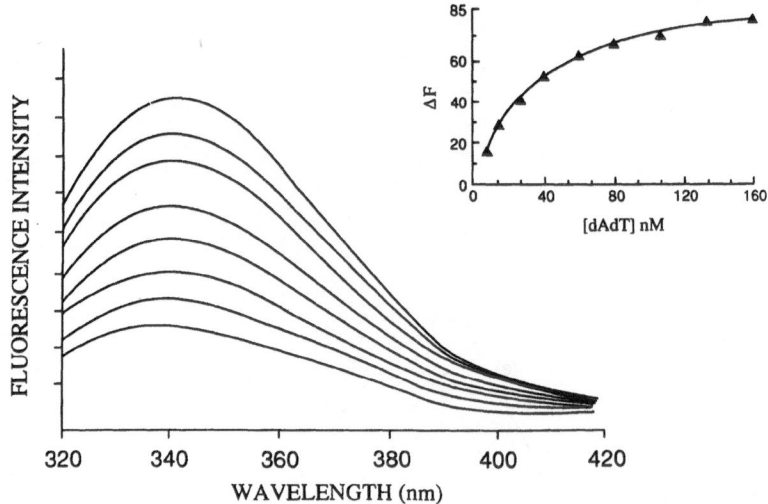

Fig. 2. Titration of HIV-1 reverse transcriptase with poly(dA)$_{500}$·p(dT)$_{12-18}$. 5 to 10 μL aliquots of a 1 mg/ml solution of the nucleic acid were added to a 5 nM solution of the enzyme in 50 mM Tris, 5 mM MgCl$_2$, pH 7.6 at 25°. Although the addition of the ligand produced a decrease in fluorescence intensity, the emission maximum and spectral bandwidth were unaffected. A saturation isotherm produced by plotting the change in fluorescence intensity monitored at 344 nm as a function of added ligand is shown in the upper right hand corner. The solid line represents the best fit of the data to equation 3.

where ΔF is the difference between the observed fluorescence and the fluorescence in the absence of nucleic acid, ΔF_{max} is the fluorescence intensity at infinite [N] and [RT]$_{tot}$ is the total protein concentration. If the total nucleic acid concentration, [N]$_{tot}$, is in large molar excess relative to [RT]$_{tot}$ then it can be assumed that [N] is approximately equal to [N]$_{tot}$. Equations 1 and 2 can then be combined to give equation 3.

$$\Delta F/\Delta F_{max} = [N]_{tot}/\{K_d + [N]_{tot}\} \qquad \{3\}$$

K_d values were determined from a nonlinear least-squares regression analysis of the titration data using equation 3. All fluorescence intensity measurements were corrected for inner filter effects arising from the addition of nucleic acid.

A typical titration is shown in Figure 2, in this case the titration of the reverse transcriptase with aliquots of poly(dA)$_{500}$·p(dT)$_{12-18}$. All of the binding studies, unless otherwise stated, were carried out in 50 mM TRIS, 5 mM MgCl$_2$, pH 7.6 and 25°. The saturation isotherm generated by plotting the change in fluorescence intensity, ΔF, as a function of added nucleic acid is shown in the upper right hand corner of Figure 2.

Apparent K_d values for binding to a series of double- and single-stranded nucleic acids are shown in Table 1. The K_d values calculated for the three commonly used homopolymeric template-primers vary by slightly less than one order of magnitude. The K_d values for poly$(rA)_{500} \cdot p(dT)_{12-18}$ and poly$(rC)_{500} \cdot p(dG)_{12-18}$ are similar to K_M values of 4.4 and 19 X 10^{-8} M, respectively, calculated in our laboratories (S. Hopkins, data not shown). This equivalency of K_d and K_M values is anticipated based on the bireactant-biproduct reaction mechanism proposed by Majumdar $et\,al.$ (1988), in which template-primer binding is described as being rapid and reversible. The K_d value of poly$(rA)_{810} \cdot p(dT)_{14}$ (Majumdar $et\,al.$, 1988) is comparable to the value calculated for the shorter homolog examined here, whereas the K_d value reported for poly$(rA)_{3000} \cdot p(dT)_{20}$ (Huber $et\,al.$, 1989) is 37-fold lower. Under the conditions utilized in this study, the enzyme will bind effectively to single-stranded DNA and RNA, as evidenced by the K_d values for poly$(rC)_{300}$ and the dT_n series. Such binding has been inferred from competition studies in which single-stranded nucleic acids have been observed to inhibit DNA synthesis on poly$(rA) \cdot$oligo(dT) template-primers (Huber $et\,al.$, 1989; Majumdar $et\,al.$, 1988). In addition, electron microscopy studies of HIV-1 reverse transcriptase have visualized the enzyme bound to single-stranded as well as double-stranded DNA (Thomas $et\,al.$, 1990). It should be noted that the K_d values in Table 1 are calculated using the concentration of nucleic acid molecules rather than the concentration of ligand binding sites. Since there is the possibility of multiple binding sites existing on both the single- and double-stranded nucleic acid molecules, the K_d values may be an overestimation of the affinity of the enzyme for the nucleic acid binding site. However, electron micrographs taken of the enzyme in the presence of single- and double-stranded DNA or DNA/RNA hybrids have shown the number of enzyme molecules bound per strand to be very low (Thomas $et\,al.$, 1990). This observation is in contrast to that made on $E.\,coli$ DNA polymerase I, which shows the enzyme to bind to nucleic acid with a high density (Griffith $et\,al.$, 1971; Kornberg, 1974).

Table 1. K_d values and Hill constants for the binding of single- and double-stranded nucleic acid by HIV-1 reverse transcriptase.

Nucleic Acid	K_d (M)	n
poly$(rA)_{500} . p(dT)_{12-18}$	1.1 ± 0.49 x 10^{-8}	0.98 ± 0.04
poly$(rC)_{500} \cdot p(dG)_{12-18}$	4.9 ± 0.17 x 10^{-8}	1.04 ± 0.08
poly$(dA)_{500} . p(dT)_{12-18}$	8.1 ± 0.05 x 10^{-8}	0.98 ± 0.03
11/20 mer[a]	4.7 ± 0.10 x 10^{-8}	1.00 ± 0.01
poly$(rA)_{12-18} . p(dT)_{10}$[b]	6.2 ± 0.16 x 10^{-7}	1.00 ± 0.03
poly$(rC)_{300}$	5.3 ± 0.21 x 10^{-8}	1.00 ± 0.06
dT_{25-30}	9.2 ± 0.37 x 10^{-7}	1.01 ± 0.08
dT_{16}	1.5 ± 0.08 x 10^{-6}	0.96 ± 0.05
dT_{10}	3.5 ± 0.11 x 10^{-6}	1.01 ± 0.09
dT_6	5.2 ± 0.21 x 10^{-6}	0.99 ± 0.06

Each value in the table represents the average of a minimum of three and a maximum of ten independent determinations.
[a]The sequence of the 11/20 mer is given in Cowart $et\,al.$ (1989).
[b]Annealing of p$(dT)_{10}$ to poly$(rA)_{12-18}$ could result in the formation of polymers of varying length.

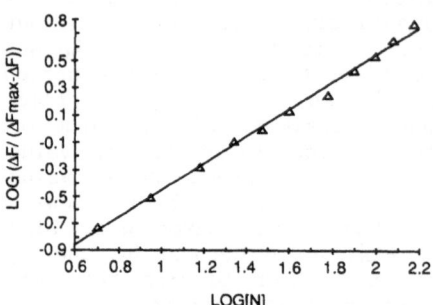
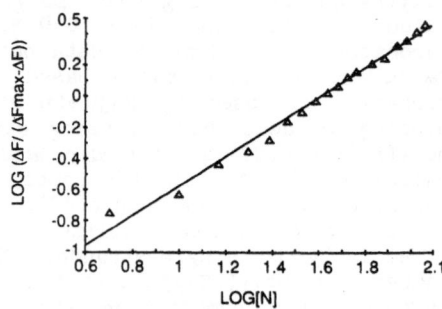

Fig. 3.   Hill plots of titration data of HIV-1 reverse transcriptase
with (a) poly(rA)$_{500}$·p(dT)$_{12-18}$ and (b) (dT)$_{10}$. The solid
lines represent a least-squares regression analysis of the
data according to the linearized form of the Hill equation

$$\log(\Delta F/(\Delta F_{max} - \Delta F)) = n\log[N] - \log K'$$

where n is the order of the binding reaction with respect to
ligand concentration (see Table 1) and K' is the concentration of nucleic acid that yields 50% of ΔF_{max} (K_d when n=1).

Hill coefficients calculated for all the ligands examined are
close to one (Table 1). There is no deviation from linearity in any of
the plots, indicating that each nucleic acid binds to a single site on
the enzyme. Representative plots are shown in Figure 3. Both the
66 kD and the 51 kD polypeptides possess those segments of the primary
sequence that make up the polymerase domain of the enzyme. Although
there are two potential nucleic acid binding sites on the heterodimer,
apparently only one is occupied. Crosslinking studies were carried out
with poly(rA)$_{12-18}$·p(dT)$_{10}$ to determine which subunit of the
heterodimer contains the nucleic acid binding site. The experimental
procedures used in the study are given in the caption to Figure 4. The
results of the study (Figure 4) show clearly that even at extremely
high concentrations of reactant, crosslinking was only observed to p66.

Analysis of the dependence of the K_d for protein-nucleic acid
interactions on the concentration of monovalent ions yields information
on the contribution of electrostatic contacts to the total binding
energy. According to the theoretical treatment done by Record and
colleagues (Record *et al.*, 1976; Lohman *et al.*, 1980), the fundamental
driving force for electrostatic association is the displacement of a
large number of cations such as Na+, K+, and Mg+2 from the nucleic acid
and perhaps the protein. The ions essentially form a screening
atmosphere around the phosphate groups on the anionic nucleic acid that
must be displaced by basic groups on the binding protein. The effect of
Na+ ions on the apparent association constant, K_{OBSD} (1/K_d), for
protein-nucleic acid interaction has been described according to the
equation

$$-\log K_{OBSD} = m'\Psi\log[Na^+] - \log K° \qquad \{4\}$$

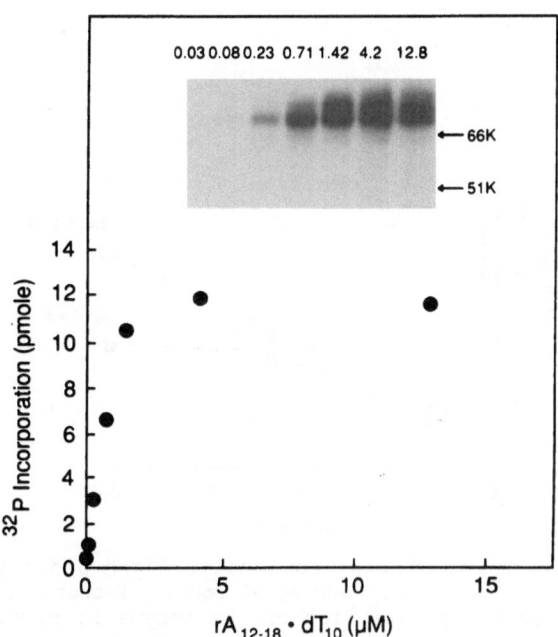

Fig. 4.   Crosslinking of HIV-1 reverse transcriptase to poly(rA)$_{12-}$
$_{18}$·p(dT)$_{10}$.   [^{32}P]-end-labeled poly(rA)$_{12-18}$.p(dT)$_{10}$ was
prepared by labeling the poly(rA)$_{12-18}$ strand using [^{32}P]ATP
and polynucleotide kinase. The labeled poly(rA)$_{12-18}$ strand
(0.1 A$_{260}$ units) was purified using a NAP5 column and
annealed to 0.16 A$_{260}$ units of (dT)$_{10}$. A standard 100 µl
crosslinking reaction contained 5 µg of heterodimeric HIV-1
reverse transcriptase, 50 mM Tris, 5 mM MgCl$_2$, 2 mM MnCl$_2$,
1 mM dithiothreitol at pH 7.6. The reaction mixture was
irradiated for 10 min with a 254 nm ultraviolet light. The
samples were then analyzed by SDS-PAGE and autoradiography
(insert). The saturation isotherm was generated by
increasing the concentration of template-primer from 26 nM to
12.8 µM and measuring the amount of ^{32}P incorporation. The
^{32}P incorporation was quantitated by cutting radiolabelled
bands from the polyacrylamide gels and Cherekov counting. A
K$_d$ value for the binary complex estimated from this data is
6.1 x 10^{-7} M. This value compares favorably to the value
calculated using the fluorescence titration technique
(Table 1).

where m' is the number of ion pairs formed in the association reaction, Ψ is the fraction of counterions bound in the thermodynamic sense to each phosphate on the nucleic acid ligand, and K° is the nonelectrostatic contribution to the association constant (the value of K_{OBSD} when the NaCl concentration is increased to 1M (Record *et al.*, 1976)). The slope of the linear log-log function (m'Ψ) is a measure of ion release in the protein-nucleic acid binding reaction.

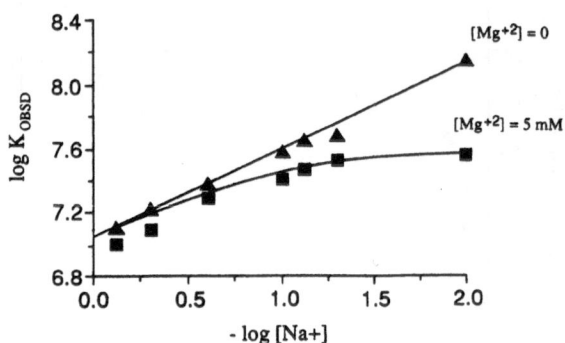

Fig. 5.   Dependence of K_{OBSD} for poly(rA)$_{500}$·p(dT)$_{12-18}$ on the concentration of Na$^+$ and Mg^{+2} ions. Titrations were carried out using a 5 nM solution of the enzyme in 50 mM Tris, pH 7.6 at 25°. Appropriate quantities of NaCl and MgCl$_2$ were added to the solution. The solid lines represent the best fit of the data to equation 4 ([Mg^{+2}] = 0) and equation 5 ([Mg^{+2}] = 5 mM) using nonlinear least-squares regression analysis.

The combined effect of Na$^+$ and Mg^{+2} ions on the thermodynamics of interaction between a charged ligand and nucleic acids has been described by Lohman *et al.* (1980) according to equation 5,

$$-\log K_{OBSD} = m'\Psi\log[Na^+] + m'\log([D]/[D_o]) - \log K^\circ \qquad \{5\}$$

where

$$[D]/[D_o] = 1/2(1 + (1 + 4K_{OBSD}^{Mg}[Mg^{+2}])^{1/2}), \qquad \{6\}$$

K_{OBSD}^{Mg} is the intrinsic association constant for the binding of Mg^{+2} to a nucleic acid site, [D] is the total nucleotide concentration, and [D$_o$] is the concentration of nucleotides having Ψ Na$^+$ ions associated with each phosphate.

The effect of increasing concentrations of NaCl on K_{OBSD} for the binding of poly(rA)$_{500}$·p(dT)$_{12-18}$ to HIV-1 reverse transcriptase in the absence and presence of 5 mM MgCl$_2$ is shown in Figure 5. The effect is minimal. K_{OBSD} changes by a factor of 10.1 when the NaCl concentration is increased from 10 mM to 750 mM. The slope (m'Ψ) calculated from the fit of equation 4 to the data gathered in the absence of Mg^{+2} is 0.62 and the intercept (log K°) is 7.08. For comparison, the values of m'Ψ and K° obtained for the nonspecific (electrostatically driven) binding of *lac* repressor to DNA are 10.0 and 1.5, respectively (deHaseth *et al.*, 1977). The addition of 5 mM MgCl$_2$ reduces K_{OBSD}. The effect is much more pronounced at low concentrations of NaCl. The explanations offered by Lohman *et al.* (1980) for the effect of Mg^{+2} on K_{OBSD} for ligand-nucleic acid interactions are (1) Mg^{+2} binds to the nucleic acid to a degree that is dependent upon the concentrations of both Mg^{+2} and Na$^+$; (2) Mg^{+2} competes with the cationic residues on the enzyme for nucleic acid phosphate groups and thereby reduces K_{OBSD}. The competition is much more effective at low concentrations of Na$^+$, where the binding density of Mg^{+2} is greater.

Discussion

The quenching of intrinsic, tryptophan fluorescence caused by the addition of nucleic acid to a solution of HIV-1 reverse transcriptase provides a straightforward method for characterizing the binding of nucleic acid to the free enzyme. The unusually large number of tryptophan residues in the enzyme allows binding assays to be performed with a high degree of sensitivity at extremely low protein concentrations. However, because the tryptophans are distributed rather uniformly throughout both domains of the 66 kD parent polypeptide and the polymerase domain of the 51 kD C-terminal fragment, interpretation of the quenching data in terms of spatial relationships is difficult. ·The approximately 50% decrease in fluorescence intensity observed upon saturation of the enzyme with nucleic acid can result from direct contact of the quenching agent with the indole side chain of tryptophan(s) at the nucleic acid binding site and/or a conformational change in the enzyme that results in a decrease in the fluorescence yield of tryptophan residues remote from the nucleic acid binding site. Regardless of the mechanism by which the fluorescence is quenched, the observation that the spectral band width and the emission maximum are unaffected by nucleic acid binding (see caption, Figure 1) indicates that the polarity of the average microenvironment experienced by the indole fluorophores remains unchanged.

The various double- and single-stranded nucleic acids examined in this study all appear to bind with a relatively high degree of affinity to a single binding site on the heterodimeric enzyme. It has been claimed that the reverse transcriptase shows a propensity to bind to a 3'-OH duplex site on a homopolymeric template-primer (Huber *et al.*, 1989). The degree to which this might be true cannot be addressed definitively from the results of this study. The number of potential protein binding sites, both single- and double-stranded, on each nucleic acid species would have to be accurately determined first. However, it is almost certain that the 11/20 mer used in this study (Table 1) contains only one, double-stranded binding site possessing a terminal 3'-OH group, while dT$_{16}$ contains only one single-stranded binding site (Table 1). Comparison of the K_d values for these two ligand binding sites indicates the enzyme prefers the duplex 3'-OH containing site by a factor of 30. Shortening of the single-stranded dT site results in a decrease in affinity (see Table 1). The problem of multiple binding sites on the large homopolymeric template-primers complicates direct comparison of their K_d values with the K_d values of nucleic acid species that possess a single protein binding site. This problem may be reflected in the higher affinity calculated for poly(rC)$_{300}$ than for dT$_{16}$. A curious observation made by Thomas *et al.* (1990) is that the enzyme, again in contrast to *E. coli* DNA polymerase I (Kornberg, 1974), shows no propensity to bind to either the 3' or the 5' terminus of single strands of DNA or RNA.

Crosslinking experiments indicate the single nucleic acid binding site to be on the 66 kD subunit of the heterodimer. Even at extremely high concentrations of poly(rA)$_{12-18}$·p(dT)$_{10}$, no crosslinking to p51 was observed (see Figure 4). Photoaffinity labeling experiments carried out in these laboratories (N. Cheng, data not shown) with 2'-deoxythymidine-5'-triphosphate gave similar results. Crosslinking is confined to p66 even at extremely high concentrations of the nucleotide. Thus it appears that the binding of both reactants in the polymerase reaction is confined to p66, and that this polypeptide is the enzymatically active portion of the heterodimer. This brings to question the role of p51. In terms of the catalytic efficiency with which the enzyme carries out the polymerase reaction, the p66/p51 heterodimer has been demonstrated to be more efficient than the p66/p66 and the p51/p51 homodimers (Lowe *et al.*, 1988). The p66/p51 heterodimer produced from the homodimer by chymotrypsin treatment has a twofold higher V_{max} than the homodimer, with no significant change in the K_m for dTTP. The p51 homodimer has been mimicked in a mutant (CTRT1; Tisdale *et al.*, 1988) and has been found to have greatly reduced reverse transcriptase activity. Lowe *et al.* (1988) suggested that p51 serves to allosterically regulate the activity of p66 and that optimal contact between the two polypeptide chains occurs after the loss of the C-terminal fragment of one of the subunits.

The major force driving association of nucleic acid with the enzyme is hydrophobic, as evidenced by the minimal effect of increasing concentrations of NaCl on the binding of poly(rA)$_{500}$·p(dT)$_{12-18}$ (Figure 5). Extrapolation of the data in Figure 5 to a NaCl concentration of 1 M, the standard state (K°) where electrostatic interactions are effectively eliminated (Manning, 1978; Record *et al.*, 1976), yields a K_d value of 8.5×10^{-8} M. The electrostatic contribution to the free energy of association of poly(rA)$_{500}$·p(dT)$_{12-18}$ with the enzyme in standard buffer can then be estimated by taking the difference in the ΔG of binding at zero NaCl ($\Delta G = -10.9$ kcal/mole) and the ΔG of binding at 1 M NaCl ($\Delta G = -9.7$ kcal/mole). The difference of 1.2 kcal/mole, which represents the stabilization of the binary complex resulting from charge-charge interaction, is only 11% of the total binding energy. In general, DNA binding can be divided functionally into two classes characterized by specific and non-specific binding. These different modes of binding are, in general, stabilized by different molecular forces. The nonspecific interaction of *lac* repressor with DNA, for example, is electrostatically driven (deHaseth *et al.*, 1977), whereas the interaction of the protein with a symmetric operator sequence is hydrophobically driven (Ha *et al.*, 1989). We felt that a limited analogy could be drawn between repressor proteins and HIV-1 reverse transcriptase insofar as the binding to a nonreactive site (either single-stranded or duplex) could be differentiated from binding to a synthesis initiation site by the forces that dictate the stability of the binary complex. However, salt-dependent binding studies carried out with poly(rC)$_{300}$ and with the 11/20 mer (Table 1), which contains a single 3'-OH duplex binding site, both gave results similar to those shown in Figure 5 (data not shown). In all three cases examined, the major force dictating the stability of the binary complex is hydrophobic.

UTILIZATION OF MOLECULAR MECHANICS AND MOLECULAR DYNAMICS TO BUILD A MODEL OF THE POLYMERASE DOMAIN OF HIV-1 REVERSE TRANSCRIPTASE

Attempts to establish a structure-function relationship for HIV-1 reverse transcriptase have been hindered, if not prohibited, by a dearth of structural information. To date, all attempts to obtain X-ray diffraction data at atomic resolution have been unsuccessful. Some investigators have suggested that it may be possible to obtain low-resolution three-dimensional models of proteins of unknown structure based on functional and primary sequence homology to proteins of known three-dimensional structure (Burnbaum *et al.*, 1990; Claessens *et al.*, 1989). The only protein having polymerase activity whose structure has been solved is the Klenow fragment of *E. coli* DNA polymerase I (Ollis *et al.*, 1985; Brick *et al.*, 1983). We therefore undertook to use the atomic coordinates of the polymerase domain of Klenow to create a hypothetical model of the polymerase domain of the viral enzyme. The immediate reason for the creation of such a model is that it will aid in the prediction of amino acid residues important in reactant recognition and binding, and thereby will serve as a guide in site-directed mutagenesis studies. Conversely, the site-directed mutagenesis studies will serve to refine the model. Burnbaum *et al.* (1990) emphasized that structural modeling of this type must be carried out cautiously, since the order and orientation of the elements of a structural motif can vary in subtle ways. However, the lack of a three-dimensional structure for the HIV-1 reverse transcriptase has severely hindered efforts to design inhibitors and/or chain terminating substrates with a higher specificity for the viral enzyme than for cellular polymerases. Given this critical need for insight into the structural motifs involved in substrate recognition and binding, we feel this approach is easily justified.

The first step in the modeling process is to align the primary sequence of the polymerase domain of HIV-1 reverse transcriptase with the primary sequence of the polymerase domain of the Klenow fragment. In addition to polymerase activity, the Klenow fragment possesses 3'-5' exonuclease activity. This activity has been ascribed to the amino terminus of the molecule and extends from residue 309 to 517 (*E. coli* DNA polymerase I numbering) (Ollis *et al.*, 1985). Since there is no functionally equivalent domain on the viral enzyme, this segment of the primary sequence was eliminated from the alignment process. The remaining 411 residue polypeptide extending from lysine 518 to histidine 928 was used. The polymerase domain of HIV-1 reverse transcriptase encompasses the 424 residue segment extending from proline 156 to the protease-sensitive cleavage site at lysine 579. This sequence was used in the alignment process. The actual alignment of the primary sequences of the two polymerase domains was carried out using a signature sequence extending from arginine 754 to methionine 768 on the Klenow fragment, and from lysine 414 to glycine 428 on HIV-1 reverse transcriptase. These segments, which show an extremely high degree of primary sequence homology, have been proposed to be the site at which 2'-deoxynucleoside-5'-triphosphates bind to the enzymes (Basu *et al.*, 1989).

In order to create the three-dimensional model, the α-carbon coordinates and residue labels of the Klenow fragment were retrieved from the Brookhaven protein databank and converted from PDB format to MacroModel (Version 2.5; Still *et al.*, 1989) format. The residue label of each carbon atom was then modified to correspond to the aligned HIV-1 reverse transcriptase residue. The relabeled MacroModel file was converted back to PDB format and used as input into the Biopolymer mode of the Sybyl program (Version 3.5 by Tripos Associates, St. Louis, MO). This mode provides a means of elaborating the α-carbon model into a full model possessing a peptide backbone and side chains. The first elaboration was to build a polyalanine backbone from the α-carbon trace using the construct backbone procedure. The alanine methyl groups were then replaced with the appropriate sidechains using the add sidechains procedure. Limited steric checking was used at this point to set the initial side chain conformation. Two regions of the Klenow fragment extending from residue 574 to residue 622 and from residue 780 to residue 787 were not resolved in the X-ray study (Ollis *et al.*, 1985). The segments of HIV-1 reverse transcriptase corresponding to the missing Klenow residues were added to the three-dimensional model using a loop searching procedure (Tripos Associates, St. Louis, MO). This procedure utilizes a protein fragment database derived from the collection of Brookhaven protein entries. This approach to the construction of a full three-dimensional model of the protein using the α-carbon positions as guidepoints follows the method outlined by Claessens *et al.* (1989).

The molecular mechanics program AMBER (Weiner *et al.*, 1984) was used to perform a molecular dynamics simulation on the structure created by the alignment. The simulation was run using a united atom model to truncate the size of the calculation. After minimization to lower the energy gradient to 0.5 rms using a nonbonded cutoff of 8Å and distance dependent dielectric, dynamics was started at 300 degrees Kelvin using a time step of 0.75 femtoseconds. A representative structure was saved at one picosecond intervals over a total of fifty picoseconds. Figures 6A and 6B show the model at times t=0 and t=34 ps, respectively, in the 50 ps dynamics time trajectory. Large-scale motion was observed in the first hundred residues starting from the amino terminus. This segment contains the large 49 residue loop that was inserted using the loop procedure (vida supra). It is possible that the motion observed in this region may be an artifact of the starting structure used to represent the missing fragment of the protein. We are presently investigating the possibility that a more stable, less mobile starting structure can be found for this region. The remainder of the molecule showed limited motion and very little change in either secondary or tertiary structure over the time course of the simulation. The molecular mechanics and dynamics calculations will be repeated using explicit water to define a solvation sphere around the molecule and a much larger nonbonded cutoff distance. It is likely that these conditions will drastically affect the dynamics of the mobile fragment since this portion of the model is highly charged.

The most prominent feature of the model is the large cleft extending down the longest axis. The approximate dimensions of this cleft are 26 Å wide, 25 Å deep and 40 Å long. Thomas *et al.* (1990) report in their electron microscopy studies of the recombinant heterodimer that a number of enzyme particles show a prominent cleft or groove running lengthwise. Initial estimates of the width and length of this cleft are 20Å by 50Å (J. Griffith, personal communication).

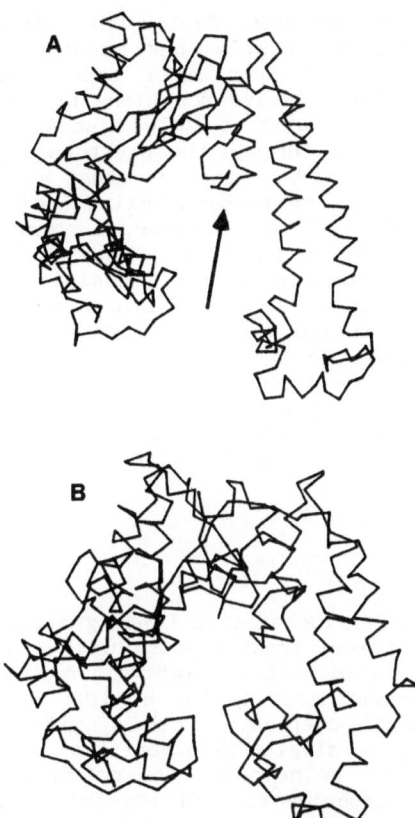

Fig. 6.   Two views of the α-carbon trace of the computer-generated
model of the polymerase domain of HIV-1 reverse transcriptase
at time (A) t=0 and (B) 34 ps in the 50 ps time trajectory.
Each view is from the front of the molecule (C terminus)
looking into the putative template-primer binding cleft.   The
arrow in panel A indicates the helix-turn-helix motif (see
text).

Thus the limited experimental data available does confirm the existence
of the cleft in the enzyme.   The analogy with the Klenow fragment of
E. coli DNA I polymerase suggests this to be the region that binds the
template primer during the course of the polymerase reaction.   This
supposition is borne out by electron micrographs of enzyme incubated
with duplex DNA, single-stranded DNA or RNA/DNA hybrids that show the
nucleic acid running along the length of the cleft (Thomas *et al.*, 1990).

Docking experiments indicate the cleft in the energy minimized
starting structure (t=0) will accommodate B-DNA after some minor
adjustments have been made in the conformation of the protein.   Docking
requires that the helix-turn-helix moiety located on the roof of the
cleft (Figure 6A) be placed in the major groove of the B-DNA.
Ollis *et al.* (1985) have speculated that α-helices J and K in the Klenow
structure serve to fix the translational position of the DNA on the

protein as the protein slides along duplex DNA. Presumably, the helix-turn-helix motif in the HIV-1 polymerase domain model would have a similar function. Docking experiments show clearly that such a construct can maintain the relative positions of the 3'-OH terminus of the nascent chain and the putative dNTP binding site as the duplex B-DNA rotates and translates through the cleft during the course of the polymerase reaction. It must be emphasized that at present there is no direct evidence that the helix-turn-helix unit pictured in the hypothetical model exists. It is, however, a common structural moiety in DNA binding proteins (Pabo and Sauer, 1984). A rather convincing feature of the present model in terms of nucleic acid binding is the asymmetric distribution of acidic and basic residues imparted by the fold. The majority of the basic residues lie within the cleft. This distribution should result in the development of a positive potential within the cleft. The development of a net positive electrostatic charge density within the cleft is consistent with this portion of the molecule binding to negatively charged nucleic acid. Calculations are presently underway to map the electrostatic potential energy surface of the polymerase model.

OVERVIEW

The development of a fluorescence binding assay has allowed us to establish that HIV-1 reverse transcriptase will bind both single- and double-stranded nucleic acid with a relatively high degree of affinity. The enzyme shows increased affinity for binding to the 3' terminus of duplex sites having a 5' overlap. Both single- and double-stranded nucleic acid bind to a single site on the heterodimeric enzyme, although there are in principal two potential binding sites. Photoaffinity labeling experiments indicate this single site to be on p66. The stability of the binary nucleic acid-protein complex is attributable to hydrophobic forces. We have developed a hypothetical model of the polymerase domain of the enzyme based on functional and sequence homology with the Klenow fragment of *E. coli* DNA polymerase I. The most prominent feature of the model is a cleft that extends down the long axis of the protein. The existence and rough dimensions of this cleft have been independently confirmed by electron microscopy studies. We are presently refining the model by including explicit water in the molecular dynamics runs and by increasing the nonbonded cutoff distance used in the calculations. We are also modifying the secondary structure of the model in specific regions based on the predictions made by several semi-empirical techniques and by studies of secondary structure content using Fourier transform infrared spectroscopy. We will use the refined version of the model in docking experiments with B-DNA to make predictions of amino acid residues important in nucleic acid binding. We will then mutate these sites and test the ability of the protein to bind nucleic acid using the fluorescence binding assay described in the text. We hope that in the absence of X-ray diffraction data this system will provide a means of establishing a detailed structure-function relationship for HIV-1 reverse transcriptase.

ACKNOWLEDGMENTS

The authors would like to thank Dr. David Barry for his support, and for many helpful discussions on the structure and function of HIV-1 reverse transcriptase.

BIBLIOGRAPHY

Basu, A., Tirumalai, R.S., and Modak, M.J., 1989, Substrate binding in human immunodeficiency virus reverse transcriptase, J. Biol. Chem., 264:8746-8752.

Brick, P., Ollis, D., and Steitz, T.A., 1983, Crystallization and 7 Å resolution electron density map of the large fragment of *Escherichia coli* DNA polymerase I, J. Molec. Biol., 166:453-456.

Burnbaum, J.J., Starzyk, R.M., and Schimmel, P., 1990, Understanding structural relationships in proteins of unsolved three-dimensional structure, Proteins: Structure, Function and Genetics, 7:99-111.

Claessens, M., Van Cutsem, E., Lasters, I., and Wodak, S., 1989, Modelling the polypeptide backbone with 'spare parts' from known protein structures, Protein Engineering, 2:335-345.

Cowart, M., Gibson, K.J., Allen, D.J., and Benkovic, S.J., 1989, DNA substrate structural requirements for the exonuclease and polymerase activities of procaryotic and phage DNA polymerases, Biochemistry, 28:1975-1983.

deHaseth, P.L., Lohman, T.M., and Record, M.T., Jr., 1977, Nonspecific interaction of lac repressor with DNA: An association reaction driven by counterion release, Biochemistry, 16:4783-4790.

Di Marzo Veronese, F., DeVico, A.L., Copeland, T.D., Oroszlan, S., Gallo, R.C., and Sarngadharan, M.G., 1986, Characterization of gp41 as the transmembrane protein coded by the HTLV-III/LAV envelope gene, Science, 229:1402-1406.

Etink, M.R. and C.A. Ghiron, 1976, Exposure of tryptophanyl residues in proteins. Quantitative determination by fluorescence quenching studies, Biochemistry, 15:672-680.

Griffith, J., Huberman, J.A., and Kornberg, A., 1971, Electron microscopy of DNA polymerase bound to DNA, J. Mol. Biol., 55:209-214.

Ha, J.-H., Spolar, R.S., and Record, M.T., Jr., 1989, Role of the hydrophobic effects in stability of site-specific protein-DNA complexes, J. Mol. Biol., 209:801-816.

Hansen, J., Schulze, T., and Moelling, K., 1987, RNase H activity associated with bacterially expressed reverse transcriptase of human T-cell lymphotropic virus III/lymphadenopathy-associated virus, J. Biol. Chem., 262:12393-12396.

Helene, C., 1971, Role of aromatic amino acid residues in the binding of enzymes and proteins to nucleic acids, Nature (London), New Biol., 234(47):120-121.

Hershberger, M.V., Lumry, R., and Verral, R., 1981, The 3-methylindole/n-butanol exciplexes: Evidence for two exciplex sites in indole compounds, Photochem. Photobiol., 33:609-617.

Hoffman, A.D., Banapour, B., and Levy, J.A., 1985, Characterization of the AIDS-associated retrovirus reverse transcriptase and optimal conditions for its detection in virions, Virology, 147:326-335.

Hopkins, S., Furman, P.A., and Painter, G.R., 1989, Investigation of the stereochemical course of DNA synthesis catalysed by human immunodeficiency virus type 1 reverse transcriptase, Biochem. Biophys. Res. Comm., 163:106-110.

Huber, H.E., McCoy, J.H., Seehra, J.S., and Richardson, C.C., 1989, Human immunodeficiency virus 1 reverse transcriptase: Template binding, processivity, strand displacement synthesis, and template switching. J. Biol. Chem.,264:4669-4678.

Johnson, M.S., McClure, M.A., Feng, D.-F., Gray, J., and Doolittle, R.F., 1986, Computer analysis of retroviral *pol* genes: Assignment of enzymatic functions to specific sequences and homologies with nonviral enzymes, Proc. Natl. Acad. Sci., USA, 83:7648-7652.

Kornberg, A., 1974, "DNA Synthesis," W.H. Freeman and Co., San Francisco.

Lakowicz, J.R., 1983, "Principles of Fluorescence Spectroscopy," Plenum Press, New York.

Larder, B., Purifoy, D., Powell, K., and Darby, G., 1987, AIDS virus reverse transcriptase defined by high level expression in *Escherichia coli*, EMBO J., 6:3133-3137.

Lohman, T.M., deHaseth, P.L., and Record, M.T., Jr., 1980, Pentalysine-deoxyribonucleic acid interactions: A model for the general effects of ion concentrations on the interactions of proteins with nucleic acids, Biochemistry, 19:3522-3530.

Lowe, D.M., Aitken, A., Bradley, C., Darby, G.K., Larder, B.A., Powell, K.L., Purifoy, D.J.M., Tisdale, M., and Stammers, D.K., 1988, HIV-1 reverse transcriptase: Crystallization and analysis of domain structure by limited proteolysis, Biochemistry, 27:8884-8889.

Majumdar, C., Abbotts, J., Broder, S., and Wilson, S.H., 1988, Studies on the mechanism of human immunodeficiency virus reverse transcriptase, J. Biol. Chem., 263:15657-15665.

Manning, G.S., 1978, The molecular theory of polyelectrolyte solutions with applications to the electorstatic properties of polynucleotides, Q. Rev. Biophys., 11:179-246.

Ollis, D.L., Brick, P., Hamlin, R., Xuong, N.G., and Steitz, T.A., 1985, Structure of large fragment of *Escherichia coli* DNA polymerase I complexed with dTMP, Nature, 313:762-766.

Pabo, C.O. and Sauer, R.T., 1984, Protein-DNA recognition, Annu. Rev. Biochem., 53:293-321.

Record, M.T., Jr., Lohman, T.M., and deHaseth, P., 1976, Ion effects on ligand-nucleic acid interactions, J. Mol. Biol., 107:145-158.

Steitz, T.A., Weber, I.T., and Matthew, J.B., 1983, Catabolite gene activator protein: Structure, homology with other proteins, and cyclic AMP and DNA binding, Cold Spring Harbor Symp. Quant. Biol. 1982, 47(1):419-426.

Still, W.C., Richards, N.G., Guida, W.C., Lipton, M., Liskamp, R., Chang, G., and Hendrickson, T. 1989, MacroModel Version 2.5, Dept. of Chemistry, Columbia University.

Thomas, D., Griffith, J., Furman, P., and Painter, G., 1990, Electron microscopic visualization of HIV-1 reverse transcriptase free and bound to DNA, J. Cellular Biology, 14D:112.

Tisdale, M., Ertl, P., Larder, B.A., Purifoy, D.J.M., Darby, G.K., and Powell, K.L., 1988, Characterization of human immunodeficiency virus type 1 reverse transcriptase by using monoclonal antibodies: Role of the C terminus in antibody reactivity and enzyme function, J. Virol., 62:3662-3667.

Walker, M.S., Bednar, T.W., and Lumry, R., 1966, Exciplex formation in the excited state of indole, J. Chem. Phys. 45:3455-3456.

Weiner, S.J., Kollman, P.A., Case, D.A., Singh, U.C., Ghio, C., Alagona, G., Profeta, S., Jr., and Weiner, P., 1984, A new force field for molecular mechanical simulation of nucleic acids and proteins, J. Am. Chem. Soc., 106:765-784.

ROUS SARCOMA VIRUS *GAG* GENE SEQUENCES

AFFECT VIRAL RNA SPLICING AND STABILITY

Mark T. McNally, George F. Barker, and Karen Beemon

Department of Biology
Johns Hopkins University
Baltimore, MD

INTRODUCTION

A feature common to the life cycles of all retroviruses is the incomplete splicing of the primary viral RNA transcript. While splicing of a subset of the primary transcripts is necessary to generate subgenomic mRNAs (including *env* mRNA and additional mRNAs in the more complex retroviruses), a major fraction of the RNA transcripts are transported to the cytoplasm in an unspliced form, where they serve both as mRNA and as genomic RNA (Coffin, 1990; Stoltzfus, 1989).

We are interested in the mechanisms by which a fraction of the viral RNA avoids being spliced, but is polyadenylated and transported to the cytoplasm intact. In the case of retroviruses such as HIV or HTLV, viral regulatory proteins (rev or rex, respectively), as well as *cis*-acting viral sequences, appear to be required for the appearance of unspliced or singly-spliced viral mRNAs in the cytoplasm (Wong-Staal, 1990; Cann and Chen, 1990). While all retroviruses must carry out this incomplete splicing, most do not appear to encode regulatory proteins necessary for this process (Coffin, 1990). Thus, it is interesting to study the different mechanisms by which the appropriate balance of unspliced to spliced viral RNAs is maintained in the cytoplasm.

We are studying the RNA processing of Rous sarcoma virus (RSV), a relatively simple virus which has three mRNAs and no known viral-encoded regulatory proteins. RSV RNA appears to be transcribed and processed entirely by cellular machinery. In contrast to most cellular pre-mRNAs, however, the primary transcript yields both spliced and unspliced cytoplasmic RNA. Subgenomic *env* and *src* mRNAs result from alternative splicing from a common 5' splice site to unique 3' splice sites. Full-length RNA serves as both mRNA for *gag* and *gag-pol* products and as virion RNA (Coffin, 1990; Stoltzfus, 1989).

The ratio of spliced/unspliced RSV RNA may be perturbed in several different ways by alteration of the viral RNA sequence. For example, insertion of an oligonucleotide 12 nts upstream of the *env* splice acceptor site results in greatly increased levels of spliced *env* mRNA (Katz et al., 1988). Revertants arising from passage of this replication-defective virus have been observed to have mutations either within the inserted sequence or in the exon region downstream of the *env* splice acceptor, suggesting that sequences favoring splicing are normally present in the exon and that the splice acceptor site is suboptimal (Katz and Skalka, 1990). Similarly, mutations in the vicinity of the *src* splice acceptor site have been observed to alter the efficiency of splicing at this site (Stoltzfus et al., 1987; Stoltzfus and Fogarty, 1989).

Advances in Molecular Biology and Targeted Treatment for AIDS
Edited by A. Kumar, Plenum Press, New York, 1991

51

We were surprised to find that deletion of certain sequences in the RSV *gag* gene (part of the intron for the subgenomic mRNAs) also increased the ratio of spliced to unspliced viral RNAs (Arrigo and Beemon, 1988). These deleted viral constructs were originally made to study effects of removal of a transcriptional enhancer sequence we had identified in the *gag* gene (Arrigo et al., 1987). Insertion of these sequences, which we have termed a Negative Regulator of Splicing (NRS), into the intron of a cellular mRNA (chicken c-*myc* intron 2) resulted in increased accumulation of the unspliced transcript, assayed by transient expression in chicken embryo fibroblasts. The NRS activity was found to be orientation dependent in both the viral and the heterologous constructs; no effect on splicing was observed when the NRS was present in the anti-sense orientation. In contrast to the unspliced retroviral RNA, the unspliced heterologous RNA was only detected in the nucleus (Arrigo and Beemon, 1988).

Arrigo and Beemon (1988) also observed that frameshift mutations in the *gag* gene of RSV were associated with a decreased steady-state level of viral unspliced RNA, and that this decrease in RNA levels was most pronounced in the cytoplasm. Since these frameshift mutations resulted in premature termination of translation within the *gag* gene, it was suggested that stability of the viral unspliced RNA may require translation to the normal termination codon. This observation has been confirmed and extended by introduction of termination codons at various sites in the RSV genome. Moreover, similar results have been observed by introduction of nonsense codons into the HIV *gag* gene (S. Arrigo, personal communication).

RESULTS AND DISCUSSION

Negative regulator of splicing within the RSV *gag* gene

We have recently extended our initial characterization of the NRS element. Using the heterologous *myc* intron assay, we have identified the minimal sequences needed for NRS activity. Maximal levels of unspliced RNA (about 70% of the total) were produced from a 300 nt NRS insert (nt 707-1006 of RSV according to sequence of Schwartz et al., 1983). Essential domains within this sequence have also been identified. Whereas a central 76 nt domain (nt 799-874) could be deleted, sequences on both sides of this domain were essential for accumulation of unspliced RNA. The minimal NRS sequence identified contains nts 777-798 and 875-930, although greater activity was observed with a larger upstream fragment containing nts 707-798, as well as the domain from nt 875-930. The domain from nt 707-798 is extremely rich in purines, whereas the region from nt 888-902 consists entirely of pyrimidines. This suggests the possibility of interaction between these two domains. Further 3' deletion of the NRS from nt 930 to nt 908 abolished its activity, suggesting that sequences downstream of the polypyrimidine tract are also required.

In addition to its absolute orientation dependence, NRS activity appears to be somewhat position dependent. Insertion of the the NRS into the *myc* intron generated unspliced RNA from that intron, without affecting splicing at a downstream intron. Further, the NRS was inactive when inserted into the upstream exon at a site 440 nt from the 3' splice site. We have found that the NRS is functional in mammalian cells (COS) as well as in avian cells. In addition, we have demonstrated activity of the NRS sequence in vivo when inserted into the adenovirus major late leader intron.

We are currently investigating the mechanism of action of the NRS. In particular, we are interested in whether it blocks spliceosome complex formation or whether it forms an inactive spliceosome. It does not appear to promote rapid nuclear transport of the unspliced RNA since unspliced heterologous RNAs containing the NRS remained in the nucleus (Arrigo and Beemon, 1988). The failure of the NRS positioned in an upstream intron to affect splicing of a downstream intron argues against its acting to compartmentalize the RNA within the nucleus. The sequence of the localized NRS suggests that it may act as a decoy splice acceptor, as proposed by Stoltzfus (1989), competing with the genuine 3' splice sites to insure production of unspliced RNA.

Nonsense codons within the RSV gag gene destabilize the unspliced viral RNA

The intracellular accumulation of the unspliced RNA of RSV was decreased when translation was prematurely terminated by the introduction of nonsense codons within its 5' proximal gene, the *gag* gene. Nonsense codons within the *pol* gene did not have this effect. The levels of spliced viral RNAs were not, however, affected by premature translation termation in the *gag* gene. These measurements were carried out by transient expression of full-length, nonpermuted viral constructs in chick embryo fibroblasts. Experiments using the transcription inhibitor actinomycin D showed that mutant unspliced RNAs were degraded more rapidly than wild-type RNA.

Mutant RNAs could be partially stabilized by coexpression of *gag* proteins in *trans*, and deletion of the nucleocapsid domain of *gag* abolished this ability. Nevertheless, unspliced viral RNA containing the same nucleocapsid deletion accumulated to wild-type levels within cells, suggesting that *gag* proteins were not required to stabilize RNAs lacking premature termination codons. Thus, it appeared that nonsense codons destabilized the RNA not because they prevented production of *gag* protein, but because they disrupted the process of translating the *gag* gene. Furthermore, analysis of double-mutant constructs containing both in-frame deletions and termination codons within *gag* suggested that RSV RNA contains *cis*-acting components which interact with the translational apparatus to influence the stability of the RNA.

Introduction of mutant viruses bearing premature termination codons in *gag* into the avian packaging cell line, Q2bn (Stoker and Bissell, 1988), resulted in less mutant RNA being packaged into virions than wild type RNA. This suggests that virion RNA, as well as viral mRNA, is subject to degradation if it cannot be completely translated.

The interaction between the RNA and ribosomes could occur in the cytoplasm, where translating ribosomes might protect *gag* RNA from nucleases. We have observed that deletion of RNA sequences downstream of the nonsense codon partially stabilizes the RNA, possibly by removing nuclease-sensitive sites. This hypothesis is similar to that proposed to explain hyper-labile RNA resulting from nonsense codons in bacterial operons (Hiraga and Yanofsky, 1972). Alternatively, the interaction between the viral RNA and the ribosomes may occur at the level of RNA transport from the nucleus to the cytoplasm or even at the level of nuclear scanning of RNAs by ribosomes, both of which have been hypothesized to explain similar results with dihydrofolate reductase and triosephosphate isomerase mRNAs (Urlaub et al., 1989; Cheng et al., 1990).

It is unexpected that coexpression of *gag* proteins should increase the intracellular accumulation of mutant unspliced RNAs. The traditional model of retroviral assembly (Coffin, 1990; Bolognesi, 1978) suggests that interaction of the unspliced RNA with the *gag* precursor occurs as the viral particle assembles at the plasma membrane. If this were the case, co-expressed *gag* proteins would be expected to package and export the mutant RNAs from the cell, thus further decreasing their accumulation. The ability of the *gag* proteins to have the opposite effect suggests that they may, instead, interact with the unspliced viral RNA at some stage prior to assembly of viral particles at the plasma membrane. It is possible, for instance, that *gag* proteins might help to export the RNA from the nucleus or might protect it from cytoplasmic degradation.

ACKNOWLEDGMENTS

This work was supported by Public Health Service grant CA48746 from the National Cancer Institute. GFB was supported by predoctoral National Research Service Award 5T32GM07231 from the Institute of General Medical Sciences and by an Owens Fellowship from the Johns Hopkins University.

REFERENCES

Arrigo, S., and Beemon, K., 1988, Regulation of Rous sarcoma virus RNA splicing and stability, Mol. Cell. Biol., 8:4858.

Arrigo, S., Yun, M. and Beemon, K., 1987, cis-acting regulatory elements within the gag genes of avian retroviruses, Mol. Cell. Biol., 7:388.

Bolognesi, D.P., Montelaro, R.C., Frank, H., and Schafer, W., 1978, Assembly of type C oncornaviruses: a model, Science, 199:183.

Cann, A.J., and Chen, I.S.Y., 1990, Human T-cell leukemia virus types I and II, in: "Fields Virology", 2nd ed., B.N. Fields and D.M. Knipe, ed., Raven Press, N.Y., N.Y.

Cheng, J., Fogel-Petrovic, M., and Maquat, L.E., 1990, Translation to near the distal end of the penultimate exon is required for normal levels of spliced triosephosphate isomerase mRNA, Mol. Cell. Biol., 10:5215.

Coffin, J.M., 1990, Retroviridae and their replication, in: "Fields Virology", 2nd ed., B.N. Fields and D.M. Knipe, ed., Raven Press, N.Y., N.Y.

Hiraga, S., and Yanofsky, C., 1972, Hyper-labile messenger RNA in polar mutants of the tryptophan operon of Escherichia coli, J. Mol. Biol., 72:103.

Katz, R. A., Kotler, M. and Skalka, A. M., 1988, cis-acting intron mutations that affect the efficiency of avian retroviral RNA splicing: implications for mechanisms of control, J. Virol., 62:2686.

Katz, R. A., and Skalka, A. M., 1990, Control of retroviral RNA splicing through maintenance of suboptimal processing signals, Mol. Cell. Biol., 10:696.

Schwartz, D.E., Tizard, R., and Gilbert,W., 1983, Nucleotide sequence of Rous sarcoma virus, Cell, 32:853.

Stoker, A.W., and Bissell, M.J., 1988, Development of avian sarcoma and leukosis virus-based vector-packaging cell lines, J. Virol., 62:1008.

Stoltzfus, C. M., 1989, Synthesis and processing of avian sarcoma retrovirus RNA, Adv. Virus. Res., 63:1.

Stoltzfus, C. M., and Fogarty, S. J., 1989, Multiple regions in the Rous sarcoma virus src gene intron act in cis to affect the accumulation of unspliced RNA, J. Virol., 63:1669.

Stoltzfus, C. M., Lorenzen, S. K., and Berberich, S. L., 1987, Noncoding region between the env and src genes of Rous sarcoma virus influences splicing efficiency at the src gene 3' splice site, J. Virol., 61:177.

Urlaub, G., Mitchell, P.J., Ciudad, C.J., and Chasin, L.A., 1989, Nonsense mutations in the dihydrofolate reductase gene affect RNA processing, Mol. Cell. Biol., 9:2868.

Wong-Staal, F., 1990, Human immunodeficiency viruses and their replication, in: "Fields Virology", 2nd ed., B.N. Fields and D.M. Knipe, ed., Raven Press, N.Y., N.Y.

RIBONUCLEASE H AND PRIMER tRNA BINDING ACTIVITIES OF

HIV REVERSE TRANSCRIPTASE AS THERAPEUTIC TARGETS

Stuart F.J. Le Grice, Oktavian Schatz[*] and Jean-Luc Darlix[+]

Div.of Infectious Diseases, CWRU School of Medicine
Cleveland, OH 44106,[*]ZFE/Bio, Hoffmann-La Roche Ltd. Basel
Switzerland,[+]Labo-retro, CRBCG du CNRS, F-31062 Toulouse
France

INTRODUCTION

Human immunodeficiency virus reverse transcriptase (RT, deoxynucleoside-triphosphate: DNA deoxynucleotidyltransferase (RNA-directed), E.C. 2.7.7.49.) has been the subject of intense investigation over the last 5 years, with the goal of developing a safe and effective therapy to stem the progression of AIDS. Although success has been achieved with inhibitors of viral replication such as AZT[1-3], the rapid emergence of tolerance to this drug[4] is a clear indication that new and improved therapies will be in constant demand for a considerable period.

In addition to its DNA- and RNA-dependent DNA polymerase activities[5,6], the C-terminal domain of HIV RT encodes a ribonuclease H (RNaseH) activity, necessary for removal of viral RNA from the DNA/RNA replication intermediate[7]. The availability of large quantities of highly pure RT through recombinant DNA technology now permits a detailed investigation into RNaseH activity. Consequently, it is not unreasonable to consider RT-associated RNaseH as another avenue for therapeutic intervention.

Retroviral replication initiates from a tRNA hybridised to a specific region near the 5' end of the viral genome, the primer binding site (pbs). Examples of this are tRNA[Trp] for Rous sarcoma virus (RSV)[8], tRNA[Pro] for Murine Lukaemia Virus (MLV)[9].As with Mouse Mammary Tumour

```
441                451                         461
Y V D G A A N R E T K L G K A G Y V T N K G R Q K V V P L T


471                      481                491
N T T N Q K T E L Q A I Y L A L Q D S G L E V N I V T D S Q
               I
               Q


501                511                521
Y A L G I I Q A Q P D K S E S E L V N Q I I E Q L I K K E K


531                      541                551
V Y L A V W P A H K G I G G N E Q V D K L V S A G I R K I L
               H
               F
```

Figure 1   Amino acid sequence of the C-terminal RNaseH domain of p66 HIV-1 RT. The sequence illustrated extends from the p51/p66 RT junction (Phe[440]/Tyr[441])[17] to Leu[560]/Phe[561], which delineates the RT and integrase (IN) domains. Amino acids sharing homology with both proposed retroviral RNaseH domains and the *E.coli* enzyme have been underlined. In addition, those amino acids altered in the present study are enclosed by boxes.

Virus (MMTV)[10], DNA sequence analysis suggests tRNA^{Lys,3} as the replication primer required by HIV[11]. In the case of RSV, a specific interaction between the replication primer and RT in the absence of viral RNA sequences has been suggested[12]. If a similar specificity could be established for HIV RT and tRNA^{Lys,3}, it is conceivable that strategies might be developed to interrupt this and inhibit the enzyme before it proceeds into the replication cycle.

This article reviews our current understanding of both the RNaseH and tRNA^{Lys,3} binding properties of HIV RT. In each case, both our results and those of other groups suggest that these properties should indeed be considered as alternative targets for anti-viral therapy.

RESULTS   [A]   RT-associated RNaseH

(i) Mutagenesis of the RNaseH domain of recombinant HIV-1 RT

Amino acid sequence comparisons have re-designated HIV RNaseH activity to the C-terminus of the 66 kDa RT polypeptide (p66)[13]. Although the putative RNaseH domains of several retroviruses share little homology with the *Escherichia coli* enzyme (Fig.1[A]), we used the structure of the latter (made available to us by Dr. K Morikawa, PERI, Japan) to introduce two amino acid substitutions, by *in vitro* site-directed mutagenesis[14], into the RNaseH domain of the HIV-1 enzyme. These mutations (Glu^{478}->Gln, His^{539}->Phe) were chosen on the basis that they (i) affected highly conserved amino acid residues, (ii) were located at helix-sheet transitions that are spatially clustered as evident from the preliminary three-dimensional structure of *E. coli* RNaseH and (iii) would not impair correct folding of the protein as predicted by the Chou-Fasman[15] or Delphi[16] algorithms. RT/RNaseH recombinant plasmids p6HRT$^{E->Q}$ and p6HRT$^{H->F}$ were constructed[17], determining synthesis of the two mutant forms of p66 RT/RNaseH. In addition, a complete HIV-1 protease expression cassette was introduced into the p66 RT-expressing plasmids, resulting in synthesis of heterodimeric p66/p51 enzymes (plasmids p6HRT$^{E->Q}$-PROT and p6HRT$^{H->F}$-PROT, respectively)[17,18]. All mutant RT/RNaseH forms, together with the wild type equivalent, were extended at their N-terminus by six histidine residues to facilitate purification by Ni^{2+}-Nitrilotriacetic acid-Sepharose (NTA-Sepharose)[19]. This procedure can be used to purify recombinant proteins directly from cell homogenates in a single step and yielded RNase- and DNase-free enzyme forms required for analysis of

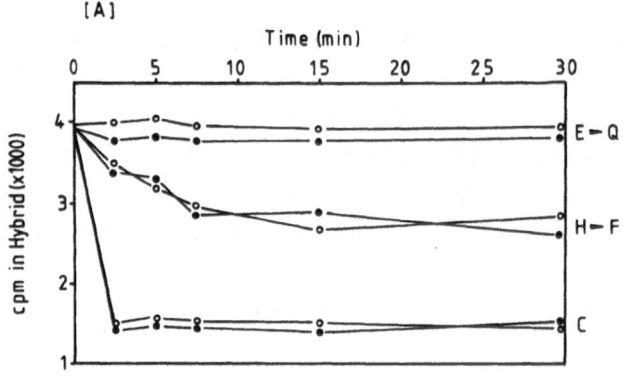

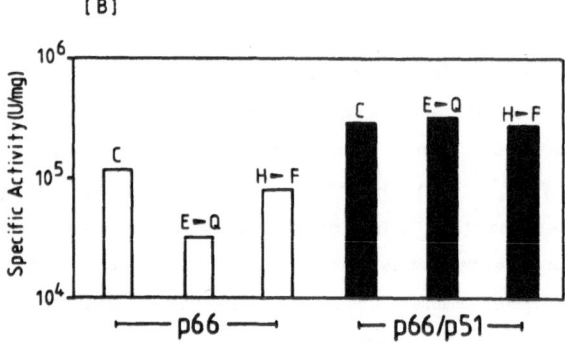

Figure 2 [A] RNaseH activity in wild-type and mutant homodimeric (p66, o) and heterodimeric (p66/p51, o) HIV RT. Activity is measured as loss of radioactivity from an RNA/DNA hybrid whose RNA component is uniformly labelled. The nature of the mutation is indicated at the right of the panel. **C**, wild type homo- and heterodimer RT. The results presented here are representative of several analyses [B] RNA-dependent DNA polymerase activity of wild typeand mutant enzymes. The nature of the mutation and enzyme form analysed is indicated above and below each column, respectively. (**C**), wild type enzyme. Specific activities (units/mg) for each enzyme form are presented and are the results of duplicate analyses. We define 1 unit of RT activity as that amount catalysing incorporation of 1 nmole precursor dGTP into polydeoxynucleotide in 10 min at 37°C, under the assay conditions specified[18].

RNaseH activity. Furthermore, comparison of non-mutant enzymes with their poly-histidine-free counterpart, indicated that the short histidine extension affected neither the polymerase nor RNaseH function of RT[19].

(ii) Ribonuclease H and polymerase characteristics of mutants

RNaseH activity of the homo- and heterodimeric enzyme preparations is illustrated in Fig. 2[A]. Substrate for these experiments was a radiolabelled RNA/DNA hybrid, prepared by *in vitro* transcription of single-stranded M13 DNA. Whereas wild type p66 and p66/p51 enzymes rapidly remove RNA from the hybrid, this function is completely eliminated from the p66 and p66/p51 E^{478}->Q mutants. Interestingly, both p66 and p66/p51 RT containing the H^{539}->F mutation exhibited partial RNaseH activity. In contrast to the data of other groups, our recent results suggest that HIV RT/RNaseH exhibits combined endonuclease and 3'->5' exonuclease activities[20]. Thus, the partial activity observed with mutant H^{539}->F may represent elimination of only one RNaseH function, although further experimentation will be neccessary to establish this.

To eliminate the possibility that mutations introduced into the C-terminal RNaseH domain did not have consequences for the overall structure of p66 RT/RNaseH, each enzyme form was analysed by protein footprinting[21] with the endoproteinases Endo ArgC, Endo LysC and Endo GluC. Following partial proteolysis, the products were fractionated by SDS/PAGE, transferred to a nylon membrane, then analysed immunologically with polyclonal rabbit antibodies against p66 RT. The similarity of proteolytic digestion patterns[22] confirmed that the structural integrity of each mutant had been preserved. In addition, the RNA-dependent DNA polymerase activity of each preparation was determined on a homopolymeric poly (rC)/oligo (dG) primer/template. As before, we determined the RT activity of both the p66 and p66/p51 form of each enzyme, the results of which are presented in Fig.2 [B]. Slight differences in specific activity were observed with the homodimeric enzymes; in contrast, heterodimeric mutants displayed no loss of activity when compared with the wild type enzyme. Since several lines of evidence now suggest that the biologically relevant form of HIV RT is the p66/p51 enzyme[18,22,23], we elected to introduce mutant E^{478}->Q, which retains full polymerase function but totally lacks RNaseH activity, into a molecular clone and determine the consequences for viral infectivity.

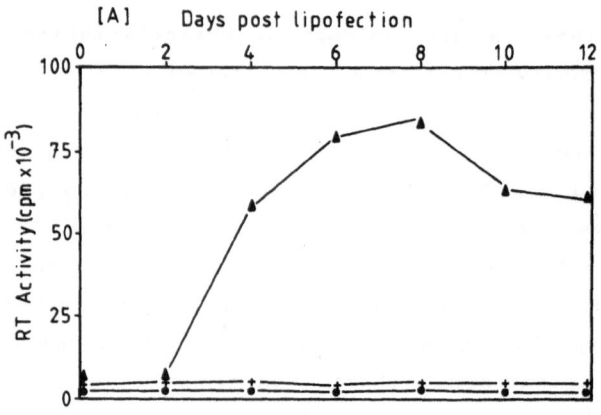

[A] Days post lipofection

[B]

Days post lipofection	2	4	6	8	10	12
wt provirus	-/+	+	+++	+++	+++	+++
E—Q mutant	-/+	-/+	-/+	-/+	-/+	-/+

Figure 3 [A] RT activity in MT-2 cultures lipofected with molecular clones of wild type (▲) or RNaseH deficient (+) HIV-1[24]. Virus pellets from 1 ml cell-free supernatants were resuspended in 40μl PBS. The solution was divided into two portions, and 20μl RT mix added to each . RT mix contained 100mM Tris/HCl, pH 8.0, 10mM DTT, 60mM $CaCl_2$, 1.2% Triton X-100, 16.6μm dGTP, 83μg/ml poly (rC)/p(dG)12-18 and 19μCi/ml [α-^{35}S]dGTP. After 1 hr, reactions were stopped by addition of 5μl 10% SDS. Mixtures were spotted onto DEAE paper and extensively washed before determination of radioactive incorporation. (o), background activity from endogenous HTLV-1 in MT-2 cell line. [B] Syncytia counts following lipofection with wild type and mutant proviral clones. +/-, <10 syncytia per well; +, 10 - 50 syncytia per well; +++, >100 syncytia per well.

(iii) RNaseH-deficient HIV is non-infectious

In order to evaluate the *in vivo* consequences of the $E^{478}->Q$ mutation, the appropriate RT portion of our recombinant plasmid was exchanged for that of an infectious molecular clone of HIV-1. The resulting full-length clone was introduced by lipofection into MT-2 cells. Following lipofection, viral infectivity was monitored either as the appearance of RT activity in culture supernatants or counting of giant multi-nucleated cells (syncytia)[24]. Fig.3 indicates the results of the analyses.

RT activity is evident in cultures of MT-2 cells within three days of lipofection with the wild-type proviral clone (Fig. 3[A]). This reaches a maximum after eight days, and thereafter declines, presumably due to cell death. The low level RT activity observed in the mock lipofection arises from endogeneous HTLV-1 present in MT-2 cells. The same level of RT activity is observed after lipofection with the mutant proviral clone for approximately two weeks after its introduction. In addition to RT activity, Fig. 3[B] illustrates the capacity of each culture for syncytia formation. Cells infected with the wild type proviral clone are virtually completely lysed six days after infection. During this time, and up to two weeks after lipofection, syncytia formation is barely observed in cultures containing the mutant provirus. The data of Figs. 3 [A] and [B] thus indicates that (a) no cellular RNaseH can complement that eliminated from HIV, and (b) elimination of virus-coded RNaseH has severely consequences for HIV infectivity.

[B] RT/tRNA interactions

(i) Gel retardation analysis of RT/tRNA complexes

Fig. 4[A], lanes 1-6 demonstrates a gel retardation assay, performed with recombinant p66/p51 HIV-1 RT and end-labelled beef liver tRNA^{Lys,3}. In the absence of RT, tRNA^{Lys,3} rapidly migrates to the bottom of the gel (lane 1). In contrast, incubation of tRNA^{Lys,3} with the HIV enzyme results in a mobility shift, indicating that all available tRNA^{Lys,3} was complexed to RT (lane 2). RT-bound tRNA^{Lys,3} could not be displaced by either tRNA^{Trp} or tRNA^{Pro}, the replication primers of RSV and MLV, respectively (Lanes 3,4). Radiolabelled tRNA^{Lys,3} could however be displaced by incubation with total virion tRNA (lane 5), indicating that this is possibly enriched for the Lys,3 species. Incubation of the

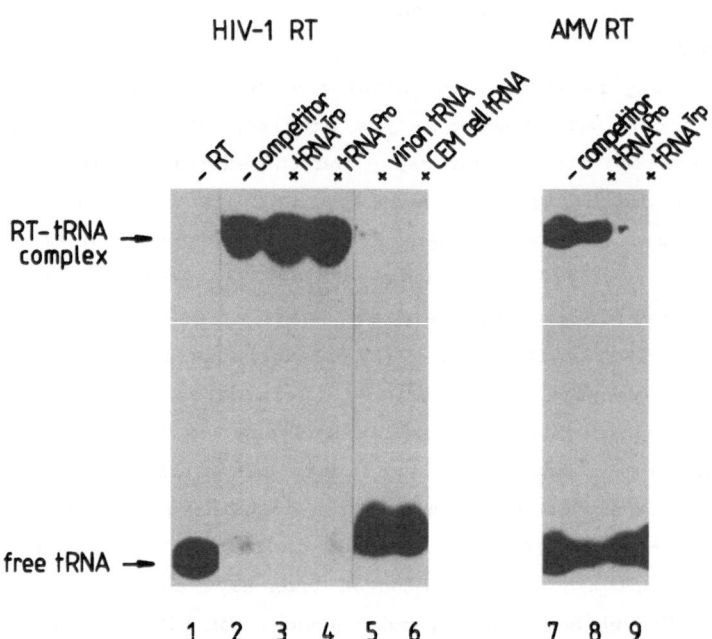

Figure 4. Gel mobility shift assays with HIV-1 RT (lanes 1 - 6) or RSV RT (lanes 7 - 9) in the presence of their replication primers, tRNA^{Lys,3} and tRNA^{Trp}, respectively[25]. Assays were performed in standard RT assay buffer, with the exception that dNTPs were omitted. Complexes of RT and radiolabelled tRNA were electrophoresed through 0.8% agarose gels in 50 mM Tris/acetate buffer, pH 7.2, containing 2 mM MgCl$_2$ and 5 mM β-mercaptoethanol for 3 hours with buffer circulation. Following electrophoresis, gels were dried and subjected to autoradiography. The position of free and complexed tRNA^{Lys,3} or tRNA^{Trp} is indicated at the side of the gel.

RT/tRNA^{Lys,3} complex in the presence of a vast excess of CEM cellular tRNA (a human acute lymphoblastic leukaemia cell line), within which tRNA^{Lys,3} is present as a sub-species, could also displace radiolabelled tRNA (lane 6). For copmparison, the same assay was performed with RSV RT and its replication primr, tRNA^{Trp}, the results of which are in lanes 7-9 The results of Fig. 4[25] thus illustrate that, both HIV and RSV RT form a specific complex with their replication primer in the absence of viral RNA sequences.

(ii) RT/tRNA^{Lys,3} cross-linking studies

To determine whether any particular region of tRNA^{Lys,3} was involved in a specific interaction with RT, complexes were treated with the cross-linking reagent *trans*-diaminedichloro-platinum (II) (*trans*-DDP)[,26]. Subsequent treatment with T1 RNase was used to removes those tRNA sequences non cross-linked to the enzyme. tRNA which had remained in contact with RT was then 5' labelled with [^{32}P]γATP, after which the complex was fractionated by SDS/PAGE and the gel subjected to autoradiography. As illustrated in Fig.5[A], radioactivity was found associated with both the 51 and 66 kDa RT polypeptides, indicating their close contact to tRNA^{Lys,3}. The small shift in apparent molecular weight furthermore suggested the presence of only a small RNA oligonucleotide. Since *trans*-DDP induced cross-links are reversible, the RT-bound oligoribonucleotide was dissociated and analysed by high resolution gel electrophoresis. Fig. 5[B] indicates that a single 12-nucleotide RNA species was obtained. Subsequent sequencing of this RNA indicated that it arose exclusively from the anticodon domain of tRNA^{Lys,3} (Fig. 6)[25]. An interesting feature of tRNA^{Lys,3} is the presence of two hyper-modified bases in its anticodon domain. Although it is somewhat speculative, the possibility must be considered that these two bases contribute to the specificity of the RT/tRNA interaction.

An noteworthy feature of Fig. 5[B] is that tRNA^{Lys,3} can be cross-linked to both the 66 and 51 kDa polypeptides of heterodimer. With the exception of a *trpE*/RT fusion[27] protein, several investigators have demonstrated that p51 RT is virtually devoid of polymerase activity and completely inactive as an RNaseH[28-30]. The likelihood that heterodimer RT is the biologically, and capacity of p51 to specifically complex with the same region of tRNA^{Lys,3} indicates that its structural integrity in heterodimer is preserved. The data here might suggest a role for p51

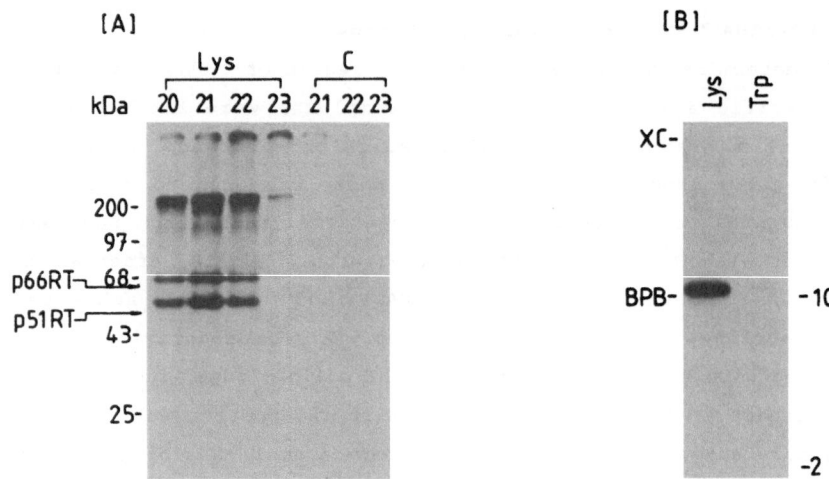

Figure 5 **[A]** Cross-linking of tRNA^{Lys,3} to p66/p51 RT. The autoradiogram of an SDS/PAGE analysis is presented. Numbers above the panel represent fractions analysed from a gel filtration column, used to separate RT-bound tRNA oligoribonucleotide complexes (following T1 RNase digestion of the complex) from other end-labelled species[25]. Migration positions of non-complexed RT are indicated on the left of the panel. The slight mobility shift in the cross-linked RT subunits represents complexes with a small oligoribonucleotide from tRNA^{Lys,3}. **[B]** High resolution gel electrophoresis of the RT bound tRNA^{Lys3} oligoribonucleotide, following reversal of the *trans*-DDP cross-link. The specificity of this interaction is indicated by the lane labelled **Trp**, which is a similar experiment between HIV RT and tRNA^{Trp}.

such as sequestration and orientation of the tRNA primer for extension by p66 of heterodimer. In order to evaluate this it will be necessary to eliminate the tRNA binding property of p66 in heterodimer and determine whether this might elongate tRNA[Lys,3] bound by the p51 component. Experiments of this nature are presently underway in our laboratory.

Discussion

In addition to DNA- and RNA-dependent DNA polymerase properties, other activities intrinsic to HIV RT might be considered as alternative therapeutic targets. This report demonstrates that virions whose RNaseH component is inactivated are non-infectious, illustrating this function as a possible candidate. In related work, it has been demonstrated that mutation of the same conserved glutamic acid residue in the putative RNaseH domain of hepatitis virus results in failure of virions to synthesise plus-strand DNA[31]. Mutagenesis experiments with *E.coli* RNaseH have also demonstrated the requirement for this residue (Glu[48] in the *E.coli* enzyme) for enzymatic activity[32]. In investigations involving the viral enzymes, an important factor was the inability of a cellular RNaseH to complement the defect in the retroviral enzyme. One possible explanation for this might be compartmentalisation of the two activities, i.e., replication of HIV in the cytoplasm, and location of cellular RNaseH in the nucleus of the host.

Although we have not analysed exactly where the block in the HIV replication cycle occurs, this is most likely at the stage following synthesis of 'strong-stop' DNA[33]. Removal of RNA from this primary replication product allows the first 'strand-jump' necessary to translocate tRNA-bound nascent DNA to the 3' end of the viral RNA genome.

In a recent communication, we demonstrated that both p66 and p66/p51 RT-associated RNaseH display combined endonuclease and 3'->5' exonuclease activities. The partial activity of RNaseH mutant His[539]->Phe may reflect inactivation of only one function, possibly suggesting two catalytic centres within the one RNaseH domain; further experimentation will however be necessary to establish this. Interestingly, the same partial inactivation is observed with *E.coli* RNaseH carrying the same mutation (His[124] in the *E.coli* enzyme). We are presently investigating this possibility using highly specific assays which differentiate between the endonuclease and exonuclease activities.

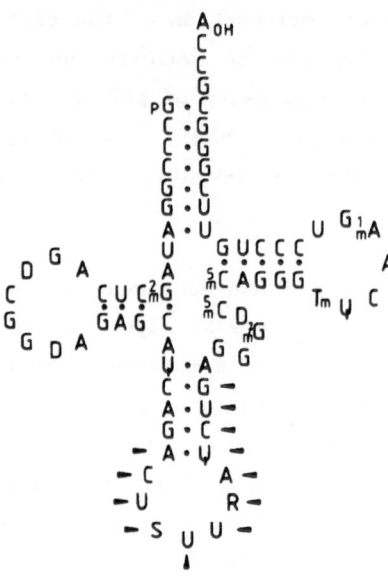

Figure 6. tRNA[Lys,3], the replication primer utilised by HIV RT. Modified bases are: **D**, dihydrouridine; Ψ, pseuoduridine; **Tm**, methyl ribothymidine. In the anticodon domain, notations **R** and **S** refer to N-[(9-b-D-ribofuranosyl-2-methylthiopurin-6-yl) carbamoyl] threonine and 2-thio-5 carboxymethyl uridine methyl eser, respectively. Structures of modified bases are indicated in the lower portion of the figure. The structure for bovine tRNA[Lys,3] is presented; sequence studies with the human equivalent are comparable with respect to both the modified and highly modifies bases (G. Keith, personal communication). Arrowheads indicate the bases of tRNA[Lys,3] anticodon domain in close contact with HIV-1 RT[25].

Acknowledgements

I wish to thank several of my former colleagues at Hoffmann-La Roche, Basel, as well as Roche Products Ltd., England, with whom fruitful discussions and collaborations resulted in certain experiments reported here. Contributions of T.Naas, F.Cromme, F.Barat, D. Lindemann, N.Borkaboti, J. Mills J.Mous, F.Grueninger-Leitch are gratefully acknowledged, as is the expert technical assistance of R.Ette, C.Udri, S.Reutener and K. Yasargil. I wish also to thank K.Morikawa, H.Heumann and J.L.Darlix for prmission to communicate data prior to publication.

References

1. H.Mitsuya, K.J.Weinhold, P.A.Furman, M.H.St.Clair, S.Nusinoff-Lehrmann R.C.Gallo, D.Bolognesi, D.W.Barry, and S.Broder, **Proc.Natl.Acad.Sci.USA 82**, 7096-7100 (1985).

2. R.Yarchoan, H. Mitsuya and S. Broder, **Scientific American, (10/1988)**, 110-119.

3. J.J.McGowan, C.Litterst and M.I.Johnston, In **'Antiviral Chemotherapy' Vol II**(Eds. J. Mills and L.Corey,) Elsevier Publishing   pp 333-345 (1989).

4. B.Larder and S.D.Kemp, **Science 246**, 1155-1157 (1989).

5. D.Baltimore, **Nature (London) 226**, 1209-1211 (1970) .

6. H.M. Temin, and S.Mizutani, **Nature (London) 226**,1211-1213 (1970).

7. K.Moelling, D.Bolognesi, H.Bauer, W.Busen, W.Plassmann, and P.Hausen, **Nature New Biol. 234**, 240-244 (1971).

8. F.Harada, R.C.Sawyer and J.E.Dahlberg, **J.Biol.Chem. 250**, 3487-3497 (1975).

9. F.Harada, G.G.Peters and J.E. Dahlberg, **J.Biol.Chem. 254**, 10979-10985 (1979).

10. G.G.Peters, and C. Glover, **J.Virol. 33**, 708-716 (1980).

11. S.Wain-Hobson, P.Sonigo, O.Danos, and M.Alizon, **Cell 40**, 9-17 (1985).

12. J.G, Levin, and J.G.Seidmann, **J.Virol.29**, 328-335 (1979).

13. M.S.Johnson, M.A.McClure, D.-F. Feng, J.Gray and R.F.Doolitle, **Proc.Natl.Acad.Sci.USA 83**, 7648-7652 (1986).

14. J.W.Taylor, J.Ott, and F.Eckstein, **Nucl.Ac.Res.24**, 8756-8785 (1985).

15. P.Y.Chou, and G.D.Fasman, **Ann.Rev.Biochem.47**, 251-276 (1978).

16. J.Garnier, D.J.Osguthorpe, and B.Robson, **J.Mol.Biol.120**, 97-120 (1978).

17. O.Schatz, F.Cromme, F.Grueninger-Leitch, and S.F.J.Le Grice, **FEBS Lett.** **257**, 311-314 (1989).

18. S.F.J.Le Grice and F.Grueninger-Leitch, **Eur.J.Biochem.187**, 307-314 (1990).

19. E.Hochuli, W.Bannwarth, H.Dobeli, R.Gentz, and D.Stueber, **BIO/TECHNOLOGY 6**, 1321-1325 (1988).

20. O.Schatz, J.Mous and S.F.J.Le Grice, **EMBO J.9**, 1171-1176 (1990).

21. H.Sheshberedaran, and L.G.Payne, **Proc.Natl.Acad.Sci.USA 85**,1. (1988).

22. V.Mizrahi,G.M.Lazarus, L.M.Miles, C.A.Meyers, and C.DeBouck, **Arch.Biochem.Biophys.273**, 347-358 (1989).

23. D.M.Lowe, A.Aitken, C.Bradley, G.K.Darby, B.A.Larder, K.L.Powell, D.J.M.Purifoy, M.Tisdale, and D.K.Stammers, **Biochemistry**, **27**, 8884-8889 (1988).

24. O.Schatz, F.Cromme, T.Naas, D.Lindemann, J.Mous, and S.F.J.Le Grice, In'**Gene Regulation and AIDS**'(T. Papas, Ed.) Portfolio Publishing Company, Texas, 304-315 (1990).

25. C.Barat, V.Lullien, O.Schatz, G.Keith, M.T.Nugeyre, F.Grueninger-Leitch, F.Barre-Sinoussi, S.F.J.Le Grice, and J.L.Darlix, **EMBO J. 8**, 3279-3285 (1989).

26. M.A.Tukalo, M.D.Kubler, D.Kern, M.Mougele, C.Ehresmann, J.P.Ebel, B.Ehresmann, and R.Giege, **Biochemistry, 26**,5200-5208 (1987).

27. V.Prasad, and S.P.Goff, **J.Biol.Chem. 264**, 16689-16693 (1989).

28. B.Larder, D.Purifoy, K.Powell, and G.Darby, **EMBO J.6**, 3133-3137 (1987).

29. A.Hizi, C.McGill, and S.H.Hughes, **Proc.Natl.Acad.Sci.USA 85**, 1218-1222 (1988).

30. J.Hansen, T.Schulze, W.Mellert, and K.Moelling, **EMBO J.7**,239-243 (1988).

31. G.Radziwill, W.Tucker, W. and H.Schaller, **J.Virol.64**, 613-620 (1990).

32. S.Kanaya, A.Kohara, Y.Miura, A.Sekiguchi, S.Iwai, H.Inoue, E.Ohtsuka, and M.Ikehara, **J. Biol. Chem. 265**, 4615-4621 (1990).

33. H.Varmus, and R.Swanstrom, In '**RNA Tumor Viruses**' (Eds. R. Weiss, N.Teich, H.Varmus and J.Coffin) Cold Spring Harbor Monograph Series, Cold Spring Harbor, N.Y. pp 369-512 (1984).

34. B.Bordier, L.Tarrago-Litvak, M.-L.Sellafranque-Andreola, D. Robert, D.Tharaud, M.Fournier,P.J.Barr,S.Litvak and L.Sarih-Cottin, **Nuc.Ac.Res.18**,429-436 (1990).

HUMAN-SPECIFIC FACTORS ARE REQUIRED FOR TAT-

MEDIATED *TRANS*-ACTIVATION OF THE HIV-1 AND HIV-2 LTRS

Michael Newstein and Peter R. Shank

Division of Biology and Medicine
Brown University
Providence, R I

Introduction

The regulation of HIV gene expression is a complex process involving the interaction of viral regulatory gene products and host cell factors. The Tat gene has been shown to play an essential role in the positive regulation of viral gene expression (6, 9). HIV-1 Tat mediates its effect through a sequence termed TAR, which has been genetically defined to reside in the R region of the LTR (19). The TAR region is present in the leader sequence of all viral encoded RNAs, as well as both ends the proviral DNA. The boundaries of HIV-1 TAR are +1 to +80 (where +1 indicates the initiation of transcription). The TAR region has the potential to fold into a stem-loop secondary structure when transcribed into RNA.

The mechanism by which Tat *trans*-activates the HIV LTR has yet to be clearly defined. However, evidence has been presented that Tat acts by increasing the rate of transcriptional initiation (5, 11, 16, 17), promoting transcriptional elongation (12, 13), and by post-transcriptional events (3, 5, 18). Recent reports have suggested a novel model of transcriptional activation in which Tat has functional interactions with the nascent viral RNA transcript and effects on upstream DNA elements within the U3 region of the LTR (1, 2, 22, 24).

Species Specificity of HIV-1 gene expression

Our group as well as other workers have shown that the HIV-1 Tat mediated *trans*-activation functions at a lower level in rodent cells as compared to human cells (14, 15, 21). Tat activity is routinely measured by using HIV-LTR-reporter gene fusions and Tat expression vectors in transient transfection experiments. We have used the human growth hormone (HGH) as a reporter construct (23). Several different human and rodent cell types were analyzed for their ability to support Tat mediated *trans*-activation on a HIV LTR-HGH reporter gene construct. As can be seen in figure 1, we have examined a variety of human cells including osteosarcoma cells (HOS), and cervical and colon carcinoma cells (HeLa and SW480 respectively). All of these cells exhibited levels of *trans*-

activation greater than 25 fold. Different rodent cell types were also assayed including mouse fibroblasts (NIH3T3), mouse kidney cells (RAG) and Chinese hamster ovary cells (CHO). All of the rodent cell types exhibited Tat *trans*-activation ratios of less than five fold. These data argue that Tat *trans*-activation is species-specific but shows relatively little specificity for different tissues within a species.

We can propose two basic models which explain the species specificity of Tat activity. The relative resistance of rodent

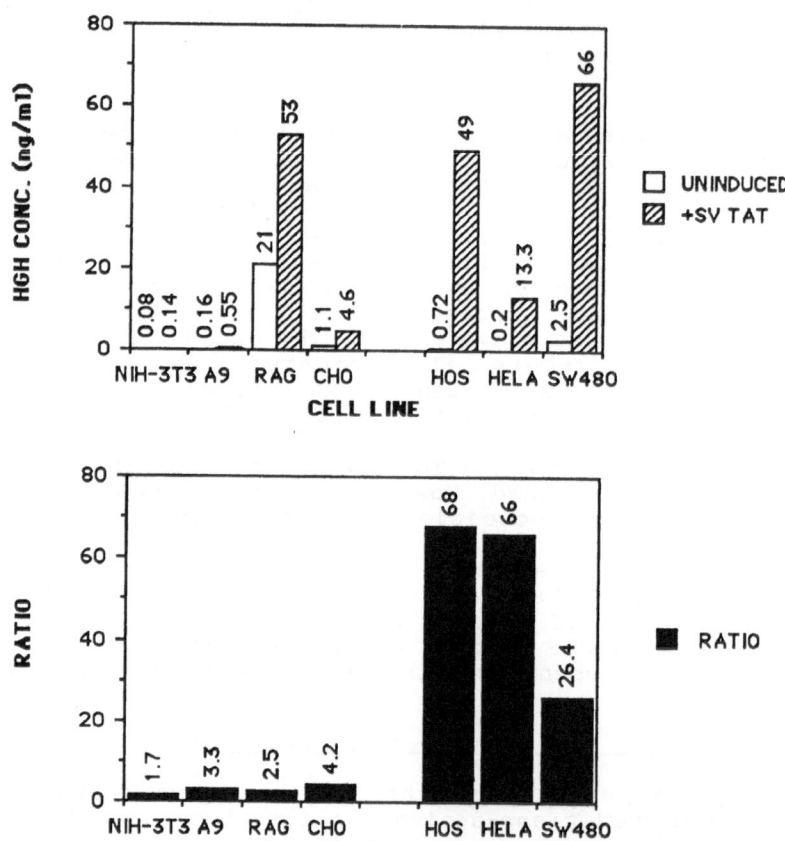

Fig. 1 Tat mediated *trans-activation* of HIV-1-HGH reporter plasmid in human and murine cells. (A) HIV-HGH expression following transient transfection. Cells were transfected by the $CaCl_2$ precipitation technique. The uninduced samples were transfected with 2ug of pHIV-1-HGH, while the plates co-transfected with pSVTAT received 2ug of pSVTAT along with 2ug of pHIV-1-HGH. All transfections were done with a total of 5ug of plasmid, with pBluescript as carrier. Culture media were assayed for HGH 72h after transfection by an immunoassay. (B) *Trans-activation* ratios of human and rodent cells. (Reproduced with permission from Reference 14).

cells to Tat *trans*-activation may be the result of a negative factor present in rodent cells. Conversely, human cells may possess a species specific Tat co-factor which is absent in rodent cells, or present in a less active form. To address this question, we have used rodent-human somatic cell and microcell hybrids in transient transfection experiments. These cells contain different compositions of human chromosomes along with the full complement of rodent chromosomes.

We have screened several rodent-human cell hybrids with different sets of human chromosomes. Some cell hybrids supported a higher level of Tat *trans*-activation as compared to the parental rodent cell lines (data not shown). This result suggests that mouse cells lack a co-factor, which is provided by the human chromosomes in the hybrid cell. This model is in agreement with other workers who have found that Tat *trans*-activation is retained at a high level in transient mouse-human heterokaryons (15).

Results of a transfection of a human-hamster hybrid containing human chromosomes 6, 8, and 11 suggested that some subset of these chromosomes was associated with an increased level of Tat mediated *trans*-activation above what is seen in the parental hamster cells (data not shown). To assess the effect of individual human chromosomes on Tat activity, we analyzed a set of mouse-human microcell hybrids. These hybrids are derived from mouse A9 fibroblasts. The experimental procedures for generating microcell hybrids, which are mouse A9 cell clones containing single human chromosomes have been published previously (20). Dominant selectable markers, e.g. the bacterial neomycin resistance gene (neo^r), are integrated into the human chromosome and provide a selective force (in the presence of the antibiotic G418) for the retention of the chromosome. Thus the human chromosomal content of the hybrids is stable when the cells are passaged in the presence of the antibiotic G418. The stability of the human chromosomes in these hybrids confers an advantage over the traditional rodent-human hybrids which may lose human chromosomes at each passage and therefore possess unequal chromosome contents.

Other workers who have screened a panel of hamster-human hybrids containing multiple human chromosomes have found that chromosome 12 is associated with a higher level of Tat activity (10). We have analyzed a set of mouse- human mono-chromosome hybrids containing single copies of chromosome 6, 8, 11, or 12. As can be seen in figure 2, the mono-chromosome hybrid containing human chromosome 12 supported the higher level of *trans*-activation, approximately ten times the level of A9 cells, but less than half of the level seen in HOS or HeLa cells. In addition, the chromosome 6 mono-chromosome hybrid supports a level of Tat activity approximately three times the level seen in the A9 cells.These data argue that a factor encoded by human chromosome 12 can complement murine cells so that they support a significantly greater level of Tat *trans*-activation. We have also determined that the chromosome 12 effect was not unique to this particular hybrid clone. Results of a transfection of a CHO-based microcell hybrid (26) containing chromosome 12 parallel those obtained with the A9 based hybrid (figure 2). Thus, a factor encoded by human chromosome 12 can complement rodent cells derived from two different species and originating from two different tissues. One could argue that the effects we see are not the result of a Tat co-factor but instead refect different levels of tat expression mediated by a factor stimulating the SV40 promoter. We believe this is not likely since we have also obtained comparable data using HIV-1 Tat expression vectors driven by the cytomegalovirus immediate-early promoter or by the HIV-1 LTR.

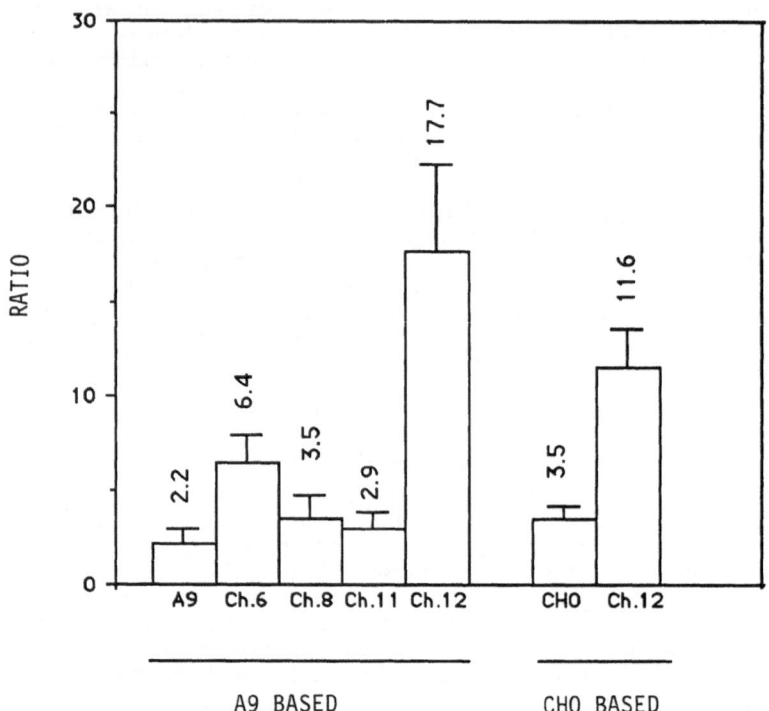

Fig. 2. Tat-mediated *trans*-activation of human-rodent microcell hybrids
Mouse A9 fibroblasts and CHO cells as well as A9 and CHO based microcell hybrids were transfected as described in the legend to Fig. 1. The ordinate represents the mean values of three independent transfections, with error bars representing the standard deviation calculated from these experiments. (Reproduced with permission from Reference 14).

Species specificity of HIV-2 gene expression

We have compared the requirements of the HIV-2 Tat protein and TAR element for human specific factors to the HIV-1 system. Interestingly, the genome of HIV-2 is more related to SIV virus of macaque monkeys than to HIV-1 (4). The Tat protein of HIV-2 shares 42% amino acid homology to the HIV-1 Tat, while the HIV-2 TAR sequence shares a 68% nucleotide sequence homology to the HIV-1 TAR region. However, the HIV-2 TAR sequence has an additional 80 bp region which is not present in HIV-1 TAR (8). This additional sequence in HIV-2 TAR has the potential to form

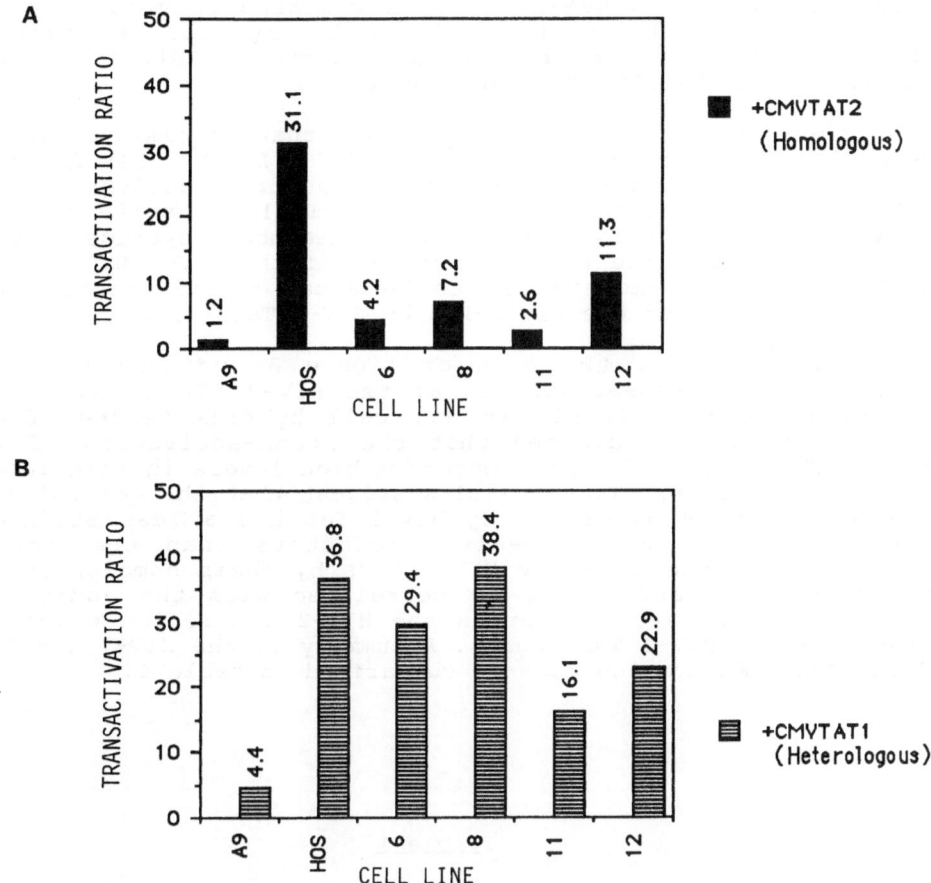

Fig. 3. Homologous and heterologous Tat *trans*-activation of HIV-2. Cyto-megalovirus immediate early promoter driven Tat expression plasmid. CMVTAT2 (pBC12 /CMV/Tat-2) was provided by Dr. Bryan Cullen. HIV-2 HGH was constructed from LTR.2CAT, which was provided by Dr. Mark Muesing. *Trans*-activation ratios were calculated following transient co-transfection as described above.

a second loop in the TAR RNA transcript. Therefore, the divergence of the HIV-2 Tat and TAR sequences may be associated with a different requirement for species-specific factors.

Figure 3 shows the results of an experiment in which we measured the *trans*-activation of HIV-2 LTR in HOS cells, A9 cells, and A9 based microcell hybrids. As is the case with the HIV-1 system, HIV-2 Tat *trans*-activates the HIV-2 LTR much more efficiently in the human HOS cells as compared to the murine A9 cells. The presence of chromosome 12 likewise appears encode a major factor which enhances HIV-2 Tat *trans*-activation of the HIV-2 LTR in rodent cells. Human chromosome 8 seems to encode an additional factor which specifically increases HIV-2

Tat *trans*-activation of the HIV-2 LTR. We have also determined that SIV Tat *trans*-activation of the SIV LTR functions much more efficiently in human cells as compared to murine cells. Furthermore, rodent-human microcell hybrids containing chromosome 12 also support a greater degree of SIV Tat *trans*-activation of the SIV LTR (data not shown).

Previous reports have demonstrated that in human cells the heterologous *trans*-activation of the HIV-1 LTR by HIV-2 Tat is 10% to 30% as efficient as the homologous *trans*-activation with HIV-1 Tat.(8, 25). We have observed similar results in both human and rodent cells as well as the microcell hybrids (figure 3). In contrast, the *trans*-activation of the HIV-2-LTR by HIV-1 Tat has been demonstrated to be equal to or greater than *trans*-activation of the HIV-2-LTR by HIV-2 Tat (8, 25).

Interestingly, the data of figure 3 indicate that the heterologous *trans*-activation of the HIV-2 LTR by HIV-1 Tat functions at high levels in all cell hybrids tested. Other experiments have indicated that the *trans*-activation of the HIV-2 LTR by HIV-1 Tat also occurs a high levels in some rodent cells (not shown). These results suggest that the heterologous *trans*-activation HIV-2 LTR by HIV-1 Tat has a less stringent requirement for human-specific cofactors than the *trans*-activation of the HIV-1 or HIV-2 LTRs by their homologous Tat proteins. This effect may be correlated with the additional sequences which are present in the HIV-2 TAR region which are absent in the HIV-1 TAR region. A summary of the HIV-1 and HIV-2 Tat *trans*-activation data is summarized in table 1.

Table 1

Summary of HIV-1 and HIV-2 Transactivation Data

			Microcell Hybrid:			
	Rodent	Human	6	8	11	12
HIV-1 LTR						
+TAT1	+	++++	++	+	+	+++
+TAT2	+	+++	+	+	+	++
HIV-2 LTR						
+TAT1	++*	++++	+++	+++	+++	+++
+TAT2	+	++++	+	++	+	+++

*Variation between rodent cell types

+	1 to 5 fold
++	5 to 10 fold
+++	10 to 20 fold
++++	> 20 fold

HUMAN CELL

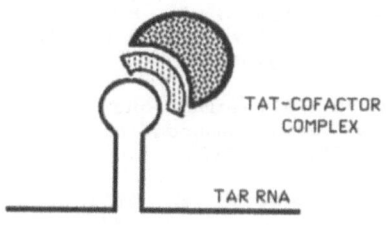

TAT-COFACTOR
COMPLEX

TAR RNA

RODENT CELL

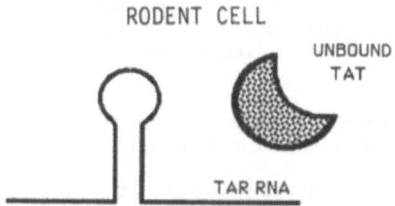

UNBOUND
TAT

TAR RNA

Fig. 4.   A model for the species specificity of HIV
gene expression

Conclusion

The regulation of HIV gene expression by Tat may represent a new type of eukaryotic gene control. The requirement of both a nascent transcript RNA structure (TAR) and upstream DNA enhancer regions for Tat *trans*-activation suggests that this process may be mediated by a protein-nucleic acid complex which includes Tat, TAR RNA, upstream enhancer binding proteins, as well as other host cell proteins. The species specificity of Tat *trans*-activation provides us with a unique opportunity to study this process. We are currently focusing on the role of human factors in the interaction of the Tat protein and TAR RNA transcripts. Preliminary results suggest that chromosome 12 may encode a protein which interacts with TAR RNA. Although there have been reports that Tat interacts directly with TAR RNA (7, 27) *in vitro*, it is likely that cellular proteins play an important role in the functional TAR RNA-TAT interaction. We have obtained preliminary data which suggests that human chromosome 12 encodes a TAR binding protein which is not detected in rodent cells. A model for the role of this protein is presented in figure 4.

Acknowledgements

We thank B. Cullen and M. Muesing for the generous gift of plamid constructs. This work was supported by grants AI-25531 and Amfar 000489.

References

1. **Berkhout, B., A. Gatignol, A. B. Rabson and K.-T. Jeang.** 1990. TAR-independent activation of the HIV-1 LTR: Evidence that tat requires specific regions of the promoter. Cell **62**: 757-767.

2. **Berkhout, B., R. H. Silverman and K.-T. Jeang.** 1989. Tat trans-activates the human immunodeficiency virus through a nascent RNA target. Cell **59**: 273-282.

3. **Braddock, M., A. Chambers, W. Wilson, M. P. Esnouf, S. E. Adams, A. J. Kingsman and S. M. Kingsman.** 1989. HIV-1 TAT "activates" presynthesized RNA in the nucleus. Cell **58**: 269-279.

4. **Clavel, F., M. Guyader, D. Guetard, M. Salle, L. Montagnier and M. Alizon.** 1986. Molecular cloning and polymorphism of the human immune deficiency virus type 2. Nature **324**: 691-695.

5. **Cullen, B. R.** 1986. Trans-activation of human immunodeficiency virus occurs via a bimodal mechanism. Cell **46**: 973-982.

6. **Dayton, A. I., J. G. Sodroski, C. A. Rosen, W. C. Goh and W. A. Haseltine.** 1986. The trans-activator gene of the human T cell lymphotropic virus type III is required for replication. Cell **44**: 941-947.

7. **Dingwall, C., I. Ernberg, M. J. Gait, S. M. Green, S. Heaphy, J. Karn, A. D. Lowe, M. Singh, M. A. Skinner and R. Valerio.** 1989. Human immunodeficiency virus 1 tat protein binds trans-activation-responsive region (TAR) RNA in vitro. Proc. Natl. Acad. Sci. U.S.A. **86**: 6925-6929.

8. **Emerman, M., L. Guyader, L. Montagnier, D. Baltimore and M. Muesing.** 1987. The specificity of the human immunodeficiency virus type 2 trans-activator is different from that of human immunodeficiency virus type 1. EMBO J. **6**: 3755-3762.

9. **Fisher, A. G., M. B. Feinberg, S. F. Josephs, M. E. Harper, L. M. Marselle, G. Reyes, M. A. Gonda, A. Aldovini, C. Debouk, R. C. Gallo and F. Wong-Staal.** 1986 . The trans-activator gene of HTLV-III is essential for virus replication. Nature **320**: 367-371.

10. **Hart, C. E., C.-Y. Ou, J. C. Galphin, J. T. Moore, J. J. Bachler, J. J. Wasmuth, S. R. Petteway and Schochetman.** 1989. Human chromosome 12 is required for HIV-1 expression in human-hamster hybrid cells. Science **246**: 488-491.

11. **Jeang, K.-T., P. R. Shank and A. Kumar.** 1988. Transcriptional activation of homologous viral long terminal repeats by the human immunodeficiency virus type 1 or the human T-cell leukemia virus type I tat proteins occurs in the absence of de novo protein synthesis. Proc. Natl. Acad. Sci. U.S.A. **85**: 8291-8295.

12. **Kao, S. Y., A. F. Calman, P. A. Luciw and B. M. Peterlin.** 1987. Anti-termination of transcription within the long terminal repeat of HIV-1 by tat gene product. Nature **330**: 489-493.

13. **Laspia, M. F., A. P. Rice and M. B. Mathews.** 1989. HIV-1 tat protein increases transcriptional initiation and stabilizes elongation. Cell **59**: 283-292.

14. **Newstein, M., E. J. Stanbridge, G. Casey and P. R. Shank.** 1990. Human chromosome 12 encodes a species-specific factor which

increases human immunodeficiency virus type 1 tat-mediated *trans*-activation in rodent cells. J. Virol. **64**: 4565-4567.

15. **Pavalakis, G. N., B. K. Felber and C. M. Wright.** 1988. A fusion assay for the detection of HIV infected cells. p. 439-466 *In* D. Bolognesi (Ed.) Human retroviruses, cancer, and AIDS. Alan R. Liss, Inc, New York.

16. **Peterlin, B. M., P. A. Luciw, P. J. Barr and M. D. Walker.** 1987. Elevated levels of mRNA can account for the *trans*-activtion of human immunodeficiency virus. Proc. Natl. Acad. Sci. USA **83**: 9734-9738.

17. **Rice, A. and M. Mathews.** 1988. Transcriptional but not translational regulation of HIV-1 by the tat gene product. Nature **332**: 551-553.

18. **Rosen, C.A., J.G. Sodroski, W.C. Goh, A.I. Dayton, J. Lippke, and W.A. Haseltine.** 1986. Post-transcriptional regulation accounts for the *trans*-activation of the human T-lymphotropic virus type III. Nature 319: 555-559.

19. **Rosen, C. A., J. G. Sodroski and W. A. Haseltine.** 1985. Location of cis-acting regulatory sequences in the human T-cell leukemia virus type I long terminal repeat. Proc. Natl. Acad. Sci. U.S.A. **82**: 6502-6506.

20. **Saxon, P. J., E. S. Srivatsan, G. V. Leipzig, J. H. Sameshima and E. J. Stanbridge.** 1985. Selective transfer of individual human chromosomes to recipient cells. Mol. Cell. Biol. **5**: 140-146.

21. **Seigel, L. J., L. Ratner, s. F. Josephs, D. Derse, M. B. Feinberg, G. R. Reyes, S. J. O'Brien and F. Wong-Staal.** 1986. Transactivation induced by human T-lymphotropic virus type III (HTLV-III) maps to a viral sequence encoding 58 amino acids and lacks tissue specificity. Virology **148**: 226-231.

22. **Selby, M. J. and B. M. Peterlin.** 1990. Trans-activation by HIV-1 tat via a heterologous RNA binding protein. Cell **62**: 769-776.

23. **Selden, R. F., K. Burke, M. E. Rowe, H. M. Goodman and D. D. Moore.** 1986. Human growth hormone as a reporter gene in regulation studies employing transient gene expression. Mol. Cell. Biol. **6**: 3173-3179.

24. **Southgate, C., M. L. Zapp and M. R. Green.** 1990. Activation of transcription by HIV-1 tat protein tethered to nascent RNA through another protein. Nature **345**: 640-642.

25. **Viglianti, G. A. and J. I. Mullins.** 1988. Functional comparison of *trans*-activation by simian immunodeficiency virus from Rhesus macaques and human immunodeficiency virus type 1. J. Virol. **62**: 4523-4532.

26. **Warburton, D., S. Gersen, M.-T. Yu, C. Jackson, B. Handelin and D. Housman.** 1990. Monochromosomal hybrids from microcell fusion of human lymphoblastiod cells containing a dominant selectable marker. Genomics **6**: 358-366.

27. **Weeks, K. M., C. Ampe, S. C. Schultz, T. A. Steitz and D. M. Crothers.** 1990. Fragments of the HIV-1 Tat protein specifically bind Tar RNA. Science **249**: 1281-1285.

ROLE OF THE TAR ELEMENT IN REGULATING HIV GENE EXPRESSION

Richard Gaynor

UCLA School of Medicine
Division of Hematology-Oncology
Los Angeles, CA 90024

The human immunodeficiency virus long terminal repeat (HIV LTR) is regulated by multiple *cis*-acting regulatory sequences.[1] Cellular factors which bind to the enhancer, SP1, TATA, and TAR regions are involved in the transcriptional regulation of HIV.[2] In addition to cellular proteins, the viral protein, *tat*, has been shown to greatly stimulate HIV gene expression from the HIV LTR.[3, 4]

Tat is an 86 amino acid protein which is highly conserved among HIV isolates.[3, 4] It contains several different domains[5], including an amino-terminal activating domain, a cysteine-rich region which functions in dimer formation, and a basic domain that functions in binding to RNA and nuclear localization.

A number of different studies have attempted to determine the role which *tat* plays in stimulating HIV gene expression.[6-10] *Tat* functions to increase both the transcription rate and the length of the synthesized transcripts.[9] Mutagenesis of the TAR region to alter the stem-loop structure in TAR RNA strongly suggests that the TAR RNA rather than the TAR DNA was critical for *tat* activation.[11-13] These studies were consistent for a role of *tat* on transcriptional initiation. However, other studies using microinjection of *tat* with pre-synthesized TAR RNA suggested that *tat* also had post-transcriptional effects.[14]

A number of studies using fusions of *tat* with either DNA or RNA binding proteins have been performed to elucidate whether *tat* interacts primarily with either DNA or RNA.[15-17] It was found that fusions of *tat* and *rev* are capable of activating the HIV LTR when the *rev* response element is placed in the position of the TAR element.[15] Likewise, when a fusion of *tat* with the procaryotic RNA binding protein R17 was created, and the R17 binding site inserted into TAR, the *tat* -R17 fusion was capable of activating the HIV LTR.[16] These results would be consistent with a role of TAR RNA acting as an anchor to bind *tat* which is then capable of interacting with the transcriptional initiation complex. However, another study in which the AP1 binding site was placed into the TAR region and fusions of *tat* and c-*jun* were cotransfected indicated that *tat* could act at the DNA level likely by interacting with the enhancer and SP1 binding domains.[17] Thus *tat* may interact with cellular proteins binding to HIV at both the DNA and RNA levels.

We have focused our attention on cellular proteins that bind to TAR RNA and DNA to determine their potential interaction with *tat*. A variety of cellular proteins have been demonstrated to bind to TAR DNA.[18, 19] These proteins include UBP1, LBP, and CTF. In addition, a variety of other less well described proteins also have been shown to

Advances in Molecular Biology and Targeted Treatment for AIDS
Edited by A. Kumar, Plenum Press, New York, 1991

bind to TAR. Cellular proteins have also been demonstrated to bind to the TAR RNA.[20-22] One of these proteins has been demonstrated to bind to the TAR loop sequences, a site which has been demonstrated to be critical for *tat* activation. In addition, it is likely that *tat* can also bind to the bulge region in TAR and this region has been demonstrated to be critical for activation of HIV gene expression.[23-26] We will describe our attempts to identify cellular proteins that bind to both TAR DNA and RNA that are important in HIV regulation. In addition, we will describe viral mutants in the TAR element and how these mutations affect viral gene expression.

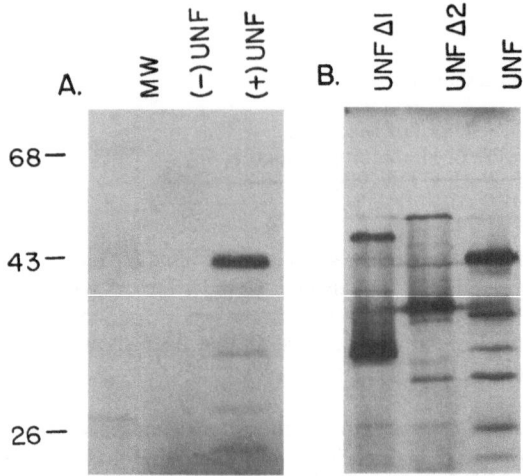

Fig. 1. *In vitro* translation of UNF RNA. (A) *In vitro* translation in either the presence or absence of UNF RNA; (B) Plasmids were truncated prior to RNA preparation and *in vitro* translation to generate truncated proteins UNFA1 and UNFA2.

Cloning of UNF; a cellular factor that binds to TAR DNA and RNA

In an attempt to identify cellular proteins that bind to TAR DNA, lambda GT11 expression cloning was performed. Ligated oligonucleotides complementary to TAR DNA extending from -18 to +25 were used to screen a lambda GT11 HeLa cDNA library and a number of positive plaques were identified. These positive phage were also screened with mutant TAR oligonucleotides and did not bind specifically. A number of these phage were found by DNA sequence analysis to encode the same cDNA and were studied in more detail.

DNA sequence analysis revealed that one of these phage contained a 3.0 kb. cDNA insert. This DNA had an open reading frame which encoded a protein of molecular weight 46 kd. A homology search revealed that there was strong homology within the putative binding domain of this protein with a ribonucleoprotein particle concensus sequence. This sequence of 60 to 80 amino acids is found in a variety of snRNP binding proteins and is involved in RNA binding. In addition, this domain is found in a variety of helix destabilizing proteins that are capable of binding single stranded DNA. There were also glycine-rich and basic domains in the carboxy-terminus of this cDNA known as UNF, or untranslated nuclear factor. Site-directed mutagenesis of each of these various regions is being performed to assay their roles on gene expression and TAR DNA binding.

UNF was cloned and expressed from bacterial expression vectors and the proteins purified by affinity chromatography. These proteins were assayed for their ability to bind to both TAR DNA and RNA using gel retardation assays. UNF bound to TAR DNA specifically and this binding was eliminated by UNF constructs containing mutations in either of the RNP concensus sequences. These results suggest that the RNP concensus sequences were critical for DNA binding. Likewise, UNF was also capable of binding to TAR RNA in gel retardation assays. Again, mutations of the RNP concensus domains eliminated this binding. These results suggest that the RNP concensus sequences were also critical for the RNA binding properties of UNF.

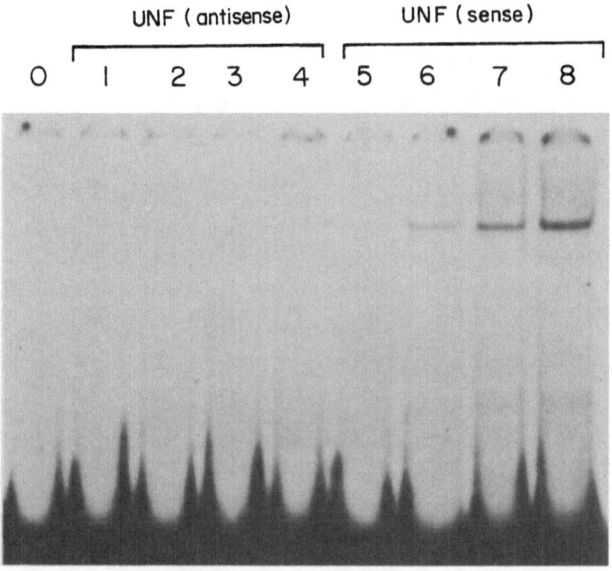

Fig. 2.   Binding of UNF to TAR DNA. Oligonucleotides corresponding to the
TAR DNA sequences were used in gel retardation with *in vitro* translated
sense and antisense UNF RNA.

To determine whether UNF was involved in regulating the gene expression of HIV, it was critical to assay its function *in vivo*. Northern analysis was performed on a variety of cell lines and the 3.0 kb. UNF transcript was identified on a variety of cell lines. However, on the murine teratocarcinoma cell line, F9, it was found that the UNF transcript was non-detectable. This provided a cell line with which transfection of an expression construct containing UNF could be assayed for its role in regulating HIV gene expression. An HIV LTR construct was fused to the chloramphenicol acetyltransferase gene (CAT) and transfected onto F9 cells in both the presence and absence of the *tat* gene. In addition, UNF expression constructs were added to these transfections. In the presence of UNF and HIV LTR CAT, there was a 7 to 10-fold increase in gene expression as compared to the absence of UNF.

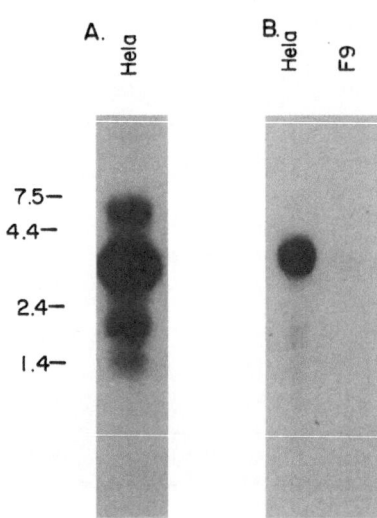

Fig. 3.   Northern analysis of UNF mRNA. Both HeLa and F9 mRNA were
used in Northern analysis with a random-primed UNF cDNA fragment.

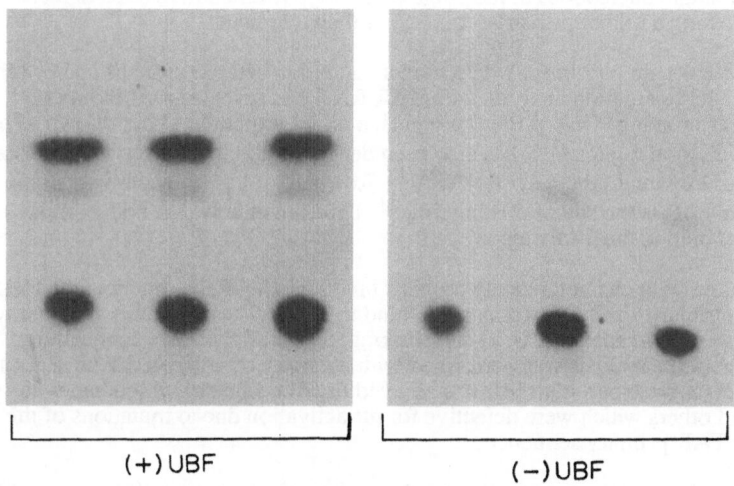

(+)UBF   (−)UBF

Fig. 4.   CAT assays of the HIV LTR in the presence and absence of UNF
         (UBF). The HIV LTR CAT plasmid was transfected into F9 cells in
         both the presence and absence of UNF expression constructs.

This suggests that UNF is a positively acting factor involved in HIV gene expression. The results of adding UNF and *tat* with the HIV LTR CAT indicated that UNF was positively acting, though some variability of the level of induction occurred. To show that the induction of UNF was specific, it was found that it did not induce expression from the construct ΔTAR-sense which contains a severely mutated TAR element. Thus, we have identified a cellular factor, UNF, that is involved in regulating HIV gene expression. Whether UNF acts primarily at the DNA or RNA level remains the subject of further investigation.

Characterization of a cellular factor URBP-180 that binds to the TAR RNA loop sequences

Studies from a number of laboratories including our own, have shown that multiple determinants were required for TAR RNA function.[11-13] These include preservation of the stem structure, the bulge sequences, the primary sequence of the loop, and in part, the primary sequence of the stem.[11-13, 24] Mutagenesis of the TAR element has revealed that disruption of either side of the stem resulted in marked defects in *tat* activation. However, restoration of stem-base-pairing resulted in complete activation. Substitution of single nucleotides in the bulge or the loop resulted in marked defects in *tat*-activation indicating a role of primary sequence of these elements.

These results are consistent with a model that either cellular proteins and/or *tat* may bind to RNA. Gel retardation analysis with TAR RNA has revealed the binding of multiple cellular proteins. One of these proteins, a 68 kd protein has been shown to bind to the TAR RNA loop region.[22] It has also been demonstrated that bacterial synthesized *tat* is capable of binding to the TAR RNA.[23, 24] Recently, it has been shown that the bulge region in TAR is the site of this binding.[24] Thus, it is likely that both cellular proteins and *tat* bind to the TAR region.

In an attempt to characterize the proteins binding to TAR, we fractionated HeLa nuclear extract to purify cellular proteins that bind to TAR. These proteins were assayed by RNA gel retardation analysis, U.V. crosslinking, and uranyl acetate footprinting. In addition, competition analysis was performed with a variety of unlabeled TAR RNAs. These TAR RNAs were part of constructs assayed *in vivo,* some of which had wild-type phenotypes and others which were defective for *tat*-activation due to mutations of the stem, loop, or TAR primary sequence.

HeLa nuclear extract was fractionated over heparin agarose, S300, mono S FPLC, and hydroxyapatite columns. Multiple gel retarded species were detected using nuclear extract. However, after further fractionation, one prominent gel retarded species was seen as shown below. To determine the molecular weight of this protein, U.V. crosslinking was performed using labeled TAR RNA. As shown below, three closely spaced species were detectable of molecular weight between 180-200 kd. Thus, either one or multiple proteins bind to TAR RNA with a molecular mass of 180-200 kd. To determine the specificity of this binding, competition analysis with unlabeled TAR RNA was performed. The wild-type RNA resulted in complete competition of the gel retarded species. A mutant which interrupted the primary sequence of the loop resulted in minimal competition of the major gel retarded species. Likewise, disruption of the stem structure also resulted in markedly decreased levels of competition. Finally, mutations of the TAR region primary sequence resulted in wild-type levels of competition. However, coupling this mutation to the loop mutation resulted in less competition than the loop mutant alone. This indicated that the primary sequence of TAR in addition to the stem structure and loop sequence influenced binding to TAR RNA.

Assay for Purification
of URBP-180

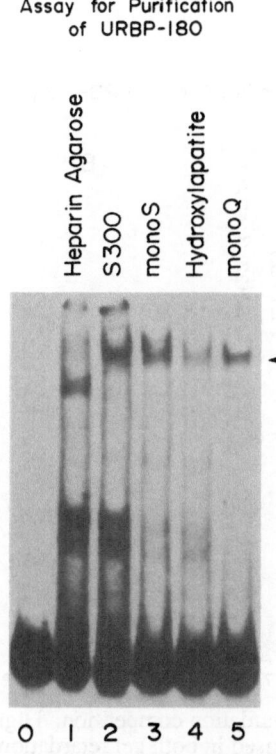

Fig. 5. RNA gel retardation of URBP-180. Cellular extracts were fractionated
on the indicated columns and active fractions which bound to TAR RNA
are indicated.

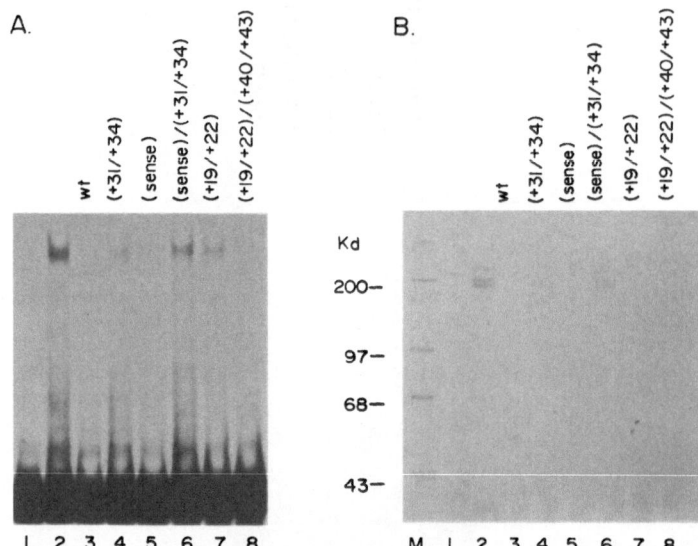

Fig. 6.   TAR RNA gel retardation competition. Highly purified fractions of
URBP-180 were used in both gel retardation (A) and UV crosslinking
(B) in the absence of competition or using a 50-fold excess of cold RNA
synthesized from wild-type, TAR loop mutant (+31/+34), TAR primary
sequence mutant (sense), a combined mutation (sense + +31/+34), TAR
stem disruption mutant (+19/+22), or a TAR stem restoration mutant
(+19/+22/+40/+43) constructs.

Thus, three major determinants of wild-type TAR function, the loop sequences, the stem structure, and the TAR primary sequence were all required for wild type binding of cellular proteins to TAR RNA. Further characterization of cellular proteins binding to TAR and their potential interactions with *tat* will be required to understand the mechanisms regulating HIV gene expression.

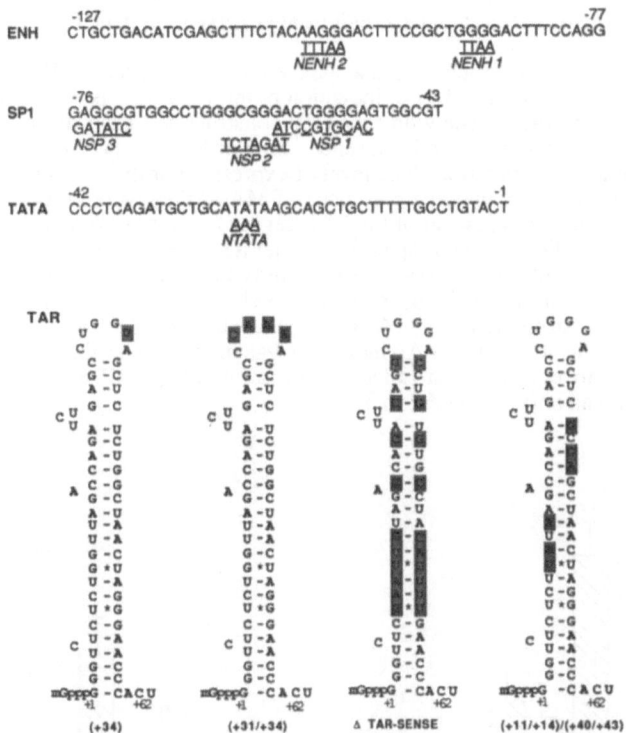

Fig. 7. Schematic of HIV LTR. Sequences in the HIV LTR mutated infectious virus in specific regulatory regions are indicated. These mutations were used to construct the mutant viruses indicated.

Regulation of TAR function in infectious viral constructs in unstimulated and PMA-stimulated Jurkat cells

To determine the role of the TAR element in the context of proviral constructs, mutated 5' and 3' LTRs containing mutations of either the TAR, enhancer, SP1, or TATA regions were inserted into infectious virus constructs. Each of the proviral constructs was introduced into Jurkat cells by electroporation. At 24 hrs post-transfection, the transfections were divided and one set was untreated and the other treated with either phorbol ester (PMA) and phytohemaglutinin (PHA). Supernatants were harvested at 48 hrs post-transfection and p24 Ag levels were quantitated.

In untreated Jurkat cells, the wild-type construct yielded approximately 8,000 pg/ml of p24 Ag antigen (Ag). Mutations of both kappa B motifs in the enhancer resulted in less than a two-fold decrease in p24 Ag levels. Mutations of either the three SP1 binding sites or the TATA element resulted in levels of p24 only slightly above detection limits. Mutations of both the loop and stem structure were also tested. A single base-pair mutation of the loop sequences (+34) resulted in approximately a 10-fold decrease in p24 Ag levels compared to the wild-type construct. A four base pair substitution in the loop sequences (+31/+34) resulted in a 35-fold decrease in p24 Ag levels compared to the wild-type construct. Mutations in the primary sequence of the TAR region (ΔTAR sense) which left stem base pairing and stem energy intact exhibited a 2.5-fold decreased in gene expression. A mutation that disrupted stem base pairing in the TAR region (+11/+14/+40+43) resulted in a 40-fold decrease in gene expression. Thus multiple upstream and downstream regulatory regions alter early viral gene expression.

Each of these constructs was also assayed in the presence of PMA. Minimal differences were noted in the level of induction of these constructs by PMA alone or PMA in combination with PHA. The wild-type HIV construct was only slightly induced in the presence of PMA as compared to its absence. Similar low levels of PMA induction were seen with the enhancer mutation. The level of expression of the SP1 and the TATA mutants were extremely low in the presence of PMA as they were in its absence. However, the results with several of the TAR region mutants were unexpected. P24 Ag levels for both mutations in the loop region, either a single base pair mutation (+34) or a four base pair substitution (+31/+34) were strongly induced by the addition of PMA. This induction ranged from 8-fold (+34) to 20-fold (+31/+34). Similarly, a mutation that disrupted the stem structure (+11/+14/+40/+43) resulted in a 15-fold level of induction in p24 Ag levels. This level of p24 Ag induction was much greater than that seen with the wild-type or enhancer mutant constructs. A mutation of the TAR region primary sequence (ΔTAR-sense) was only induced 2 to 3-fold.

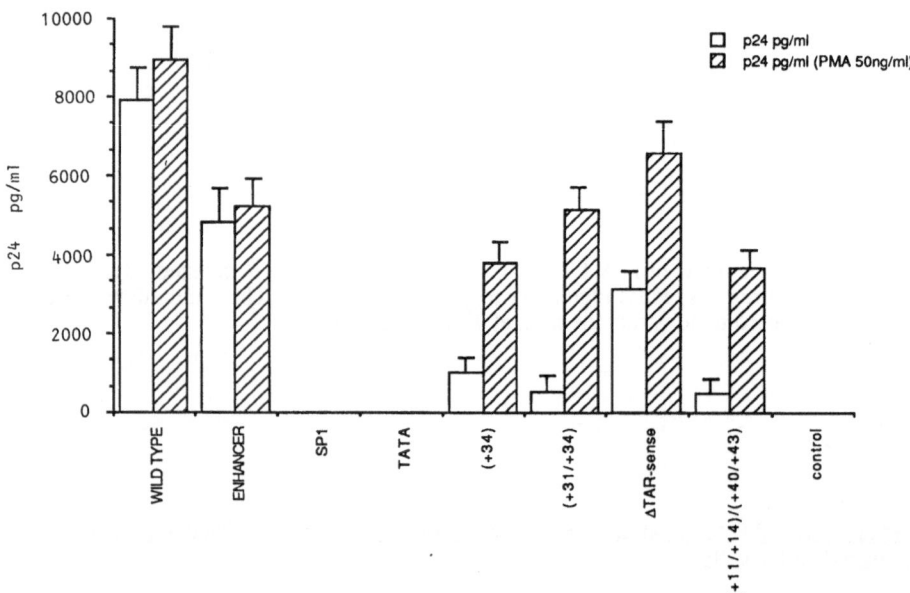

Fig. 8.   p24Ag expression of HIV mutants in unstimulated and PMA-stimulated Jurkat cells. Infectious viral constructs either wild-type or containing mutations in the enhancer, SP1, TATA, TAR loop (+34), TAR loop (+31/+34), TAR sense, and TAR stem disruption (+11/+14/+40/+43) were electroporated into Jurkat cells in either the presence or absence of PMA and p24Ag assays were performed 48 hrs post-transfection.

It was important to determine whether the *tat* protein was required for high levels of gene expression with HIV TAR mutants in PMA-treated Jurkat cells. A deletion of the cysteine domains in the *tat* gene was constructed which resulted in a protein defective for transactivation. Constructs containing this *tat* mutation included the wild-type construct, the enhancer mutant, the loop substitution mutant (+31/+34), and the stem disruption mutant (+11/+14/+40+43). As shown in Figure 4, the wild-type construct gave high levels of expression in Jurkat cells in both the presence and absence of PMA. However, constructs containing *tat* deletions were extremely defective in both the presence and the absence of PMA as compared to the wild-type construct.

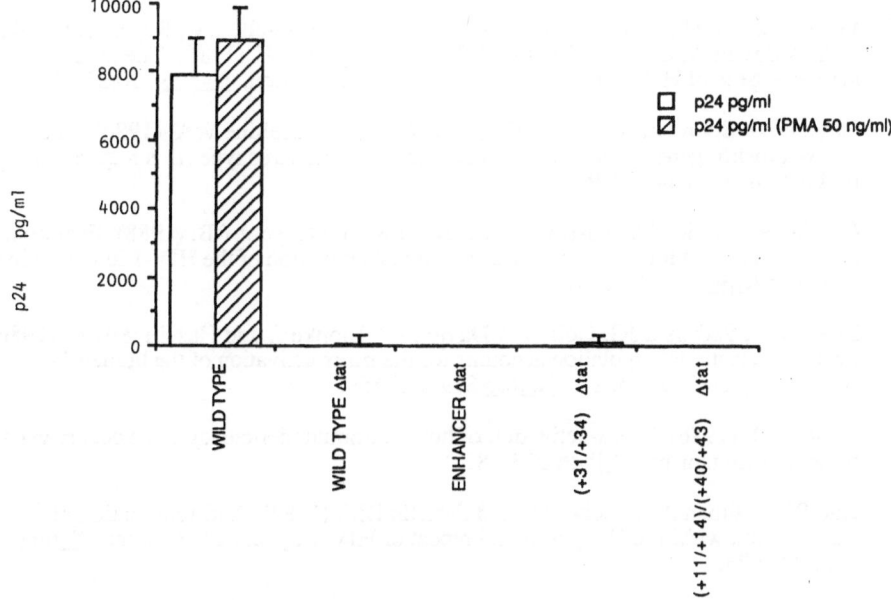

Fig. 9. p24 expression of HIV *tat* mutants in unstimulated and PMA-stimulated Jurkat cells. Infectious viral constructs either the wild-type or containing additional mutants in *tat* (wild-type, enhancer, TAR loop (+31/+34), or TAR stem disruption (+11/+14/+40/+43) were electroporated into Jurkat cells in both the presence or absence of PMA and p24Ag levels measured at 48 hrs post-transfection.

In activated Jurkat cells, two critical determinants of TAR function, the loop sequences and the stem secondary structure, were much less critical for gene expression than in unstimulated Jurkat cells. This was demonstrated by the fact that in the absence of PMA, loop and stem mutants were 10 to 35-fold less active than wild-type, while in the presence of PMA these same mutants are only 2 to 3-fold decreased compared to wild-type. The wild-type construct was only minimally induced by PMA as was the enhancer mutant. Furthermore, *tat* was required for high levels of gene expression with both the wild-type construct and with TAR mutants in activated Jurkat cells. These results suggest that in activated Jurkat cells, *tat* is able to interact with upstream promoter elements to increase HIV gene expression in a TAR-independent manner.

These results suggest that the TAR element can interact with multiple cellular proteins binding to both DNA and RNA. It is possible that *tat* is able to interact with these proteins. The effect of subtle mutations of the TAR element in viral gene expression reinforces the critical role of this region in viral replication. Further cloning and *in vitro* studies of cellular proteins and *tat* will help to understand the mechanism of HIV gene expression.

REFERENCES

1. Rosen CA, Sodroski JG, and Haseltine WA. (1985) The location of cis-acting regulatory sequences in the human T cell lymphotropic virus type III (HTLV-III/LAV) long terminal repeat. Cell 41:813-823.

2. Garcia JA, Harrich P, Pearson L, Mitsuyasu R, and Gaynor RB. (1987) Interactions of cellular proteins involved in the transcriptional regulation of the human immunodeficiency virus. EMBO J. 6:3761-3770.

3. Fisher AG, Feinberg MB, Josephs SF, Harper ME, Marselle LM, Reyes G, Gonda MA, Aldovini A, Debouk C, Gallo RC, and Wong-Staal F. (1986) The trans-activator gene of HTLV-III is essential for virus replication. Nature 320:367-371.

4. Dayton AI, Sodroski JG, Rosen CA, Goh WC, and Haseltine WA. (1986) The trans-activator gene of the human T cell lymphotropic virus type III is requred for replication. Cell 44:941-947.

5. Garcia JA, Harrich D, Pearson L, Mitsuyasu R, and Gaynor RB. (1988) Functional domains required for tat-induced transcriptional activation of the HIV-1 long terminal repeat. EMBO J. 7:3143-3147.

6. Rosen CA, Sodroski JG, Goh WC, Dayton AI, Lippke J, and Haseltine WA. (1986) Post-transcriptional regulation accounts for the trans-activation of the human T-lymphotropic-virus type III. Nature 319:555-559.

7. Cullen BR. (1986) Trans-activation of human immunodeficiency virus occurs via a bimodal mechanism. Cell 46:973-982.

8. Kao SY, Calman AF, Luciw PA, and Peterlin BM. (1987) Anti-termination of transcription within the long terminal repeat of HIV-1 by tat gene product. Nature 330:489-493.

9. Laspia MF, Rice AP, and Matthews MB. (1989) HIV-1 tat protein increases transcriptional initiation and stabilizes elongation. Cell 59:283-292.

10. Berkhout B, Silverman RH, and Jeang KT. (1989) Tat trans-activates the human immunodeficiency virus through a nascent RNA target. Cell 59:273-282.

11. Feng S, and Holland EC. (1988) HIV-1 tat trans-activation requires the loop sequence within tar. Nature 334:165-167.

12. Garcia JA, Harrich D, Soultanakis E, Wu F, Mitsuyasui R, and Gaynor RB. (1989) Human immunodeficiency virus type 1 LTR TATA and TAR region sequences required for transcriptinal regulation. EMBO J 8:765-778.

13. Selby MJ, Bain ES, Luciw PA, and Peterlin BM. (1989) Structure, sequence, and position of the stem-loop in tar determine transcriptional elongation by tat through the HIV-1 long terminal repeat. Genes & Dev. 3:547-558.

14. Braddock M, Chambers A, Wilson W, Esnouf MP, Adams SE, Kingsman AH, and Kingsman SM. (1989) HIV-1 tat "activates" presynthesized RNA in the nucleus. Cell 58:269-279.

15. Southgate C, Zapp ML, and Green ML. (1990) Activation of transcription by HIV-1 tat protein tethered to nascent RNA through another protein. Nature 345:640-642.

16. Selby MJ, and Peterlin BM. (1990) Trans-activation by HIV-1 *tat* via a heterologous RNA binding protein. Cell 62:769-776.

17. Berkhout B, Gatignol A, Rabson AB, and Jeang KT. (1990) TAR-independent activation of the HIV-1 LTR; evidence that *tat* requires specific regions of the promoter. Cell 62:757-767.

18. Wu FK, Garcia JA, Harrich D, and Gaynor RB. (1988) Purification of the human immunodeficiency virus type I enhancer and TAR binding proteins EBP-1 and UBP-1. EMBO J. 7:2117-2129.

19. Jones KA, Luciw PA, and Duchange N. (1988) Structural arrangements of transcription control domains within the 5' untranslated leader regions of the HIV-1 and HIV-2 promoter. Genes Dev. 2:1101-1114.

20. Gaynor R, Soultanakis E, Kuwabara M, Garcia J,. and Sigman DS. (1989) Specific binding of a HeLa cell nuclear protein to RNA sequences in the human immunodeficiency virus transactivating region. Proc. Natl. Acad. Sci. *USA* 86:4858-4862.

21. Gatignol A, Kumar A, Rabson A, and Jeang KT. (1989) Identification of cellular proteins that bind to the human immunodeficiency virus type 1 trans-activaton-responsive TAR element RNA. Proc. Natl. Acad. Sci. *USA* 86:7828-7832.

22. Marciniak RA, Garcia-Blanco MA, and Sharp PA. (1990) Identification and characterization of a HeLa nuclear protein that specifically binds to the trans-activation-response (TAR) element of human immunodeficiency virus. Proc. Natl. Acad. Sci. USA 87:3624-3628.

23. Berkhout B, and Jeang KT. (1989) *Trans*-activation of human immunodeficiency virus type 1 is sequence specific for both the single-stranded bulge and loop of the *rans*-acting-responsive hairpin: a quantitative analysis. J. Virol. 63:5501-5504.

24. Dingwall C, Ernberg I, Gait MJ, Green SM, Heaphy S, Karn J, Lowe AD, Singh M, Skinner MA, and Vallerio R. (1989) Human immunodeficiency virus 1 tat protein binds *trans*-activation responsive region (TAR) RNA in vitro. Proc. Natl. Acad. Sci. USA 86:6925-6929.

25. Roy S, Parkin C, Rosen J, Itovitch J, and Sonenberg N. (1990) Structural requirements for trans-activation of human immunodeficiency virus type 1 long terminal repeat-directed gene expression by *tat*: Improtance of base pairing, loop sequence and bulges in the *tat*-responsive sequence. J. Virol. 64:1402-1406.

26. Roy S, Delling U, Chen CH, Rosen CA, and Sonnenberg N. (1990) A bulge structure in HIV-1 TAR RNA is required for *tat* binding and *tat*-mediated trans-activation. Genes & Dev. 4:1365-1373.

REGULATION OF HIV-1 GENE EXPRESSION BY THE

TAT PROTEIN AND THE TAR REGION

Michael F. Laspia, Shobha Gunnery, Mark Kessler, Andrew P. Rice
and Michael B. Mathews

Cold Spring Harbor Laboratory
Cold Spring Harbor, New York 11724

INTRODUCTION

In addition to the structural genes gag, pol and env, the HIV family of human retroviruses encodes a number of novel regulatory genes (reviewed by Varmus, 1988; Cullen and Green, 1989; Pavalkis and Felber, 1990). One of these, the tat gene, is essential for viability and encodes a trans-acting regulator of HIV gene expression (Arya et al., 1985; Sodroski et al., 1985b). A cis-acting element in the long terminal repeat (LTR), called TAR, that is located downstream of the site of transcription initiation is required for Tat transactivation (Rosen et al., 1985; Jakobovits et al., 1988; Hauber and Cullen, 1988; Selby et al., 1989). The untranslated leader formed by transcription of TAR and present at the 5' end of all HIV-1 mRNAs is capable of forming a stem and loop structure (Muesing et al., 1987). Recent studies indicate that at least one role for TAR RNA in gene regulation is to provide a binding site for Tat in the vicinity of the HIV-1 promoter (Southgate et al., 1990; Selby and Peterlin, 1990; Berkhout et al., 1990). The exact mechanism of Tat stimulation of HIV-1 gene expression remains controversial (Sharp and Marciniak, 1989). While the primary effect appears to be transcriptional, evidence has also been provided for additional or even predominant posttranscriptional effects.

We have studied the mechanism of transactivation of HIV-1 gene expression by Tat and found that Tat interacts with TAR to increase HIV-1 transcription bimodally, by stimulating transcriptional initiation and also by increasing the processivity of elongating RNA polymerases (Laspia et al., 1989; Laspia et al., 1990). We have also obtained evidence for an additional regulatory role for the HIV-1 leader. RNAs possessing the HIV-1 leader sequence at their 5' end are capable of inhibiting the double-stranded (ds) RNA-activated protein synthesis inhibitor DAI (Gunnery et al., 1990). This suggests that TAR may also regulate translation.

RESULTS

Transcriptional Regulation of HIV-1 Gene Expression

A role for Tat in transcriptional regulation of HIV-1 gene expression was initially suggested by the observation that Tat dramatically increases LTR-promoted RNA levels (Cullen, 1986; Peterlin et al., 1986; Wright et al., 1986; Hauber et al., 1987; Muesing et al., 1987; Jakobovits et al., 1988; Rice and Mathews, 1988). This was confirmed by the finding that rates of reporter gene transcription are elevated by Tat (Hauber et al., 1987; Kao et al., 1987; Rice and Mathews, 1988; Jeang et al., 1988;

Advances in Molecular Biology and Targeted Treatment for AIDS
Edited by A. Kumar, Plenum Press, New York, 1991

Jakobovits et al., 1988; Sadaie et al., 1988). A model for transcriptional regulation has been proposed which suggests that Tat does not increase initiation of transcription, but instead acts as an antiterminator permitting elongation of RNA polymerases that would otherwise terminate transcription within TAR (Kao et al., 1987; Selby et al., 1989).

To investigate the molecular mechanism underlying Tat transactivation we have employed a recombinant adenovirus vector, containing an HIV-1 LTR-directed reporter gene, chloramphenicol acetyltransferase (CAT) to infect HeLa cells that stably express Tat. Using this model system, we found earlier (Rice and Mathews, 1988) that Tat promotes a large increase in the accumulation and rate of synthesis of HIV-1-directed CAT RNA, but does not affect the translation of CAT RNA. We have extended our investigation of Tat regulation of HIV-1 transcription with the recombinant adenovirus system. The effects of Tat on HIV-1 transcription rates and on the accumulation of cytoplasmic levels of LTR-directed RNA have been analyzed and compared with transactivation by the general transcriptional activator, the adenovirus E1A protein (Berk, 1986; Jones et al., 1988). Tat and general transcriptional activators synergistically stimulate HIV-1 transcription and we have also examined the mechanism underlying this effect.

Differential regulation of HIV-1 cytoplasmic RNA levels by Tat and E1A. Analysis of accumulation of HIV-1 promoted cytoplasmic RNA levels by ribonuclease protection assay has indicated that there are two classes of correctly initiated transcripts (Figure 1A). One class contains long, predominantly polyA+, RNA and corresponds to full length transcripts. The other class contains short, exclusively polyA-, RNA, approximately 55 to 59 nucleotides in length, and corresponds to termi-

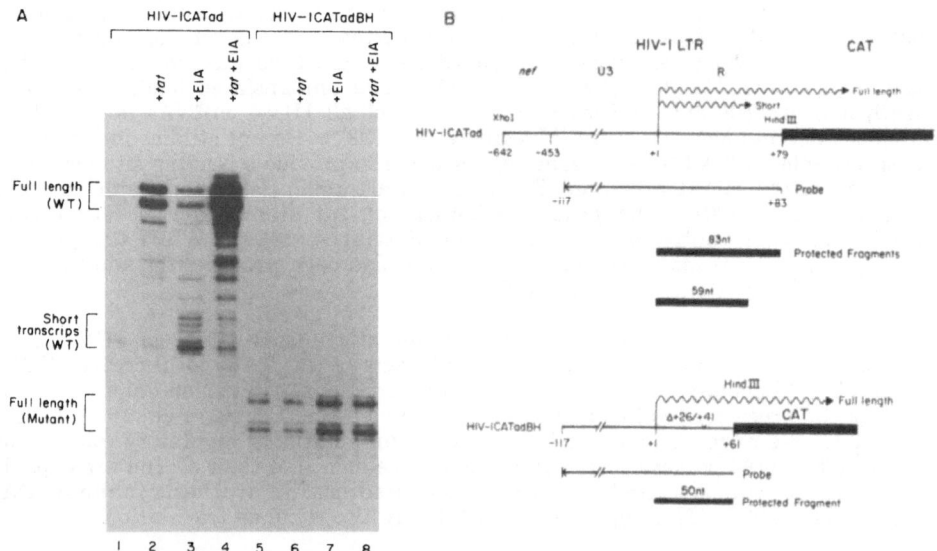

Fig. 1. Analysis of HIV-1-directed cytoplasmic RNA levels by RNase protection assay. (A) HeLa cells (lane 1,5), HeLa/tat cells (lane 2,6), and HeLa cells (lane 3,7) or HeLa/tat cells coinfected with dl309 (lane 4,8) were infected with HIV-1 CATad (lanes 1-4) or HIV-1CATadBH. Recombinant adenovirus infection was at an moi of 100; cytoplasmic RNA was isolated 24 hrs post-infection. RNase protections were performed as described (Laspia et al., 1989). (B) Schematic of the recombinant adenovirus and antisense riboprobe. HIV-1CATad contains the HIV-1 LTR sequence -642 to +83; HIV-1CATadBH -127 to +60 with a deletion of +26 to +41. Protected probe fragments corresponding to full length RNA and short transcripts are marked.

nated, possibly processed RNAs. Tat stimulates the level of full length RNA, but does not increase the level of the short transcripts. E1A, on the other hand, increases the level of both the full length and the short transcripts. A recombinant adenovirus containing a TAR deletion is not transactivated by Tat, as expected, but is by E1A (Figure 1A) indicating that TAR is not an essential promoter element. These results are consistent with Tat, as well as E1A, acting to increase HIV-1- promoted transcription. However, since Tat increases the level of full length RNA without increasing the short transcripts while E1A increases both full length and short transcript levels, it appears that Tat does not act solely to increase transcriptional initiation. On the other hand, since the total amount of RNA is increased by Tat, it seems that Tat is not acting solely to suppress transcription termination.

Tat interacts with TAR to increase transcriptional initiation and to stabilize elongation. A direct measurement of RNA synthesis has been performed using nuclear run-on assays (Greenberg and Ziff, 1984) and short single-stranded DNA probes corresponding to the HIV-1 leader (fragment I) and neighboring portions of the CAT gene (fragments II and III) (Figure 2). Analysis of transcription rates in various portions of a gene by nuclear run-on assay with very short pulse labeling intervals provides an estimate of RNA polymerase distribution over that region of the gene. In the absence of Tat, transcription in the promoter proximal region is low and decreases approximately 10-fold in the CAT gene with a 2 minute pulse label (Figure 2A). This polar effect on transcription suggests that RNA polymerase density along the template declines with increasing distance from the promoter. Tat increases transcription rates (15-fold) in the HIV-1 leader and increases transcription in the CAT gene sequences as well. Transcriptional polarity is partially suppressed in the presence of Tat; 30% of the RNA polymerases in the leader region transcribe into the CAT gene while only 10% do so without Tat. Thus, Tat appears to have two effects: one is to increase promoter proximal RNA polymerase distribution and the other is to suppress transcriptional polarity. E1A also produces a large increase in promoter proximal transcription and increases transcription in the CAT gene sequences as well. However, in the presence of E1A, transcriptional polarity is seen in fragments II and III that is quantitatively similar to that in HeLa cells. Therefore, both Tat and E1A increase promoter proximal transcription, and Tat, in addition, is capable of suppressing transcriptional polarity.

The large increase in promoter proximal transcription produced by Tat suggests that Tat acts to increase transcriptional initiation. However, an alternative explanation for this effect is that Tat acts to suppress a block to elongation in TAR giving the misleading appearance of increased initiation. To evaluate this alternative, we examined transcription rates in the immediate vicinity of the promoter (Figure 2B). Both Tat and E1A increase transcription rates in a subfragment of probe I corresponding to the first 24 nucleotides of the HIV-1 leader. Thus Tat (and E1A) increase RNA polymerase density very near the transcription start site, supporting the model that Tat acts to increase the level of transcriptional initiation.

To determine the relative contributions of initiation and elongation to Tat transactivation, we generated additional single-stranded DNA probes corresponding to the remainder of the LTR-directed CAT transcription unit. The relative increase in transcription rate at the 3' end of the transcription unit is a measure of the stimulation of overall transcription of the gene and reflects the maximum potential contribution of transcription to increased gene expression. The relative contribution of elongation is calculated by dividing the increase in overall transcription by the increase in initiation. In the absence of transactivators, polarity reduces transcription in the 3' end of the gene by 95% (Figure 3). Tat increases promoter proximal transcription 9-fold. In addition, Tat suppresses polarity such that 40% of the initiating RNA polymerases transcribe to the end of the gene. This results in an 80-fold increase in overall transcription rates due to Tat. Therefore, in this experiment Tat stimulates transcription rates 9-fold due to its ability to stabilize elongation and 9-fold due to its ability to stimulate initiation. E1A increases promoter proximal transcription 4-fold and also produces a smaller, but reproducible, stimulation of elongation suggesting that E1A is also exerts a small stabilizing influence on transcriptional elongation.

Transcription rates have also been analyzed in mutants containing deletions within the TAR region (Figure 2C). Tat does not stimulate transcription of TAR deletion mutants, confirming the requirement of TAR for the Tat stimulatory effect on HIV-1 transcription. E1A is capable of stimulating transcription of the TAR mutants, however, confirming that TAR is not required for high levels of HIV-1 transcription in all cases. Furthermore, since the basal level of transcription is neither increased nor decreased by deletions in TAR, TAR is not a negative element (e.g., a terminator) nor a positive element required for basal levels of transcription.

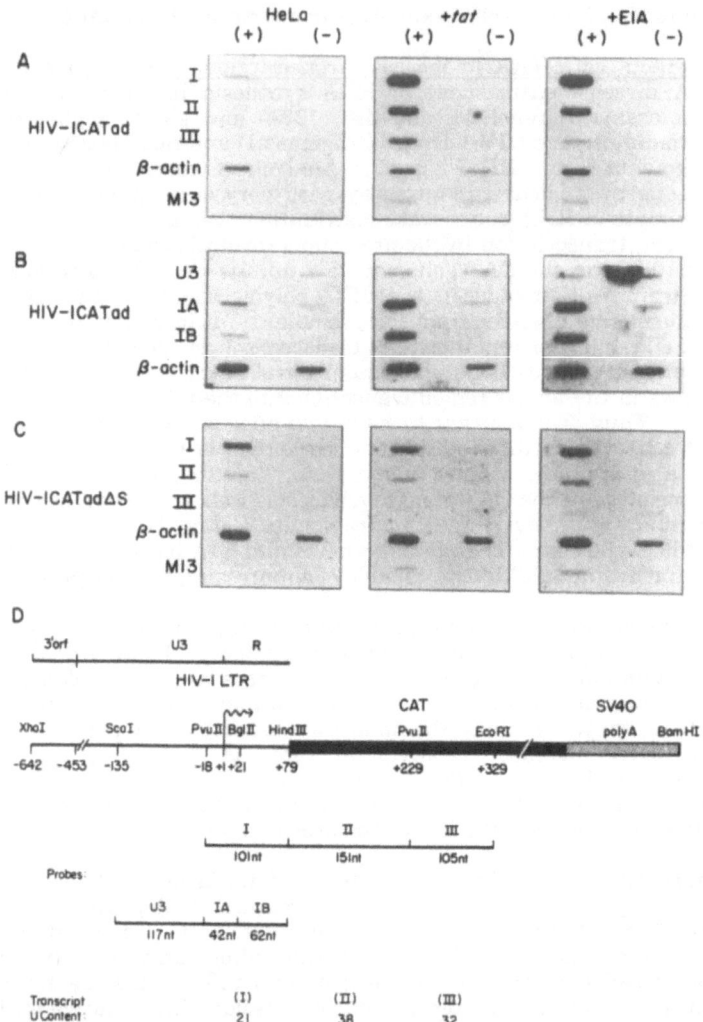

Fig. 2. Analysis of HIV-1 transcription by nuclear run-on assay. Nuclear run-on transcription assays in HIV-1CATad- (A, B) or HIV-1CATΔS- (C) infected HeLa cells (HeLa), HeLa/tat cells (⁺tat), or HeLa cells infected with dl309 (⁺E1A). HIV-1CATad is described in Figure 1. HIV-1CATadΔS contains the HIV-1 LTR sequence ⁻117 to ⁺83 with a deletion of nucleotides ⁺35 to ⁺38. Infections were carried out at an moi of 100; nuclei were isolated at 24 hours post-infection. Nascent transcripts were pulse-labeled with [α-^{32}P]UTP as described (Laspia et al., 1989) for 2 minutes in (A, B) or 5 minutes in (C). Transcription of β-actin was measured to control for the recovery of nascent transcripts in the nuclear run-on assay. (D) Schematic of the DNA probes used in the nuclear run-on assay. Adapted from Laspia et al (1989) with permission.

<u>Transcriptional synergy between Tat and E1A.</u> The interaction between Tat and other transcriptional activators may be involved in the transition from low basal transcription during latency to high level expression in the active stages of viral growth (Cullen and Greene, 1989; Pomerantz et al., 1990). As an example of such interactions, we examined the combined effect of Tat and the general transcriptional activator, E1A, on LTR-directed gene expression (Laspia et al., 1990). Simultaneous expression of Tat and E1A stimulates LTR-promoted CAT activity to more than 4-times the extent predicted if they were acting additively. Increased CAT enzyme expression is accompanied by an increased accumulation of LTR-directed cytoplasmic RNA (Figure 1A). Quantitation of RNA levels by RNase protection assay indicates that Tat and E1A individually stimulate the level of full length transcript 16-fold and 4-fold, respectively. Together, Tat and E1A produce a 4-fold larger increase than is predicted if they acted additively. As expected, synergy between Tat and E1A is abolished in a TAR deletion mutant. These results indicate that Tat and E1A cooperate at the HIV-1 promoter to produce a synergistic stimulation in the accumulation of full length LTR-promoted RNA.

Analysis of the effect of simultaneous expression of Tat and E1A on HIV-1-promoted transcription rates showed, quite unexpectedly, that Tat + E1A elicit only a slight increase in promoter proximal transcription compared Tat alone, indicating

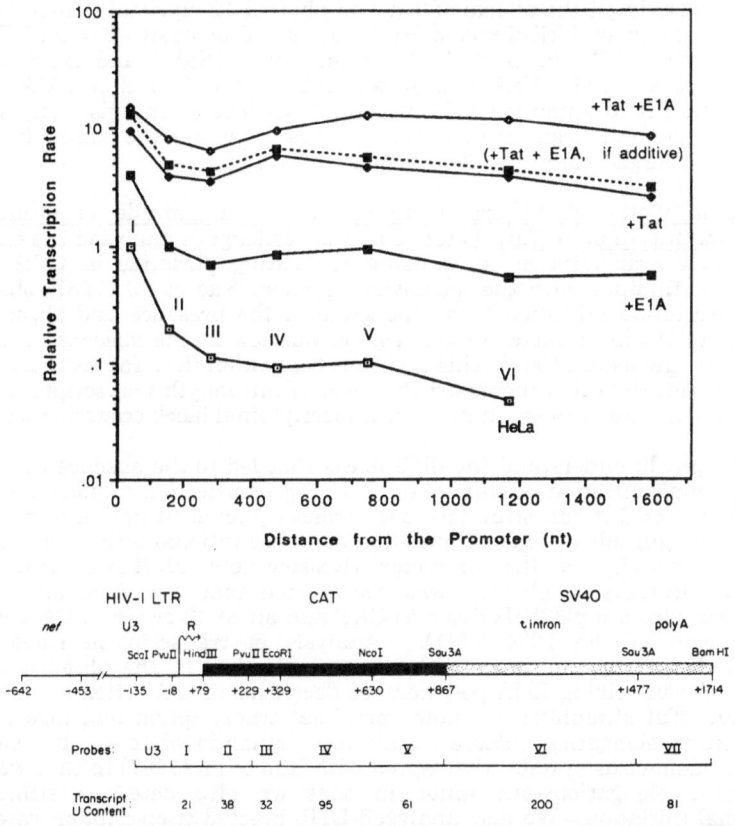

Fig. 3 Relative transcription rate throughout the HIV-1-promoted transcription unit. The nuclear run-on assay was quantified by measuring hybridization to probe fragments bound to nitrocellulose with a Betagen Scanner. Counts were normalized for transcript uridine content and to the β-actin transcription rate (to correct for RNA recovery). The transcription rate for each fragment relative to transcription in fragment I in HeLa cells is plotted against the midpoint of each fragment. A schematic of probes used in nuclear run-on assay is shown.

that synergy is not due to increased   rates of transcriptional initiation. Quantitation revealed that Tat + E1A produce a roughly additive increase in promoter proximal transcription. Likewise, in the adjacent region of the CAT gene (fragments II and III), Tat + E1A again did not elicit a more than additive increase in transcription rates. However, Tat + E1A produce an increase over E1A alone indicating that when transcription initiation is elevated by E1A, Tat suppresses transcriptional polarity in fragments II and III just as it does in the absence of E1A. Therefore, when the level of transcriptional initiation is elevated (by E1A in this case) the primary effect of Tat is to increase elongation. This may afford an explanation for the lack of an effect of Tat on initiation in some systems (Kao et al., 1987, and see below).

Examination of the effects of Tat + E1A on transcription rates in the 3' end of the gene indicates that together they dramatically suppress polarity leading to a large increase in HIV-1-promoted overall transcription (250-fold). This stimulation in overall transcription is much larger than can be explained if the two transactivators acted additively and is due to suppression of transcriptional polarity since 76% of the initiating RNA polymerases transcribe to the end of the gene in the presence of Tat + E1A whereas only 40% do so with Tat alone. Thus, Tat + E1A act in an additive manner to stimulate transcriptional initiation, but it is increased elongation that principally accounts for transcriptional synergy.

We also analyzed the effects of Tat and phorbol 12-myristate 13-acetate (PMA) on the accumulation of LTR-directed RNA levels. Previously, Tat and PMA were shown to stimulate LTR-directed CAT enzyme levels (Nabel and Baltimore, 1987; Tong-Starken et al., 1987; Seikevitz et al., 1987). We find that PMA, like E1A, cooperates with Tat to stimulate LTR-directed RNA levels synergistically suggesting that Tat and general transcriptional activators may operate through a similar mechanism to synergistically stimulate HIV-1 transcription.

Position of the SV40 origin influences the magnitude of stimulation of transcriptional initiation by Tat. Peterlin and his colleagues analyzed the effect of Tat on LTR-directed transcription rates using replicating plasmids in COS cells. In contrast to our findings with the adenovirus system, Kao et al. (1987) observed the rates of transcription initiation to be the same in the presence and absence of Tat. While transcription in promoter distal regions was low in the absence of Tat, it was increased in the presence of Tat. This and the observation that Tat decreased the level of the short transcripts and increased the level of full length transcripts led them to conclude that Tat is an antiterminator of a transcriptional block occurring within TAR.

We sought to understand the differences that led to the absence of a Tat effect on transcriptional initiation in this system. Using transfection of plasmids into COS cells we found that Tat can stimulate both transcriptional initiation and elongation. However, the magnitude of Tat's stimulation of transcriptional initiation is dependent on the basal activity of the promoter (Kessler and Mathews, manuscript in preparation). Initially, a plasmid was constructed that contained an HIV-1 LTR sequences from plasmid pU3RIII fused to CAT and an SV40 origin of DNA replication cloned upstream of the LTR (pH1). Analysis of transcription rates following transfection of this plasmid into COS cells reveals that in the absence of Tat the distribution of transcribing RNA polymerases decreases as a function of distance from the promoter. Tat stimulates promoter proximal transcription and also produces a small increase in elongation. These results are similar to what we observe with the recombinant adenovirus system; they agree with Kao et al. (1987) in that we find that Tat stimulates elongation but differ in that we also detect a stimulation of transcriptional initiation. We also analyzed LTR-directed transcription rates using a plasmid with LTR sequences from the HIVSF2 isolate and the SV40 origin cloned downstream of the CAT cassette (pSt5). Interestingly, the basal level (without Tat) of promoter proximal transcription with this construct is several fold higher than with pH1. Correspondingly, the basal level of CAT enzyme activity with pSt5 is 8-fold higher than with pH1. Tat does not cause an appreciable increase in the level of promoter proximal transcription with pSt5, but, like pH1, promoter distal transcription is dramatically increased with the addition of Tat. These results indicate that Tat's ability to stimulate transcriptional initiation is reduced when the basal level of transcription initiation is elevated. Under these circumstances the primary effect of Tat is to increase transcriptional elongation. This is in agreement with our finding that Tat's effect on initiation is small and its primary effect is to increase elongation when initiation is increased by E1A .

Analysis of Tat's effect on cytoplasmic RNA levels indicates that transfection of COS cells with pH1 produces a pattern of HIV-1-directed RNA levels similar to that following infection of HeLa cells with the recombinant adenovirus. Without Tat, approximately equimolar amounts of both full length and short transcripts are present in very low levels. Tat produces a large increase in the full length RNA without increasing the level of the short RNAs. In contrast, when pSt5 is transfected without Tat, full length RNA levels are low and short RNA accumulates to a high level. Tat causes a decrease in the short transcripts and a large increase in the full length RNA. These results are qualitatively similar to those of Kao et al. (1987) using a similar plasmid. However, since Tat increases the total amount of HIV RNA (i.e, the decrease in short RNA is not enough to account for the increase in full length RNA) Tat does not appear to be acting to antiterminate the short transcripts.

To investigate the basis for this elevated level of transcriptional initiation we determined whether it was due to sequence differences between the pH1 and pSt5 LTRs. The pH1 vector contains an LTR from pU3RIII ($^-$642 to $^+$83), obtained from the HXB2 clone (Sodroski et al., 1985a) while pSt5 contains LTR sequences derived from the SF2 isolate (Sanchez-Pescador et al., 1985) and includes additional sequences corresponding to the remainder of the R region and the U5 region. The DNA sequences of these two isolates are about 10% different within the U3 region. We have shown that neither the additional sequences unique to pSt5 nor the sequence differences in the U3 and R regions of the two LTRs is responsible for the increased basal level of promoter proximal transcription in this plasmid. Rather the basal level of gene expression is influenced by the position of the SV40 origin. Plasmid pH2 is identical to pH1 except that the SV40 origin is located downstream of the HIV-1 CAT transcription unit, as in pSt5. With pH2 the basal level of CAT enzyme activity is 25-fold higher than pH1. We are currently performing nuclear run-on analysis to determine whether the position of the SV40 origin influences the basal level of transcription initiation. The prediction is that, like pSt5, the basal level of transcription initiation with pH2 will be high and not appreciably increased by Tat. Pointing in this direction, analysis of cytoplasmic RNA levels of the full length and the short transcripts indicates that pH2 behaves identically to pSt5. Determination of plasmid DNA levels following Hirt extraction indicates that the transcriptional differences are not due to differences in copy number of plasmid DNA. Thus, our data support the hypothesis that the position of the SV40 origin region relative to the HIV-1 LTR-promoted CAT cassette influences the basal level of transcriptional initiation and the relative contributions of initiation and elongation to transactivation by Tat.

The HIV-1 Leader Inhibits Activation of DAI

Paradoxically, while a positive role for TAR in HIV-1 transcription is clearly established, evidence has also been presented that TAR RNA exerts a negative effect on mRNA translation. Secondary structure in the 5′ untranslated leader RNA has been shown to inhibit the translation of mRNA with the leader in cis (Parkin et al., 1988). In cell extracts, the stem structure of TAR RNA is also reported to inhibit initiation of translation in trans by activating the dsRNA-activated inhibitor of protein synthesis, DAI (Edery et al., 1989; SenGupta and Silverman, 1989). DAI (also known as dsI and p68) is an interferon-induced protein kinase that when activated by dsRNA undergoes autophosphorylation and subsequently phosphorylates eIF-2. This sequesters an essential second initiation factor (eIF-2B or GEF) and blocks the initiation of translation (Proud, 1986).

We examined the effect of TAR RNA on the activation of DAI in vitro. An RNA corresponding to nucleotides $^+$1 to $^+$82 of the HIV-1 leader was synthesized using a T7 expression system and purified by electrophoresis through a denaturing polyacrylamide gel (stage 1 RNA). Partially purified stage 1 RNA activates DAI resulting in autophosphorylation of DAI and phosphorylation of eIF-2 (Figure 4A), in agreement with previous results (Edery et al., 1989; SenGupta and Silverman, 1989). Further analysis indicated that the activity present in stage 1 preparation is not TAR RNA itself, however. TAR RNA, being an imperfect stem-loop structure, is digested by the single-stranded RNA specific enzyme RNase T_1. RNase III, on the other hand, digests perfectly duplexed dsRNA longer than 20 base pairs (Robertson and Hunter, 1975), and is unable to act upon imperfectly paired duplexes, such as in TAR RNA. The capability of the partially purified preparation of TAR RNA to activate DAI is unaffected by incubation with RNase T_1 but is lost after digestion with RNase III indicating that TAR RNA itself is not responsible for activation of DAI. Furthermore, the activation of DAI by partially purified RNA is probably due to a dsRNA contaminant in the preparation. Based on these results we purified TAR RNA further. Partially purified TAR RNA (stage 1), was subjected to electrophoresis through a non-denaturing gel (stage 2 RNA), and then to chromatography over a cellulose column under conditions designed to separate single- and double-stranded RNA (stage 3 RNA). Stage 2 and stage 3 purified TAR RNA exhibited little or no ability to cause DAI autophosphorylation and phosphorylation of eIF-2 (Figure 4A). Stage 3 RNA tested over a wide range of concentration did not activate DAI. Thus, activation of DAI is not an intrinsic property of TAR RNA.

Some short double stranded or highly structured RNA species can interact with DAI in such a way as to prevent its activation. The best studied of these is a small, highly structured, adenovirus encoded RNA, called VA RNA_I, that protects protein synthesis from the consequences of DAI activation by dsRNA made late in adenovirus infection (Kitajewski et al., 1986; O'Malley et al., 1986; Maran and Mathews, 1988). To test whether TAR RNA could function to block activation of DAI by dsRNA, reovirus dsRNA and stage 3 TAR RNA were incubated with DAI. We found that TAR RNA blocks activation of DAI by dsRNA (Figure 4B). This inhibition is due to TAR RNA, since the RNase sensitivity of the inhibitory component matches that of TAR RNA: both were eliminated by RNase T1, but neither was affected by RNase III. Thus, like VA RNA_I (Mellits et al., 1990), the inhibition is an intrinsic property of TAR RNA. Moreover, an anti-sense version of TAR RNA failed to block DAI activation by ds RNA. These results indicate that the DAI activation property associated with TAR RNA transcribed in vitro is not due to TAR RNA per se; on the contrary, purified TAR RNA, like adenovirus VA RNA, prevents the activation of DAI by dsRNA in vitro. Whether it also exerts this positive effect in vivo remains to be seen.

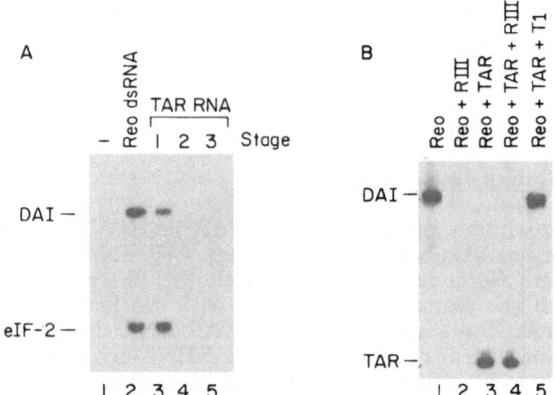

Fig. 4   Effect of TAR RNA on the protein kinase DAI. (A) DAI activation by TAR RNA synthesized in vitro.   DAI activation assays   contained no RNA (lane 1), 10 ng/ml reovirus dsRNA (lane 2), or stage 1 TAR RNA (lane 3), stage II TAR RNA (lane 4) or stage III RNA (lane 5) at 200 ng/ml. The position of DAI and the α-subunit of eIF-2 is marked.   (B) Inhibition of dsRNA activation of DAI by purified TAR RNA. All lanes contained reovirus dsRNA (10 ng/ml), alone (lane 1) or preincubated with RNase III (lane 2). Lanes 3-5 also contained stage 3 TAR RNA (25 µg/ml), undigested (lane 5), preincubated with RNase III (lane 4) or RNase T1 (lane 5). Adapted from Gunnery et al. (1990).

DISCUSSION

These studies with an HIV-1 recombinant adenovirus have provided evidence that Tat interacts with TAR to elevate transcriptional initiation. In addition, Tat also appears to stabilize transcriptional elongation through the HIV-1 leader and downstream sequences. The combination of these two effects leads to a large overall increase in transcription. Tat stimulates transcription following binding to TAR RNA (Roy et al. 1990; Weeks et al., 1990; Southgate et al., 1990; Selby and Peterlin, 1990; Dingwall et al. in press). Thus, TAR RNA appears to function as a scaffold for Tat immediately downstream of the HIV-1 promoter and is analogous to cis-regulatory DNA elements in that it provides a transcription factor binding site. We propose that Tat binds to nascent HIV-1 transcripts via TAR to mediate an increase in initiation by causing the efficient formation at the HIV-1 promoter of an RNA polymerase complex that is capable of stable transcriptional elongation. In other words, Tat represents a novel transcriptional activator in its ability to mediate increased transcriptional initiation by binding to nascent RNA transcripts.

The synergy between Tat and E1A is likely to be a paradigm for the interaction between Tat and general transcriptional activators since PMA also cooperates with Tat to stimulate HIV-1 transcription synergistically. Cooperation between Tat and general activators, resulting in synergistic stimulation of gene expression, may be involved in the transition from transcriptional quiescence during viral latency to active expression during productive infection (Cullen and Greene, 1989; Pomerantz et al., 1990). Our results indicate that Tat and E1A act to promote the formation of a more highly processive transcriptional complex than is formed in the presence of Tat alone. Thus, Tat is able to cooperate with transcription factors that act through upstream promoter elements to stimulate HIV-1 transcription rates synergistically and this may serve as a molecular trigger which induces HIV-1 to transit from transcriptional latency to active gene expression.

The effect of Tat on initiation is variable in magnitude. In recombinant adenovirus-infected cells, Tat fails to stimulate initiation from the E1A-induced

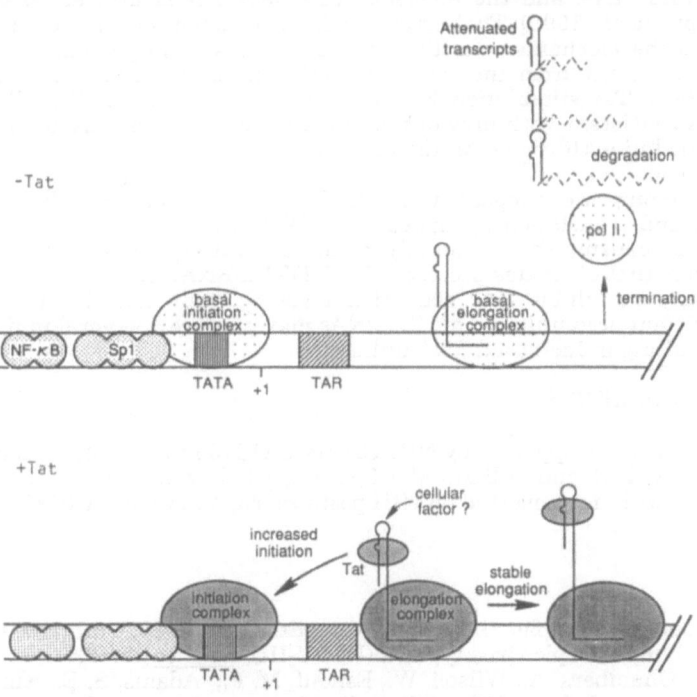

Fig. 5 Model for transactivation of HIV-1 gene expression by Tat.

promoter to the extent that it stimulates   the basal level of initiation from the HIV-1 promoter. Likewise, in plasmid transfection assays in COS cells with constructs that exhibit a high basal level of transcriptional initiation, Tat does not produce a further stimulation in initiation. These finding indicate that the magnitude of Tat's effect on transcription is low when the basal activity of the promoter is high and that, under these circumstances, the primary effect of Tat is to stabilize elongation. This observation could have implications for the development of systems which seek to analyze the stimulation of transcription by Tat in vitro since the magnitude of the initiation effect may be markedly influenced by the strength of the promoter.

While Tat's role as a transcriptional regulator of HIV-1 gene expression is clearly established, its role in translational control remains controversial. Several studies have concluded that Tat and TAR cooperate to stimulate translation, either as well as, or instead of transcription (Cullen, 1986; Feinberg et al., 1986; Rosen et al., 1986; Wright et al., 1989; Braddock et al., 1989) while other studies failed to discern an effect at the translational level (Peterlin et al., 1986; Jeang et al., 1988; Rice and Mathews, 1988; Jakobvits et al., 1988). The findings described here may provide a basis for resolving this controversy. We find that highly purified preparations of TAR RNA, like adenovirus VA RNA$_I$, inhibit activation of the cellular protein kinase, DAI in vitro. While VA RNA is abundant in adenovirus infected cells it is not known whether transcripts containing the TAR sequence accumulate sufficiently in HIV-1-infected cells to block DAI activation. However, TAR sequences are present in an ideal location in HIV mRNAs to spare them from the inhibitory effects of DAI activation on polypeptide chain initiation. It has been postulated that VA RNA ensures the translation of mRNAs with which it specifically associates (O'Malley et al., 1989); similarly TAR may act in cis to increase the translation of HIV-1 mRNAs.

A number of investigators have noticed that Tat stimulates reporter enzyme activity to a greater extent than reporter RNA levels and have ascribed this effect to posttranscriptional regulation by Tat (Cullen, 1986; Feinberg et al.,1986; Rosen et al., 1986; Wright et al., 1986). We have found that both E1A and Tat increased the utilization of TAR-CAT RNA and that a correlation exists between the level of cytoplasmic CAT RNA and the efficiency with which it is utilized to produce CAT enzyme (Laspia et al., 1990). The correlation is independent of the presence or absence of Tat. While the mechanism for the effect is unknown, we postulate that increased utilization may result from increased protection from the inhibitory effects of DAI activation due to Tat stimulation HIV-1-directed cytoplasmic TAR-CAT RNA levels. This model is consistent with previous data indicating that Tat has no direct effect on CAT RNA translation (Rice and Mathews, 1988).

In summary we propose that Tat/TAR interactions coordinately regulate transcription and translation by increasing HIV-1 transcription rates. Tat elevates transcription by causing the efficient formation of an RNA polymerase complex at the HIV-1 promoter that elongates processively. HIV-1 mRNAs bearing the TAR structure may act locally to inhibit DAI activation favoring their selective translation. In addition, sufficient accumulation of HIV RNAs may produce a general inhibition of DAI activation resulting in increased RNA utilization.

ACKNOWLEDGEMENTS

This work was supported by NIH Grants CA13106 to M.B.M., AI25308 to A.P.R. and AI27270 to A.P.R. and M.B.M. M.F.L is supported by an AmFAR research scholar award 700148. M.K. is supported an NIH postdoctoral fellowship AI07877.

REFERENCES

Arya, S. K., Guo, C., Josephs, S. J., and Wong-Staal, F., 1985, Trans-activator gene of human T-lymphotropic virus type III (HTLV-III), Science 229:69.
Braddock, M., Chambers, A., Wilson, W., Esnouf, M. P., Adams, S. E., Kingsman, A. J. and Kingsman, S. M., 1989, HIV-1 Tat "activates" presynthesized RNA in the nucleus, Cell 58:269.

Berk, A. J., 1986, Adenovirus promoters and E1A transactivation. Ann. Rev. Genet. 20:45.

Berkhout, B., Gatignol, A., Rabson, A. B., and Jeang, K.-T., 1990, TAR-independant activation of the HIV-1 LTR: evidence that Tat requires specific regions of the promoter, Cell 62:757.

Cullen, B. R., 1986, Trans-activation of human immunodeficiency virus occurs via a bimodel mechanism, Cell 46:973.

Cullen, B. R., and Greene, W.C., 1989, Regulatory pathways governing HIV-1 replication, Cell 58:423.

Dingwall, C. Ernberg, I., Gait, M. J., Green, S. M., Heaphy, S., Karn, J., Lowe, A.D., Singh, M., and Skinner, M. A., HIV-1 tat protein stimulates transcription by binding to a U-rich bulge in the stem of the TAR RNA structure, EMBO J., in press.

Edery, I., Petryshyn, R., and Sonenberg, N., 1989, Activation of double-stranded RNA-dependent kinase (dsI) by the TAR region of HIV-1 mRNA: a novel translational control mechanism, Cell 56:303.

Feinberg, M. B., Jarrett, R. F., Aldovini, A., Gallo, R. C., and Wong-Staal, F., 1986, HTLV-111 expression and production involve complex regulation at the levels of splicing and translation of viral RNA, Cell 46:807.

Greenberg, M. E., and Ziff, E. B., 1984, Stimulation of 3T3 cells induces transcription of the c-fos proto-oncogene, Nature 311: 433.

Gunnery, S., Rice, A. P., Robertson, H. D., and Mathews, M. B., 1990, HIV-1 TAR RNA can prevent activation of the protein kinase DAI, Proc. Natl. Acad. Sci. USA 87:in press.

Hauber, J., Perkin, A. Heimer, E. P., and Cullen, B. R, 1987, Trans-activation of human immunodeficiency virus gene expression is mediated by nuclear events, Proc. Natl. Acad. Sci. USA 84:6364.

Hauber, J., and Cullen, B. R., 1988, Mutational analysis of the trans-activation-responsive region of the human immunodeficiency virus type I long terminal repeat. J. Virol. 62:673.

Jakobovits, A., Smith, D. H., Jakobovits, E. B., and Capon, D. J., 1988, A discrete element 3' of human immunodeficiency virus 1 (HIV-1) and HIV-2 mRNA initiation sites mediates transcriptional activation by an HIV-1 trans activator, Mol. Cell. Biol. 8:2555.

Jeang K-T, Shank, P. R., and Kumar, A., 1988, Transcriptional activation of homologous viral long terminal repeats by the human immunodeficiency virus type 1 or the human T-cell leukemia virus type 1 tat proteins occurs in the absence of de novo protein synthesis, Proc. Natl. Acad. Sci. USA 85:8291.

Jones, N. C., Rigby, P. J., and Ziff, E. B., 1988, Trans-acting protein factors and the regulation of eukaryotic transcription: lessons from studies on DNA tumor viruses, Genes and Devel. 2:267.

Kao, S.-Y., Calman, A. F., Luciw, P. A., and Peterlin, B. M., 1987, Anti-termination of transcription within the long terminal repeat of HIV by the tat gene product, Nature 330:489.

Kessler, M., and Mathews, M. B., The mechanism of transactivation by Tat of HIV-1-directed transcription is determined by the basal promoter activity, manuscript in preparation.

Kitajewski, J., Schneider, R. J., Safer, B, Munemitsu, S. M, Samuel, C. E., Thimmappaya, B., and Shenk, T., 1986, Adenovirus of VA1 RNA antagonizes the antiviral action of interferon by preventing activation of the interferon-induced eIF-2α kinase, Cell 45:195.

Laspia, M. F., Rice, A. P., and Mathews, M. B., 1989, HIV-1 Tat protein increases transcriptional initiation and stabilizes elongation, Cell 59:283.

Laspia, M.F., Rice, A.P., and Mathews, M.B., 1990, Synergy between HIV-1 Tat and adenovirus E1A is principally due to stabilization of transcriptional elongation, Genes and Development, 12:in press.

Maran, A., and Mathews, M. B., 1988, Characterization of the double-stranded RNA implicated in the inhibition of protein synthesis in cells infected with a mutant adenovirus defective for VA RNA$_1$, Virol 164:106.

Mellits K. H., and Mathews, M. B., 1988, Effects of mutations in stem and loop regions on the structure of adenovirus VA RNA$_1$, EMBO J. 7:2849.

Muesing, M. A., Smith, D. H., and Capon, D. J., 1987, Regulation of mRNA accumulation by human immunodeficiency virus trans-activator protein, Cell 48:691.

Nabel, G., and Baltimore, D., 1987, An inducible transcription factor activates expression of human immunodeficiency virus in T-cells, Nature 326:711.

O'Malley, R. P., Mariano, T. M., Siekierta, J., and Mathews, M. B., 1986, A mechanism for the control of protein synthesis by adenovirus VA RNA$_1$, Cell 44:391.

O'Malley, R. P., Duncan, R. F., Hershey, J. B. and Mathews, M. B., 1989, Modification of protein synthesis initiation factors and the shut-off of host protein synthesis in adenovirus-infected cells, Virol. 168:112.

Parkin, N. T., Cohen, E. A., Darveau, A., Rosen, C., Haseltine, W. and Sonenberg, N., 1988, Mutational analysis of the 5' non-coding region of human immunodeficiency virus type1: effects of secondary structure on translation, EMBO J. 7:2831.

Pavlakis G. N., and Felber, B. K., 1990, Regulation of expression of human immunodeficiency virus, New Biol. 2:20.

Peterlin, B. M., Luciw, P. A., Barr, P. J., and Walker, M. D., 1986, Elevated levels of mRNA can account for the trans-activation of human immunodeficiency virus, Proc. Natl. Acad. Sci. USA 83:9734.

Pomerantz, R. J., Trono, D., Feinberg, M. B., and Baltimore, D., 1990, Cells nonproductively infected with HIV-1 exhibit an aberrant pattern of viral RNA expression: a molecular model for latency, Cell 61:1271.

Proud, C., G., 1986, Guanine nucleotides, protein phosphorylation and the control of translation, TIBS 11:73.

Rice, A. P., and Mathews, M. B., 1988, Transcriptional but not translational regulation of HIV-1 by the tat gene product, Nature 332:551.

Rice, A. P., and Mathews, M. B., 1988b, Trans-activation of the human immunodeficiency virus long terminal repeat sequences, expressed in an adenovirus vector, by the adenovirus E1A 13S protein, Proc. Natl. Acad. Sci. USA 85:4200.

Robertson, H. D. and Hunter, T., 1975, Sensitive methods for the detection and characterization of double helical ribonucleic acid, JBC 250:418.

Rosen, C. A., Sodroski, J. G., and Haseltine, W. A., 1985, The location of cis-acting regulatory sequences in the human T cell lymphotropic virus type III (HTLV-III/LAV) long terminal repeat, Cell 41:813.

Rosen, C. A., Sodroski, J. G., Goh, W. C., Dayton, A. I., Lippe, J., and Haseltine, W. A., 1986, Post-transcriptional regulation accounts for the trans-activation of the human T-lymphotropic virus type III, Nature 319:555.

Roy, S., U. Delling, Chen, C.-H., Rosen, C. A., and Sonnenberg, N., 1990, A bulge structure in HIV-1 TAR RNA is required for Tat binding and Tat-mediated trans-activation, Genes and Devel. 12:in press.

Sadaie, M. R., Benter, T., and Wong-Staal, F., 1988, Site-directed mutagenesis of two trans-regulatory genes (tat-III, trs) of HIV-1, Science 239:910.

Sanchez-Pescador, R., Power, M. D., Barr, P. J., Steimer, K. S., Stempein M. M., Brown-Shimer, S. L., Gee, W. W., Renard, A., Randolph, A., Levy, J. A., Dina, D., and Luciw, P. A., 1985, Nucleotide sequence and expression of an AIDS-associated retrovirus (ARV-2), Science 227:484.

Selby, M. J., Bain, E. S., Luciw, P. A., and Peterlin, B. M., 1989, Structure, sequence, and position of the stem-loop in tar determine transcriptional elongation by tat through the HIV-1 long terminal repeat, Genes and Devel. 3:547.

Selby, M. J., and Peterlin, B. M., 1990, Trans-activation by HIV-1 Tat via a heterologous RNA binding protein, Cell 62:769.

Siekevitz, M., Josephs, S. F., Dukovich, M., Peffer, N., Wong-Staal, F. W., and Greene, W. C., 1987, Activation of the HIV-1 LTR by T cell mitogens and the trans-activator protein of HTLV-1. Science238:1575.

SenGupta, D. N., and Silverman, R. H., 1989, Activation of interferon-regulated dsRNA-dependent enzymes by human immunodeficiency virus-1 leader RNA, Nucl. Acid Res. 17:969.

Sharp P. A., and Marciniak, R. A., 1989, HIV TAR: An RNA enhancer? Cell 59:229.

Sodroski, J., Rosen, C., Wong-Staal, F., Salahuddin, S. K., Popovic, M., Arya, S., Gallo, R. C., and Haseltine, W. A., 1985a, Trans-acting transcriptional regulation of human T-cell leukemia virus type III long terminal repeat, Science 227:171.

Sodroski, J. G., Patarca, R., Rosen, C., Wong-Staal, F., and Haseltine, W., 1985b, Location of the trans-activation region on the genome of human T-cell lymphotropic virus type III, Science 229:74.

Southgate, C., Zapp, M. L., and Green, M. R., 1990, Activation of transcription by HIV-1 Tat protein tethered to nascent RNA through another protein, Nature 345:640.

Tong-Starksen, S.E., Luciw, P.A., and Peterlin, B.M., 1987, Human immunodeficiency virus long terminal repeat responds to T-cell activation signals. Proc. Natl. Acad. Sci. USA 84:6845.

Varmus, H., 1988, Regulation of HIV and HTLV gene expression, Genes and Devel. 2:1055.

Weeks, K. M., Ampe, C., Schultz, S. C., Steitz, T. A., and Crothers, D. M., 1990, Fragments of the HIV-1 Tat protein specifically bind TAR RNA, Science 249, 1281-1285.

Wright, C. M., Felber, B. K., Paskalis, H., and Pavlakis, G. N., 1986, Expression and characterization of the trans-activator of HTLV-III/LAV virus, Science 234:988.

HIV-1 TAT: A TRANSCRIPTIONAL ACTIVATOR THAT RECOGNIZES A STRUCTURED RNA TARGET

Ben Berkhout and Kuan-Teh Jeang

Laboratory of Molecular Microbiology
National Institute of Allergy and Infectious
Diseases
National Institutes of Health
Bethesda, MD 20892

INTRODUCTION

The human and simian immunodeficiency viruses (HIVs and SIVs) are unique among retroviruses in encoding for many small regulatory proteins in addition to Gag, Pol and Env (reviewed in 1). Although counterpart regulatory proteins are apparently absent or yet to be described for avian and murine retroviruses, HIV greatly depends on the functions of some of these proteins (Tat, Rev, Vif) for its viability (reviewed in 2). Other non-structural proteins (Nef, Vpr, Vpu, and Vpx for HIV-2) have moderating effects and are not absolutely essential for viral propagation in tissue culture.

Tat is one intensely studied regulatory protein. Although it is known to be essential for HIV viability, its exact role remains somewhat controversial. Tat arguably works at a transcriptional and/or a post-transcriptional level. Wide ranging experimental findings have variously implicated Tat in transcriptional initiation, transcriptional anti-termination/elongation and translation. How, mechanistically, this small protein (14 to 21kDa, depending on the particular isolate) accomplishes these pleiotropic activities is unclear. In part, the interaction of Tat with RNA, DNA, and cellular proteins has been invoked to explain its biological role.

In this report we review some of the recent findings on Tat mechanism. We will discuss the interactions between Tat and LTR RNA and DNA sequences, and the contributory role of cellular proteins in *trans*-activation. We will speculate on a model to explain the role of Tat in the induction of transcription from the LTR promoter.

Tat is synthesized from two coding exons (Figure 1). The first exon codes for amino acids 1-72; the second codes for amino acids 73-86 (or 73-101, depending on the isolate). Within the first exon portion of Tat is a stretch of basic amino acids (position 49 to 57, domain 4) that functions as a nuclear-nucleolar localization signal and that plays an important role in Tat *trans*-activation (ref. 3-7). Another function has been attributed to this region of Tat. It has been suggested that the amino acids between positions 49 to 57 are responsible for specific binding by Tat to TAR RNA (8-10). Alternatively, these basic amino acids function primarily to direct Tat into the nucleus. Recent experiments replacing this Tat domain with a basic domain from the HIV-1 Rev protein (11) or with the nuclear localization signal of adenovirus E1a (Jeang, unpublished observation) have found a maintainance of *trans*-activation function. These results, while compatible with a subcellular localization role, questions the exact amino acid sequence within Tat for specificity of binding to TAR RNA.

Amino acids in the C-terminus of Tat are not essential for *trans*-activation since a truncated 56 amino acid version of Tat is substantially active in transfection experiments (4,12,13). Addition of large unrelated protein domains (e.g. β-galactosidase) to this short version of Tat also does not affect activity (5,6). However, the conservation of the C-terminal portion of the Tat open reading frame, including exon-intron borders, does strongly suggest that these sequences are important for the virus. The recent identification of a cell adhesion domain in the second coding exon of Tat might be one such function (14-16).

A region of Tat that is highly conserved between distantly related viral isolates is the cysteine-rich domain (residues 22 to 37, domain 2). This region is important for *trans*-activation and has been proposed as a metal-chelating dimerization domain (17). Individual mutations in six out of the seven cysteine residues abolish Tat function when assayed in transfection experiments (4,6,13). Whether this is due to a disturbance in dimerization or in overall protein folding is not clear. The identity of several non-cysteine residues in this region is also very important for Tat activity (18). It is not known whether the cysteine-domain serves a structural role or that it is actually involved in functional interactions of Tat with other proteins.

The extreme N-terminus of Tat (domain 1) has an acidic domain of 13 amino acids with amphipathic characteristics. Mutations that changed the acidic nature of this region affected *trans*-activation (19). This result suggests, but does not prove, that this region might serve as an acidic promoter activating domain as has been described for many other *trans*-activator proteins (20).

Tat dramatically increases viral RNA and protein synthesis by approximately two orders of magnitude over basal levels. This is thought to be due to an increase in the rate of initiation and/or elongation of transcription. Although such an effect might *a priori* be expected to act through a *trans*-activation response element (TAR) in the viral DNA, it is known that Tat uses a TAR RNA (21,22). TAR maps to position +19 to +42 of HIV-1 transcripts and has a secondary structure containing an extended hairpin (see Figure 2, refs 22-25). It is this RNA structure, in combination with specific sequence elements in the single-stranded bulge and loop domains that are important for *trans*-activation (21,26,27). A very similar set of RNA sequence and structure elements is found in the TAR elements of the HIV-2 and SIV viruses (26, 28, 29). These elements are likely to be the contact points for Tat (8-10) and cellular proteins (30-32). Consistent with this idea, overproduction of TAR RNA results in inhibition of Tat function (33), probably through competition for such proteins.

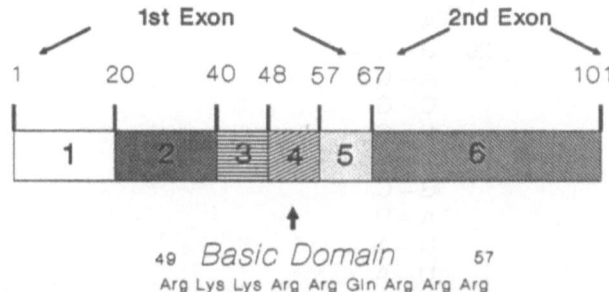

Fig 1. **Functional domains of Tat.** The Tat protein of HIV-1 (strain SF2) is 101 amino acids long. Based on mutational analysis (13), Tat was divided into six domains. Domains 1 (N-terminus), 2 (cysteine-rich region) and 4 (basic region), which are essential for *trans*-activation, are discussed in the text.

Tat binds to the bulged nucleotides of TAR RNA *in vitro* (8-10). This finding does not rule out that cellular co-factors are involved in Tat-mediated *trans*-activation or binding within the cell. In fact, it has been known that Tat is very poorly active in certain genetic backgrounds such as murine cells (12). An indication that this defect is due to the absence of a particular protein function in rodent cells comes from complementation studies using hamster-human cell hybrids. Such experiments showed that the presence of human chromosome 12 restores high levels of *trans*-activation to hamster cells (34,35). This observation is consistent with the identification of several TAR RNA

binding proteins (30-32) and a Tat binding protein (36) in
human cells. Since Tat binds only to TAR RNAs containing a
wild-type bulge sequence (9,10), it is conceivable that
cellular TAR-binding factors recognize important sequence
elements in the loop of the hairpin. While these binding
studies are illustrative, complete understanding of the
molecular nature of Tat action awaits the development of an
in vitro transcription system dependent on reconstituted Tat
and cellular proteins.

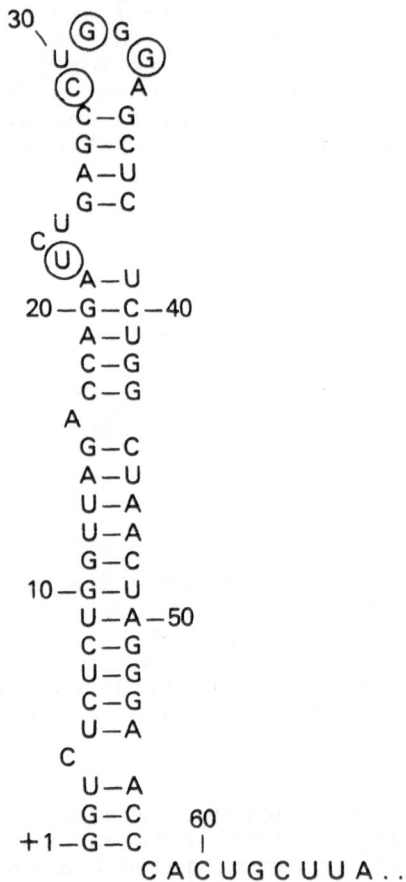

Fig 2. **Secondary structure of HIV-1 TAR RNA (strain HXB-2).**
The hairpin structure was experimentally determined
(22,23) and is consistent with phylogenetic data (29).
The minimal TAR element represents the upper portion of
the hairpin, from position +17 to +43 (22,24,25).
Important sequence-elements are located in the single-
stranded bulge- and loop-domains. Mutation of the
nucleotides circled dramatically interfered with Tat
action (21,26,27). The bulge element has been reported
to be the binding site for Tat (9,10).

Several models have been suggested to explain how Tat can stimulate transcription through a target within the viral transcript. These proposals are schematically depicted in Figure 3 and are described below.

A. The "classical" anti-termination model (Figure 3A). In this model (37,38), the rate of transcriptional initiation at the LTR promoter is constant whether in the absence or presence of Tat. Without Tat, premature transcriptional terminations lead to the production of short promoter-proximal RNAs. In the presence of Tat, a protein complex is assembled on the nascent TAR RNA, including Tat and TRBP (cellular TAR RNA binding protein). This complex modifies the elongating RNA polymerase (indicated by Pol* in Figure 3A) in such a way that it can efficiently continue into the structural genes. This anti-termination model resembles in great detail the proposed mechanism of action of the anti-terminator N protein of phage lambda (39,40). Mechanistically, the model is also reminiscent of the action of the bacterial terminator protein rho (41) and the process of transcriptional attenuation as described for several operons concerned with amino acid biosynthesis (42). In all these procaryotic cases the nascent RNA is recognized by proteins that determine the fate of the elongating RNA polymerase.

The short transcripts synthesized in the absence of Tat have been detected in several experimental systems (37,43,44) and form the essence of this model. These approximately 59-base long RNAs were suggested to be the result of specific transcriptional termination in the promoter-proximal region (37). Because deletion of TAR does not increase expression, no transcriptional termination signal can be present in TAR. Instead, direct measurements of the density of elongating RNA polymerases have indicated that termination is a gradual process (43). Transcriptional polarity, producing short transcripts that are heterogeneous in length (38,43,44), is one explanation for this gradual fall off.

B. A modified version of the anti-termination model (Figure 3B) shows an interaction of the nascent TAR RNA-Tat-TRBP complex with RNA polymerases present at the upstream promoter (45). Similar to model A, the frequency of initiation is not affected by Tat, but the initiating RNA polymerase is modified from termination-prone (Pol) to termination-resistant (Pol*). Although mechanistically different from model A, both models have similar net results, only short transcript in the absence of Tat, and only full-length ones in the presence of Tat. Note that unlike the short transcripts in model A, the ones in model B are not the actual precursors for the longer transcripts.

C. A third model is the initiation model (Figure 3C), which is based on the demonstration that trans-activation leads to increased transcriptional initiations (25,43,46,47). Model C is similar to model A and B with respect to the assembly of a functional TAR-Tat-TRBP complex

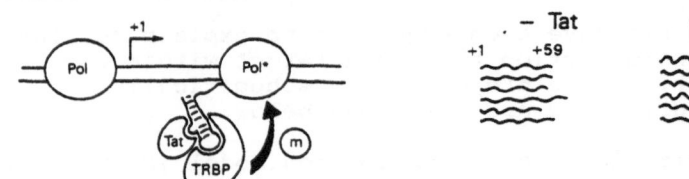

A. "Classical" Anti-termination

Transcripts Produced

− Tat   + Tat

B. Modified Anti-termination

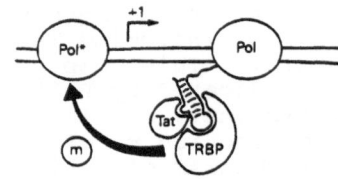

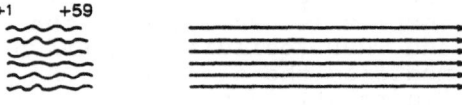

C. Initiation

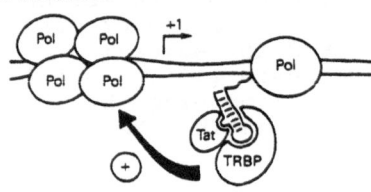

D. Initiation plus Modified Anti-termination

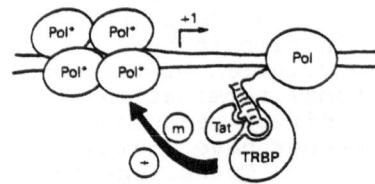

Fig 3. **Proposed models for Tat *trans*-activation.** A schematic representation of Tat action is shown on the left, the quantity and length of transcripts synthesized in the absence or presence of Tat are shown on the right. The models show two types of RNA polymerases, one initiating at the promoter sequences upstream of position +1, the other engaged in transcription of the HIV sequences. In all models, Tat and a cellular TAR-binding protein (TRBP) attach to a nascently transcribed TAR RNA which is folded into a typical hairpin structure. At this point the models differ. In model A Tat modifies (m) the elongating RNA polymerase such that it becomes termination-resistant (indicated by Pol*). Thus, the short transcripts seen in the absence of Tat become extended. In model B, the anti-termination signal modifies the initiating RNA polymerase. Note that the short transcripts are not the actual precursors for the long transcripts. In model C, Tat increases (+) the frequency of transcriptional initiations. Model D incorporates both the initiation component (model C) and the modification component (model B) such that Tat induces the synthesis of more and full-length transcripts.

on the nascent transcript. In model C, however, Tat
increases the rate of transcriptional initiation at the
upstream promoter. Although RNA as transcriptional enhancer
has not been reported before in any biological system,
several lines of evidence favor this model (48).

     D. Model D incorporates the features of both models B
and C (Figure 3D, ref. 43). In the absence of Tat, there is
both low basal level of transcriptional initiation and
polarity during transcriptional elongation. The TAR-Tat-
TRBP complex triggers new initiations at the upstream
promoter, but at the same time renders these RNA polymerases
termination-resistant (43).

     One important question raised by all models is whether
there is sufficient time for the assembly of the TAR-Tat-
TRBP complex. In model A the anti-termination signal has to
reach the elongating RNA polymerase before it actually
terminates prematurely and dissociates from the DNA
template. In model C the initiation signal from the TAR
element to the upstream promoter is expected to become less
efficient during the course of elongation which distances
TAR from the LTR promoter. Thus, formation of the TAR
hairpin structure and binding of Tat and TRBP proteins has
to be extremely rapid or, alternatively, the elongation of
the RNA polymerase has to be slow. From procaryotic
examples where recognition of nascent RNA strands is
involved, pausing of the transcriptional machinery has been
reported (41,42). We suggest that pausing plays a role in
HIV *trans*-activation. One result compatible with this idea
is the finding that constructs that transiently form a TAR
hairpin, that is not maintained in the mature RNA, do allow
for efficient Tat *trans*-activation (22).

     It seems rather straightforward to devise experiments
that distinguish between termination/anti-termination versus
initiation. In particular, the observation of short
transcripts in the absence of Tat seems to support a
termination/anti-termination model. However, as happens so
often in biological systems, the outcome of such experiments
varies greatly depending on the type of plasmids and/or cell
lines used. Abundant short transcripts were described in
transient transfection experiments using plasmids that
replicate to high copy-numbers in COS cells (37,38) and in
in vitro transcription assays with sarcosyl-treated cellular
extracts (44). In contrast, we and others have reported a
dramatic Tat-mediated increase in RNA synthesis without
detecting appreciable levels of short transcripts in the
absence of Tat (25,43,46,47,49). The fact that our plasmids
do not replicate might explain the difference in phenotype.
A separate experimental protocol, using adenovirus-based
vectors, showed some short transcripts, but the major effect
of Tat was to increase the rate of initiation of
transcription (43). Overall, we believe that the
transcriptional induction function of Tat can be explained
without invoking short transcripts and/or the suppression of
premature terminations. It is possible that the observed
short transcripts reflect transcriptional pausing in the
promoter-proximal region.

Other results argue against the "classical" anti-termination model. The observation that TAR rapidly loses activity when its position is moved further downstream in the transcript (38 and Berkhout, unpublished observations), suggests that spacing between TAR and the promoter is important. This is consistent with all the models, except model A. In addition, Tat function becomes very inefficient when TAR is coupled to heterologous promoters (23,29,49-51). We found that the removal of both the upstream NF-kB and Sp-1 motifs severely reduced the Tat response (49). The requirement for specific HIV regulatory DNA sequences is not consistent with model A and is most compatible with an initiation model.

TRANS-ACTIVATION IN THE ABSENCE OF TAR

A schematic representation of a nascent TAR-Tat complex interacting with upstream promoter elements is shown in Figure 4A. The molecular details of the simultaneous interactions of Tat-TRBP with the RNA binding site (TAR) and with the DNA activation site (LTR promoter) remain to be solved. The drawn model only proposes that one function of TAR is to form a scaffold for the assembly of a *trans*-activation complex. This does not resolve whether Tat itself or a cellular co-factor is the actual activator protein.

Some recent results suggest that Tat is the transcriptional activator. For example, Southgate et al. (47) showed that Tat can efficiently activate transcription when directed to transcripts through an RNA-binding domain of an unrelated protein (the HIV-1 Rev protein, see Figure 4B). In this context, with the Tat-Rev fusion protein bound to the Rev target (RRE RNA), TAR is dispensable. Similar results were obtained by Selby and Peterlin using a fusion between Tat and an RNA-binding bacteriophage coat protein, in combination with the phage RNA operator (Figure 4B, ref.45). These experiments showed that one function of TAR is to locate Tat to the proximity of the promoter. The fact that Tat can efficiently function in the absence of any TAR sequences suggests qualitatively that cellular TAR-binding proteins may not be directly involved in the induction of the upstream LTR promoter. Not totally clear is whether cellular factors synergize the activity of Tat. It remains possible that cellular factors are essential for the functional binding of Tat to the natural TAR binding site. These experiments do not address the role of cellular Tat-binding proteins (36).

We showed that Tat remains active when tethered to the promoter as a DNA-binding fusion protein (Figure 4C, 49). The DNA-binding domain of the Jun protein directs Tat to the AP-1 DNA binding sites that substitute the TAR sequences. As in the previous experiments, TAR RNA becomes dispensable in this context. More importantly, the observation that Tat can *trans*-activate when attached to DNA strongly supports the initiation model for Tat action. In fact, the Tat-Jun protein resembles generic transcriptional activators, in that it has a DNA-binding domain (Jun) and a second activation domain (Tat, perhaps the acidic N-terminus).

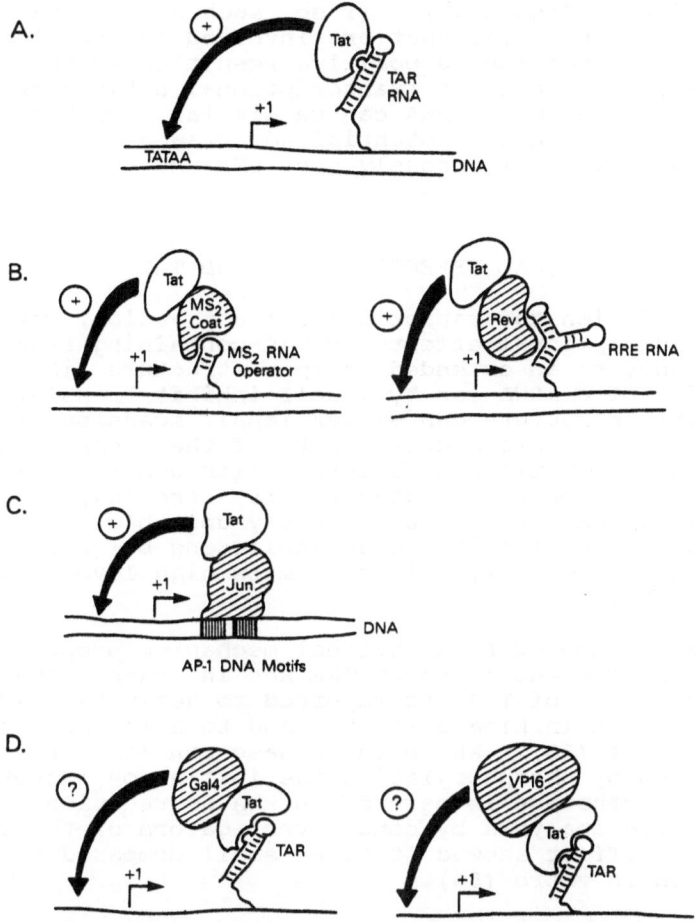

Fig 4. *Trans*-**activation through non-TAR RNA and DNA targets.** Wild-type *trans*-activation is schematically depicted in panel A. The TAR target can be functionally replaced by unrelated RNA-binding sites when Tat is fused to a protein that will recognize this new RNA structure (panel B). Efficient *trans*-activation was reported for Tat fused to bacteriophage MS2 coat (45) or the HIV-1 Rev protein (47). Tat also remained active when presented to the LTR promoter in the form of a DNA-binding protein (panel C, 49). It remains to be tested whether the activation domain of Tat can be substituted for by that of transcriptional activator proteins like Gal4 or VP16 (panel D).

The data presented above clearly indicate that the nucleic acid binding moiety of Tat can be substituted. Alternatively, it will be of interest to test whether the "acidic activation domains" of transcription factors like Gal4 or VP16 can function when coupled to the RNA-binding domain of Tat (Figure 4D). If so, such hybrids can define the amino acids in Tat that are involved in specific RNA binding. In addition, a positive result would suggest that the molecular details of transcriptional activation from a position in the DNA or RNA can be similar. We have speculated (49) on the potential advantages of a Tat-mechanism that simultaneously uses DNA and RNA targets for HIV.

POST-TRANSCRIPTIONAL EFFECTS OF TAR AND TAT

In addition to transcriptional activation, Tat may have post-transcriptional effects on TAR-containing transcripts. The presence of an extended hairpin structure like TAR in the leader of a mRNA can by itself inhibit translation (52). This stable structure can either impair scanning of the ribosome or decrease accessibility of the 5'cap structure to translational initiation factors. Although this *cis*-inhibition can be rather dramatic *in vitro* (53,54), the TAR structure appears to be sufficiently unstable, due perhaps to the presence of multiple destabilizing bulge elements in the hairpin (Figure 2). This permits high level translation *in vivo* (49).

A more complex translational mechanism proposes the inhibition of translation by TAR RNA in *trans*. The double-stranded nature of TAR was reported to activate dsRNA-dependent protein kinase (55,56) and to a lesser extent 2-5A-synthetase (55). While the kinase can inhibit translation by phosphorylating the initiation factor eIF2-a, the 2-5A-synthetase is part of an RNA degradation pathway. Quantitative analysis by some investigators of the *trans*-inhibitory effect showed it to be small compared to *cis*-inhibition *in vitro* (54).

In some experimental systems, the Tat-induced increase in mRNA is insufficient to account for the increase in protein synthesized from that mRNA (57). Another demonstration of a post-transcriptional Tat effect has been reported in oocyte injection experiments (58). Expression of pre-synthesized TAR-CAT mRNA was enhanced when it was co-injected with the Tat protein. In addition, purified Tat specifically enhances the translation of TAR-containing mRNAs *in vitro* (54). The structural and sequence requirements for TAR in this post-transcriptional mechanism(s) are yet to be clarified. The post-transcriptional components are believed to be a lesser component of Tat-induced gene expression.

REFERENCES

1. Coffin, J.M. (1990) Retroviridae and their replication. In *Virology* (ed. Fields, B.N. and Knipe, D.M.), pp 1437-1500. Raven press, New York.

2. Cullen, B.R. and Greene, W.C. (1990) Functions of the auxiliary gene products of the human immunodeficiency virus type 1. *Virology* 178, 1-5.

3. Dang, C.V. and Lee, W.M.F. (1989) Nuclear and nucleolar targeting sequences of c-erb-A, c-myb, N-myc, p53, HSP70, and HIV Tat proteins. *J. Biol. Chem.* 264, 18019-18023.

4. Garcia, J.A., Harrich, D., Pearson, L., Mitsuyasu, R. and Gaynor, R.B. (1988) Functional domains required for tat-induced transcriptional activation of the HIV-1 long terminal repeat. *EMBO J.* 7, 3143-3147.

5. Hauber, J., Malim, M.H. and Cullen, B.R. (1989) Mutational analysis of the conserved basic domain of human immunodeficiency virus tat protein. *J. Virol.* 63, 1181-1187.

6. Ruben, S., Perkins, A., Purcell, R., Joung, K., Sia, R., Burghoff, R., Haseltine, W.A. and Rosen, C.A. (1989) Structural and functional characterization of human immunodeficiency virus tat protein. *J. Virol.* 63, 1-8.

7. Siomi, H., Shida, H., Maki, M. and Hatanaka, M. (1990) Effect of a highly basic region of human immunodeficiency virus Tat protein on nucleolar localization. *J. Virol.* 64, 1803-1807.

8. Dingwall, C., Ernberg, I., Gait, M.J., Green, S.M., Heaphy, S., Karn, J., Lowe, A.D., Singh, M., Skinner, M.A. and Valerio, R. (1989) Human immunodeficiency virus 1 Tat protein binds *trans*-activation-responsive region (TAR) RNA in *vitro*. *Proc. Natl. Acad. Sci. USA* 86, 6925-6929.

9. Roy, S., Delling, U., Chen, C-H., Rosen, C.A. and Sonenberg, N. (1990) A bulge structure in HIV-1 TAR RNA is required for Tat binding and Tat-mediated trans-activation. *Genes Dev.* 4, 1365-1373.

10. Weeks, K.M., Ampe, C., Schultz, S.C., Steitz, T.A. and Crothers, D.M. (1990). Fragments of the HIV-1 Tat protein specifically bind TAR RNA. *Science* 249, 1281-1285.

11. Subramanian, T., Kuppuswamy, M., Venkatesh, L., Srinivasan, A. and Chinnadurai, G. (1990) Functional substitution of the basic domain of the HIV-1 trans-activator, Tat, with the basic domain of the functionally heterologous Rev. *Virology* 176, 178-183.

12. Seigel, L.J., Ratner, L., Josephs, S.F., Derse, D., Feinberg, M.B., Reyes, G.A., O'Brien, S.J. and Wong-Staal, F. (1986) *Trans*-activation. induced by human T-lymphotropic virus type III (HTLV-III) maps to a viral sequence encoding 58 amino acids and lacks tissue specificity. *Virology* 148, 226-231.

13. Kuppuswamy, M., Subramanian, T., Srinivasan, A. and Chinnadurai, G. (1989) Multiple functional domains of Tat, the transactivator of HIV-1, defined by mutational analysis. *Nucl. Acids Res.* 17, 3551-3561.

14. Frankel, A.D. and Pabo, C.O. (1988) Cellular uptake of the Tat protein from the human immunodeficiency virus. *Cell* 55, 1189-1193.

15. Brake, D.A., Debouck, C. and Biesecker, G. (1990) Identification of an Arg-Gly-Asp (RGD) cell adhesion site in human immunodeficiency virus type 1 transactivator protein, Tat. *J. Cell Biol.* 111, 1275-1281.

16. Ensoli, B., Barillari, G., Salahuddin, S.Z., Gallo, R.C. and Wong-Staal, F. (1990) Tat protein of HIV-1 stimulates growth of cells derived from Kaposi's sarcoma lesions of AIDS patients. *Nature* 345, 84-86.

17. Frankel, A.D., Bredt, D.S. and Pabo, C.O. (1988) Tat protein from human immunodeficiency virus forms a metal-linked dimer. *Science* 240, 70-73.

18. Rice, A.P. and Charlotti, F. (1990) Mutational analysis of the conserved cysteine-rich region of the human immunodeficiency virus type 1 Tat protein. *J. Virol.* 64, 1864-1868.

19. Rappaport, J., Lee, S-J., Khalili, K. and Wong-Staal, F. (1989) The acidic amino-terminal region of HIV-1 Tat protein constitutes an essential activating domain. *New Biol.* 1, 101-110.

20. Ptashne, M. (1988) How eucaryotic transcriptional activators work. *Nature* 335, 683-689.

21. Feng, S. and Holland, E.C. (1988) HIV-1 tat *trans*-activation requires the loop sequence within TAR. *Nature* 334, 165-167.

22. Berkhout, B., Silverman, R.H. and Jeang, K.T. (1989) Tat trans-activates the human immunodeficiency virus through a nascent RNA target. *Cell* 59, 273-282.

23. Muesing, M.A., Smith, D.H., and Capon, D.J. (1987) Regulation of mRNA accumulation by a human immunodeficiency virus *trans*-activator protein. *Cell* 48, 691-701.

24. Hauber, J. and Cullen, B.R. (1988) Mutational analysis of the *trans*-activation-responsive region of the human immunodeficiency virus type I long terminal repeat. *J. Virol.* 62, 673-679.

25. Jakobovits, A., Smith, D.H., Jakobovits, E.B. and Capon, D.J. (1988) A discrete element 3′ of human immunodeficiency virus (HIV-1) and HIV-2 mRNA initiation sites mediates transcriptional activation by an HIV *trans*-activator. *Mol. Cell. Biol.* 8, 2555-2561.

26. Berkhout, B. and Jeang, K.T. (1989) Trans-activation of human immunodeficiency virus type 1 is sequence specific for both the single-stranded bulge and loop of the trans-acting-responsive hairpin: a quantitative analysis. *J. Virol.* 63, 5501-5504.

27. Roy, S., Parkin, N.T., Rosen, C., Itovitch, J. and Sonenberg, N. (1990) Structural requirements for *trans*-activation of human immunodeficiency virus type 1 long terminal repeat-directed gene expression by Tat: importance of base pairing, loop sequence, and bulges in the Tat-responsive sequence. *J. Virol.* 64, 1402-1406.

28. Emerman, M., Guyader, M., Montagnier, L., Baltimore, D. and Muesing, M.A. (1987) The specificity of the human immunodeficiency virus type 2 *trans*-activator is different from that of human immunodeficiency virus type 1. *EMBO J.* 6, 3755-3760.

29. Berkhout, B., Gatignol, A., Silver, J. and Jeang, K.T. (1990) Efficient *trans*-activation by the HIV-2 Tat protein requires a duplicated TAR RNA structure. *Nucl. Acids Res.* 18, 1839-1846.

30. Gatignol, A., Kumar, A., Rabson, A. and Jeang, K.T. (1989) Identification of cellular proteins that bind to the human immunodeficiency virus type 1 *trans*-activation-responsive TAR element RNA. *Proc. Natl. Acad. Sci. USA* 86, 7828-7832.

31. Gaynor, R., Soultanakis, E., Kuwabara, M., Garcia, J. and Sigman, D.S. (1989) Specific binding of a Hela cell nuclear protein to RNA sequences in the human immunodeficiency virus *trans*-activating region. *Proc. Natl. Acad. Sci. USA* 86, 4858-4862.

32. Marciniak, R.A., Garcia-Blanco, M.A. and Sharp, P.A. (1990) Identification and characterization of a HeLa nuclear protein that specifically binds to the *trans*-activation-response (TAR) element of human immunodeficiency virus. *Proc. Natl. Acad. Sci. USA* 87, 3624-3628.

33. Graham, G.J. and Maio, J.J. (1990) RNA transcripts of the human immunodeficiency virus transactivation response element can inhibit action of the viral transactivator. *Proc. Natl. Acad. Sci. USA* 87, 5817-5821.

34. Hart, C.E., Ou, C-Y., Galphin, J.C., Moore, J., Bachler, L.T., Wasmuth, J.J., Petteway, S.R. and Schochetman, G. (1989) Human chromosome 12 is required for HIV-1 expression in human-hamster hybrid cells. *Science* 246, 488-491.

35. Newstein, M., Stanbridge, E.J., Casey, G. and Shank, P.R. (1990) Human chromosome 12 encodes a species-specific factor,which increases human immunodeficiency

virus type 1 Tat-mediated trans activation in rodent cells. *J. Virol.* 64, 4565-4567.

36. Nelbock, P., Dillon, P.J., Perkins, A. and Rosen, C.A. (1990) A cDNA for a protein that interacts with the human immunodeficiency virus Tat transactivator. Science 248, 1650-1653.

37. Kao, S.Y., Calman, A.F., Luciw, P.A. and Peterlin, B.M. (1987) Anti-termination of transcription within the long terminal repeat of HIV-1 by tat gene product. *Nature* 330, 489-493.

38. Selby, M.J., Bain, E.S., Luciw, P.A. and Peterlin, B.M. (1989) Structure, sequence, and position of the stem-loop in TAR determine transcriptional elongation by tat through the HIV-1 long terminal repeat. *Genes Dev.* 3, 547-558.

39. Barik, S., Ghosh, B., Whalen, W., Lazinski, D. and Das, A. (1987) An antitermination protein engages the elongating transcription apparatus at a promoter-proximal recognition site. *Cell* 50, 885-899.

40. Horwitz, R.J., Li, J. and Greenblatt, J. (1987) An elongation control particle containing the N gene transcriptional antiterminator protein of bacteriophage lambda. *Cell* 51, 631-641.

41. Platt, T. (1986) Transcription termination and the regulation of gene expression. *Annu. Rev. Biochem.* 55, 3390-372.

42. Landick, R. and Yanofsky, C. (1987) Transcriptional attenuation. In *Escherichia coli and Salmonella typhimurium: Cellular and molecular biology* (ed. F.C. Neihardt, J.L. Ingraham, K.B. Low, B. Magasanik, M. Schaechter and H.E. Umbarger) pp. 1276-1301. American Society for Microbiology, Washington, D.C.

43. Laspia, M.F., Rice, A.P. and Matthews, M.B. (1989) HIV-1 Tat protein increases transcriptional initiation and stabilizes elongation. *Cell* 59, 283-292.

44. Toohey, M.G. and Jones, K.A. (1989) In vitro formation of short RNA polymerase II transcripts that terminate within the HIV-1 and HIV-2 promoter-proximal downstream regions. *Genes Dev.* 3, 265-282.

45. Selby, M.J. and Peterlin, B.M. (1990) Trans-activation by HIV-1 Tat via a heterologous RNA binding protein. Cell 62, 769-776.

46. Jeang, K.T., Shank, P.R. and Kumar, A. (1988) Transcriptional activation of homologous viral long terminal repeats by the human immunodeficiency virus type 1 or the human T-cell leukemia virus type 1 tat proteins occurs in the absence of *de novo* protein synthesis. *Proc. Natl. Acad. Sci. USA* 85, 8291-8295.

47. Southgate, C., Zapp, M.L. and Green, M.R. (1990)
    Activation of transcription by HIV-1 Tat protein
    tethered to nascent RNA through another protein.
    Nature 345, 640-642.

48. Sharp, P.A. and Marciniak, R.A. (1989) HIV TAR: an RNA
    enhancer? *Cell* 59, 229-230.

49. Berkhout, B., Gatignol, A., Rabson, A.B. and Jeang,
    K.T. (1990) TAR-independent activation of the HIV-1
    LTR: evidence that Tat requires specific regions of the
    promoter. *Cell* 62, 7257-767.

50. Rosen, C.A., Sodroski, J.G. and Haseltine, W.A. (1985)
    The location of cis-acting regulatory sequences in the
    human T cell lymphotropic virus type III (HTLV-III/LAV)
    long terminal repeat. *Cell* 41, 813-823.

51. Peterlin, B.M., Luciw, P.A., Barr, P.J. and Walker,
    M.D. (1986) Elevated levels of mRNA can account for
    the *trans*-activation of human immunodeficiency virus
    (HIV). *Proc. Natl. Acad. Sci. USA* 83, 9734-9738.

52. Kozak, M. (1989) The scanning model for translation:
    an update. *J. Cell. Biol.* 108, 229-241.

53. Parkin, N.T., Cohen, E.A., Darveau, A., Rosen, C.,
    Haseltine, W. and Sonenberg, N. (1988) Mutational
    analysis of the 5' noncoding region of human
    immnnodeficiency virus type 1: effects of secondary
    structure on translation. *EMBO J.* 7, 2831-2837.

54. SenGupta, D.N., Berkhout, B., Gatignol, A., Zhou, A.
    and Silverman, R.H. (1990) Direct evidence for
    translational regulation by HIV-1 leader RNA and Tat
    protein. *Proc. Natl. Acad. Sci. USA*, in press.

55. SenGupta, D.N. and Silverman, R.H. (1989) Activation
    of interferon-regulated, dsRNA-dependent enzymes by
    HIV-1 leader RNA. *Nucl. Acids Res.* 17, 969-978.

56. Edery, I., Petryshyn, R. and Sonenberg, N. (1989)
    Activation of double-stranded RNA-dependent kinase
    (dsI) by the TAR region of HIV-1 mRNA: a novel
    translational control mechanism. *Cell* 56, 303-312.

57. Cullen, B.R. (1986) *Trans*-activation of human
    immunodeficiency virus occurs via a bimodal mechanism.
    Cell 46, 973-982.

58. Braddock, M., Chambers, A., Wilson, W., Esnouf, M.P.,
    Adams, S.E., Kingsman, A.J. and Kingsman, S.M. (1989)
    HIV-1 Tat "activates" presynthesized RNA in the
    nucleus. *Cell* 58, 269-279.

HIV-1 Tat *trans*-activates in the Absence of its Target

Mark J. Selby
B. Matija Peterlin

Howard Hughes Medical Institute and
Departments of Medicine, Microbiology and Immunology
University of California at San Francisco
San Francisco, CA. 94143-0724

The human immunodeficiency virus type 1 (HIV-1) is the etiologic agent of the acquired immunodeficiency disease syndrome (AIDS). Transcription of HIV-1 is regulated by the virally encoded *trans*-activator, Tat (Varmus, 1988; Cullen and Greene, 1989; Jones 1989; Sharp and Marciniak, 1989; Pavlakis and Felber, 1990). Tat acts at the 5' LTR to greatly increase levels of viral RNA and proteins (Rosen et al, 1985; Peterlin et al, 1986; Muessing et al., 1987). Tat has been shown to be essential for efficient viral replication and cytopathology (Dayton et al., 1986; Fisher et al., 1986).

The target site for Tat, known as TAR, was localized to just 3' of the site of HIV-1 transcription initiation (Rosen et al., 1985; Wright et al., 1986; Muessing et al, 1987). TAR functions in a position- and orientation-dependent manner and is therefore unlike a transcriptional enhancer (Peterlin et al., 1986; Muessing et al., 1987). Transcription through TAR yields a RNA that can assume a stem-loop structure in vitro (Muessing et al., 1987). Mutational analyses of TAR demonstrated that Tat recognizes TAR as RNA and not DNA and that nucleotides from positions +19 to +45 are critical for responsiveness to Tat (Selby et al., 1989). Furthermore, using the TAR stem-loop and competing secondary structures, Berkhout et al. (1989) demonstrated conclusively that Tat recognizes TAR as nascent RNA.

Since Tat action leads to an increase in the level of expression at the RNA level, does it do so like a transcriptional activator? A short review of transcriptional activators is warranted so as to note features of experiments with these factors that may be applied to understanding Tat

function.   Figure 1 shows some basic elements of
transcription. First, polymerase, which is a generic term for
several factors, interacts with, what is in most cases, the
TATA box.   Polymerase does not function in a vacuum; it can be
and is often modified by factors binding to adjacent sites on
the DNA template.   Those that augment transcription are called
transcription activators.   The activators are typically
bifunctional proteins, consisting of a domain that recognizes
the target DNA (binding domain) and another that probably
interacts with polymerase or an associated factor to increase
expression (activation domain).   The fusion of activators or
activation domains to other DNA-binding proteins confers a
new DNA-binding specificity upon the activator and can thus
"target" the activator to a different transcription unit
(Figure 2). Dissection of such fusions proteins has allowed
for the assignment of activation and DNA binding domains.
Since Tat functions at the level of transcription, does it do
so in a fashion analogous to a DNA binding transcription
factor?

Our goal was to use the aforementioned scheme to determine
if Tat can function when provided with a DNA target site
specificity in the absence of TAR.   Therefore, the entire Tat
protein was fused to amino acid 87 of LexA,   a bacterial DNA-
binding protein.   LexA is an ideal fusion partner because 1.
it has no function in eukaryotic cells and thus background is
not an issue; 2. it has a well-characterized binding site and
binds with high-affinity; 3. fusion of some transcription
activators to LexA have been shown to be functional (Godowski
et al, 1988 and references therein) 4. LexA has no known
affinity for RNA.   To "target" LexA-Tat to a transcription
unit, the LexA binding site was introduced 5' of the HIV-1 cap
site.   As shown in Table 1, Tat and Lex-Tat $trans$-activated an
HIV-1 template that had an intact TAR element.   Co-
transfections with these two Tat's and the wild-type LTR-CAT
(HIVTARCAT) showed numerically equivalent $trans$-activations,
indicating the Tat moiety of the fusion was functionally
intact.   Wild-type Tat and the fusion, as expected, failed to
$trans$-activate a TAR-deletion template (HIVΔTAR-CAT).   Both
$trans$-activators were equally active on a HIV-1 template that
had a lexA binding site engineered between the NF$_K$B and Sp1
binding sites.   The basal level expression of the Lex-A
binding site template did not differ from the HIV-1 parent
recombinant, demonstrating that the introduction of the LexA
site did not interfere with expression.   However, both Tat and
the Lex-Tat fusion were also unable to $trans$-activate the ΔTAR
derivative of the LexA binding site template. Furthermore, the
LexA-Tat fusion also was unable to $trans$-activate templates
that had LexA binding sites within the TAR region that were
unresponsive to Tat.   To ensure that the lexA-containing
templates were responsive to activation by a DNA-binding
transcription activator, all the templates were evaluated by
co-transfection with the lex-glucocorticoid receptor fusion
(Godowiski et al,1988).   In the presence of the inducer
dexamathasone, all the lexA-bearing templates responded with
increased levels of the reporter gene product.   Thus, while
the lex-Tat fusion was non-functional, a similar fusion of
LexA 1-86 to the GR was functional on the same templates.
These observations suggests that Tat cannot be "presented" to
the transcription apparatus via a DNA-binding protein and
retain the ability to $trans$-activate.

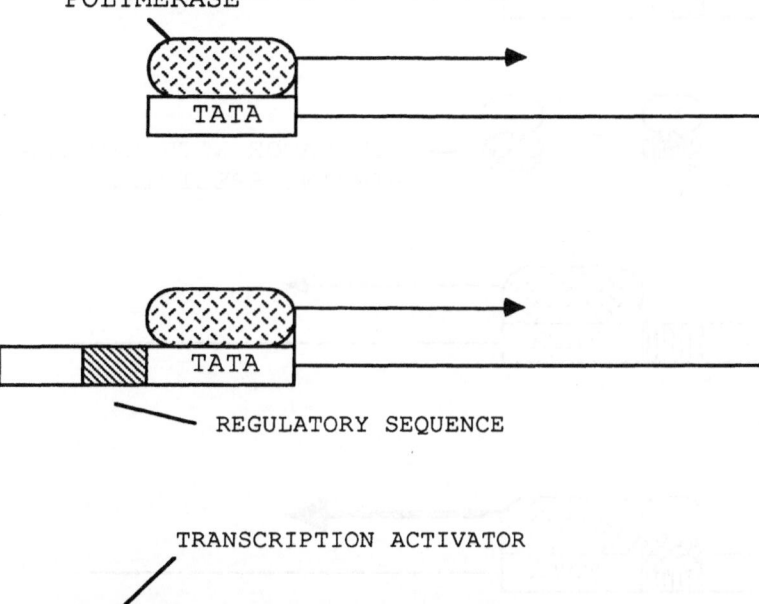

POLYMERASE

TATA

REGULATORY SEQUENCE

TRANSCRIPTION ACTIVATOR

TATA

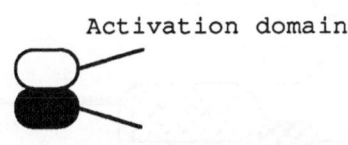

Activation domain

DNA binding domain

Figure 1. A schematic depiction of transcription
activation. Initiations of transcription
are depicted as arrows and their strengths
are denoted by arrows of different
thickness. The bifunctional character of
many transcription activators, which often
bind to regulatory sequences, is also shown.

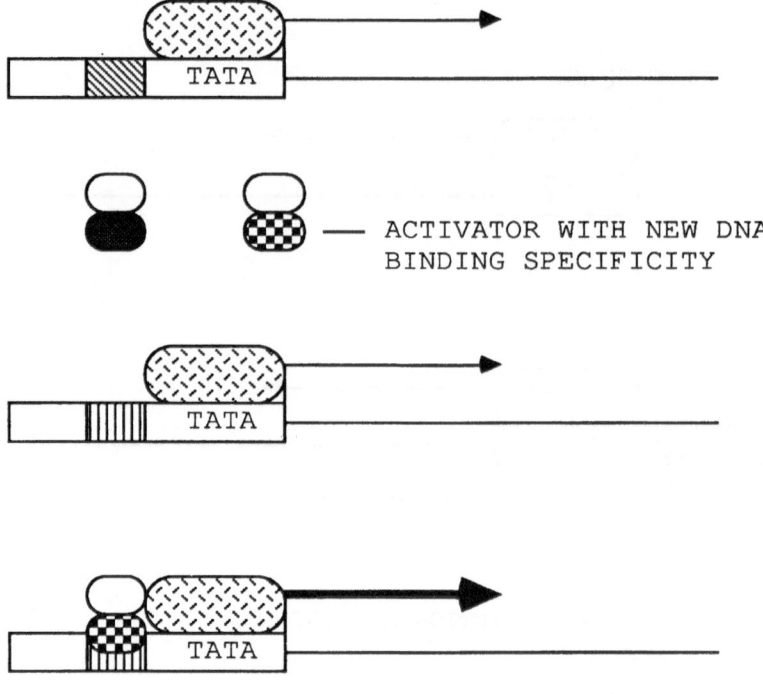

ACTIVATOR WITH NEW DNA
BINDING SPECIFICITY

CAN TAT, WHEN "PRESENTED" BY A DNA-BINDING
PROTEIN, INCREASE THE LEVEL OF EXPRESSION?

TAT

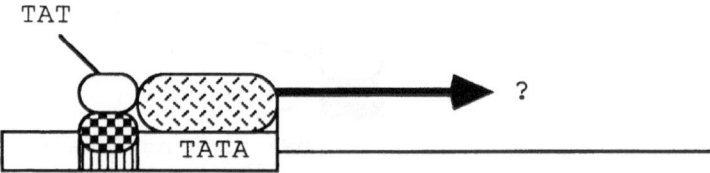

Figure 2. A schematic depiction of how the specificity
of a transcription activator can be altered by
fusion. The different binding specificities
as well as target sites are denoted by
different patterns. The lower part of this
figure illustrates how Tat can be "targeted"
to a transcription unit via a DNA-binding
protein.

Since several laboratories have demonstrated that Tat recognizes TAR as RNA, we opted to make a fusion of Tat to an RNA binding protein and evaluate its function when the cognate binding site is introduced in place of TAR. We selected the coat protein of bacteriophage MS2 to fuse to Tat. The coat protein (CP) was selected because it's binding characteristics have been extensively studied and a molecular clone is available (Romaniuk et al., 1987; Peabody, 1990). The coat protein was fused to amino acid 67 of Tat to generate the Tat-coat protein fusion (pTCP). The fusion was evaluated by co-transfection with a variety of DNA templates. We observed that both Tat and TCP *trans*-activated a wild-type HIV-1 template efficiently (Table 2a). The level of *trans*-activation of TCP was about 50% of wild-type Tat, but identical to a Tat truncated after amino acid 67. Thus, we can state that there is no loss of Tat function upon fusion to coat protein. Finally, we transfected TCP with a HIV-1 template where the TAR element was completely deleted and replaced with the binding site for coat protein. Incidentally, the coat protein binding site is also a RNA stem-loop structure. Significant *trans*-activation of this template by the Tat-coat protein fusion was observed (Table 2b). Tat, either full-length or truncated to amino acid 87, had no effect on the MS2 target site template. As an additional control, we demonstrated that coat protein itself had no effect on this template. Collectively, these results show that Tat, when fused to a protein of different RNA binding specificity, can function in absence of its natural target. Thus, in contrast to the Tat-DNA binding protein fusion, Tat functions well when tethered to a novel RNA binding protein.

Table 1. *Trans*-activation of HIV templates by TAT and Lex-TAT recombinants

template	*trans*-activator	TA
HIVTARCAT	TAT	+
"	Lex-TAT	+
HIVΔTARCAT	TAT	−
"	Lex-TAT	−
HIV-lex-TAR-CAT	TAT	+
"	Lex-TAT	+
HIV-lex-ΔTARCAT	TAT	−
"	Lex-TAT	−

The parental template is HIVTARCAT where CAT has been fused to position +80 in the LTR. Derivatives bearing the LexA binding site between the NF$_K$B and Sp1 sites are denoted with lex in their plasmid designations. Derivatives lacking the TAR region are denoted as ΔTAR recombinants. The ability of a *trans*-activator to *trans*-activate (TA) is given as + or −.

Table 2a. *Trans*-activations of wild-type HIV templates by the Tat-coat protein fusion

Template[a]	Trans-Activator[b]	fold-TA (%of TAT)[c]
HIV-TAR-CAT	TAT(1-100)	100
"	TAT(1-67)	54.2
"	TCP(1-67/CP)	57.3
"	CP	<1

Trans-activation of a wild-type TAR-template (a) by TAT and TAT-coat protein fusion. The length of TAT in each effector is shown parenthetically (b). Fold-*trans*-activation as a % of TAT fold-*trans*-activation is presented (c). TCP represents the TAT-coat protein fusion recombinant. CP denotes coat protein driven by the SV-40 early promoter.

Table 2b. *Trans*-activations of templates containing the MS2 Coat Protein Binding Site or TAR

Template[a]	*Trans*-activator[b]	fold-TA (% of TCP)[c]
HIV-TAR-CAT	TCP	100
"	TAT	172
pLTR-MS2	TCP	15.6
"	TAT	<1

The ability of TAT and the fusion (b) to *trans*-activate are evaluated on both TAR- and MS2-bearing templates (a). Fold-*trans*-activation is presented as a % of TAT-coat protein *trans*-activation on a TAR-template.

To demonstrate that the coat protein moiety of the TCP fusion was responsible for targeting Tat to the target RNA, we engineered several different mutations into the coat protein binding site. Mutations in the binding site that have been shown to reduce binding in vitro (Carey et al., 1983; Lowary and Uhlenbeck, 1987) would be expected to show corresponding decreases in *trans*-activations. Co-transfections of TCP with templates bearing point-mutations that decreased binding affinities showed reduced *trans*-activations as predicted. Furthermore, deletions of 1 or 2 nucleotides in the binding site completely eliminated *trans*-activation. These data demonstrate that coat protein targets Tat to the transcription unit.

The utility of this fusion is that it allows us to examine mutants in Tat without regard to mutating the region responsible for RNA binding and thus simplifies the assignment of the activation and RNA binding regions. To identify these regions, we made several deletions in the Tat moiety. Specifically, we removed the N-terminal region which has some alpha-helical character (Rappaport et al., 1989), the cysteine-rich region which might be involved in metal-linked dimerization (Frankel et al., 1988), and the basic region which consists of 9 basic amino acids (Endo et al., 1989). Table 3 shows that deletion of the N-terminal and cysteine-rich regions result in a non-functional *trans*-activator on both TAR- and MS2-target templates. Thus, these regions constitute an indispensable region for *trans*-activation. Whether they are required for contact with polymerase or an associated factor and/or are required for proper folding of the molecule remains to be elucidated. The most interesting deletion was that which removed only the 9 basic amino acids. This deletion recombinant was inactive on a wid-type TAR-containing template. However, not only was *trans*-activation of the MS2 template retained with this deletion, it is even greater than that observed with the control Tat-coat protein fusion. This may have resulted from the removal of an interfering RNA binding motif or alternatively from a more favorable folding of the deleted molecule relative to the parental fusion. *Trans*-activation by the basic region deletion of the MS2 template and not the TAR template, indicates this region is required for the recognition of TAR. Because coat protein provides the RNA targeting function in the fusion, the basic region is dispensable for function on the MS2 template, but not on the TAR template. This is the first functional demonstration that the basic domain targets Tat to TAR.

Table 3. *Trans*-activations by deletion mutants in the Tat moiety of the fusion

Trans-activator[a]	% conversion[b]	fold-TA[c]
TCP	37.3	218
TCP-ΔCYS	0.169	1.0
TCP-ΔNT	0.261	1.53
TCP-ΔB	54.8	320
Coat protein	0.171	1.0

The fusion *trans*-activator and derivatives are shown in the first column (a). Percent conversions of CAT activity (b) and fold-*trans*-activations (fold-TA)(c) by deletions of the cysteine-rich region (ΔCYS), the N-terminus (ΔNT) and the basic region (ΔB) are presented.

We have shown that Tat can be tethered to a RNA binding protein and function, while a fusion to a DNA protein was non-functional. Our results demonstrate that Tat, unlike all known transcriptional modifiers, can only function in the context of nascent RNA targets. Our investigations have also shown that coat protein serves to target Tat to the target site and that the ability to *trans*-activate lies solely in the Tat moiety. Our results also show that the TAR element itself is dispensable for *trans*-activation as it can be replaced with an unrelated sequence and function. Furthermore, these data suggest that factors which bind 3' of the cap site are also dispensable as *trans*-activation results in spite of their absence. These observations are consistent with the hypothesis that TAR is a positive-acting element that simply presents Tat to the transcription apparatus. This makes intuitive sense given that deletion of TAR does not result in increased basal expression, but is rendered Tat non-responsive. Future experiments will 1. need to identify the cellular factor(s) that is required to mediate Tat function and 2 clarify the mechanism of Tat *trans*-activation in terms how expression is enhanced.

The functional Tat-coat protein chimera is the first successful fusion of a factor that modifies transcription to a RNA binding protein. Other fusions with transcription activators are sure to follow. The ability of other fusions to function or lack thereof will determine the uniqueness or universality of Tat function.

References

Berkhout, B., Silverman, R.H., and Jeang, K.-T. (1989). Tat *trans*-activates the human immunodeficiency virus through a nascent RNA target. Cell 59, 273-282.

Carey, J., Lowary, P.T., and Uhlenbeck, O.C. (1983). Interaction of R17 coat protein with synthetic variants of its ribonucleic acid binding site. Biochemistry 22, 4730-4737.

Cullen, B.R. and Greene, W.C. (1989). Regulatory pathways governing HIV-1 replication. Cell 44, 941-947.

Dayton, A.I., Sodroski, J.G., Rosen, C.A., Goh, W.C., and Haseltine, W.A. (1986). The *trans*-activator gene of the human T cell lymphotropic virus type III is required for replication. Cell 44, 941-947.

Endo, S., Kubota, S., Siomi, H., Adachi, A., Oroszlan, S., Maki, M., and Hatanaka, M. (1989). A region of basic amino-acid cluster in HIV-1 Tat protein is essential for *trans*-acting activity and nucleolar localization. Virus Genes 3, 99-110.

Fisher, A.G., Feinberg, M.B., Hosephs, S.F., Harper, M.E., Marselle, L.M., Reyes, G., Gonda, M.A., Aldovini, A., Debouk, C., Gallo, R.C., and Wong-Staal, F. (1986). The *trans*-activator gene of HTLV-III is essential for virus replication. Nature 320, 367-371.

Frankel, A.D., Bredt, D.S., and Pabo, C.O. (1988). Tat protein from human immunodeficiency virus forms a metal-linked dimer. Science 240, 70-73.

Godowski, P.J., Picard, D., and Yamamoto, K.R. (1988). Signal transduction and transcriptional regulation by glucocorticoid receptor-LexA fusion proteins. Science 241, 812-816.

Jones, K.A. (1989). HIV trans-activation and transcription control mechanisms. New Biol. 1, 127-135.

Lowary, P.T. and Uhlenbeck, O.C. (1987). An RNA mutation that increases the affinity of an RNA-protein interaction. Nucl. Acids Res. 15, 10483-10493.

Muessing, M., Smith, D.H. and Capon, D.J. (1987). Regulation of mRNA accumulation by a human immunodeficiency virus trans-activator protein. Cell 48, 691-701.

Pavlakis, G.N. and Felber, B.K. (1990). Regulation of expression of human immunodeficiency virus. New Biol. 2, 20-31.

Peabody, D.S.(1990) Translational repression by bacteriophage MS2 coat protein expressed from a plasmid. A system for genetic analysis of a protein-RNA interaction. J.Biol. Chem. 265, 5684-5689.

Peterlin, B.M., Luciw, P.A., Barr, P.J. and Walker, M.D. (1986). Elevated levels of mRNA can account for the transactivation of human immunodeficiency virus (HIV). Proc. Natl. Acad. Sci. USA 83, 9734-9738.

Rappaport, J., Lee, S.-J., Khalili, K. and Wong-Staal, F. (1989). The acidic amino-terminal region of HIV-1 Tat protein constitutes an essential activating domain. New Biol. 1, 101-110.

Romaniuk, P.J., Lowary, P., Wu, H.-N., Stormo, G. and Uhlenbeck, O.C. (1987). RNA binding site of R17 coat protein. Biochemistry 26, 1563-1568.

Selby, M.J., Bain, E.S., Luciw, P.A. and Peterlin, B.M. (1989). Structure, sequence, and position of the stem-loop in tar determine transcriptional elongation by tat through the HIV-1 LTR. Genes Dev., 3, 547-558.

Sharp., P.A. and Marciniak, R.A. (1989). HIV TAR: an RNA enhancer? Cell 59, 229-230.

Varmus, H.E. (1988). Regulation of HIV and HTLV gene expression. Genes Dev. 2, 1055-1062.

Wright, S.M., Felber, B.K., Paskalis, H. and Pavlakis, G.N. (1986). Expression and characterization of the trans-activator of HTLV-III/LAV virus. Science 234, 988-992.

TRANS–ACTIVATION REQUIRES THE BINDING OF THE HIV-1 *TAT* PROTEIN TO TAR RNA

Colin Dingwall, Ingemar Ernberg [1], Michael J. Gait, Shaun Heaphy, Jonathan Karn and Michael Skinner [2]

Medical Research Council Laboratory of Molecular Biology, Hills Road Cambridge, CB2 2QH, England

Introduction

The viral *trans*-activator protein, *tat*, stimulates HIV transcription by recognizing the *trans*-activation-responsive region (TAR). Genetic experiments have shown that TAR is a *cis*-acting element which is located immediately downstream of the transcription start site, between nucleotides +1 and +59 (Dayton *et al.*, 1986; Rosen *et al.*, 1986; Fisher *et al.*, 1986; Hauber *et al.*, 1987; Muesing *et al.*, 1987).

Tat-dependent transcription requires only the presence of the TAR element. Transcription from viral LTRs truncated upstream of the Sp1 sites (Jakobovits *et al.*, 1990) or fused to heterologous promoters (Peterlin *et al.*, 1986; Cullen, 1986; Muesing *et al.*, 1987; Berkhout *et al.*, 1990b) is stimulated strongly by *tat*. However, TAR is functional only when placed in the correct orientation and 3' to the cap site (Muesing *et al.*, 1987; Selby *et al.*, 1989; Hauber & Cullen, 1988; Jakobovits *et al.*, 1988; Peterlin *et al.*, 1986). The position of TAR relative to the transcription start site is also important; when the distance between the 5'-end of TAR and the cap site is increased, a progressive reduction in *trans*-activation levels is observed (Hauber & Cullen, 1988; Selby *et al.*, 1989).

There is now strong evidence that TAR RNA rather than TAR DNA is the active element (reviewed in Sharp & Marciniak, 1989). Transcripts containing the TAR RNA sequence form a highly stable, nuclease-resistant, stem-loop structure (Muesing *et al.*, 1987; Berkhout *et al.*, 1989). Point mutations which destabilize the TAR stem by disrupting base-pairing usually abolish *tat*-stimulated transcription. However, it is the stability of the TAR stem rather than its sequence that is crucial, for as long as Watson-Crick base pairing is maintained, nearly every base-pair of the stem may be altered without affecting *trans*-activation (Feng & Holland, 1988; Berkhout & Jeang, 1989; Garcia *et al.*, 1989; Selby *et al.*, 1989; Roy *et al.*, 1990a). The TAR element must be transcribed in the nucleus and the RNA product correctly folded in order for *trans*-activation to occur. Berkhout *et al.* (1989) introduced anti-sense sequences, designed to destabilize formation of the TAR RNA hairpin-loop structure in nascent transcripts, in positions either 5' or 3' to TAR. Placement of the antisense sequences on the 5'-side of TAR prevents the wild-type hairpin from forming and leads to a block in *trans*-activation, but placement of the antisense sequence on the 3'-side of TAR allows for the formation of the normal TAR RNA stem-loop structure and efficient *trans*-activation.

Several recent reports suggest strongly that *tat* is presented to the transcription machinery by binding directly to nascent transcripts carrying the TAR RNA element. First, our previous studies have shown that HIV-1 *tat* protein expressed in *E. coli* is an RNA binding protein with

---

1 Present address: Department of Tumour Biology, Karolinska Institutet, Box 60400, S-0401, Stockholm, Sweden

2 Present address: A.F.R.C. Institute for Animal Health, Houghton Labortory, Cambridge PE17 2DA, UK.

Advances in Molecular Biology and Targeted Treatment for AIDS
Edited by A. Kumar, Plenum Press, New York, 1991

a high affinity for TAR RNA (Dingwall et al., 1989). Direct binding of tat to TAR has also been reported by Roy et al. (1990b) and by Cordingley et al., (1990). Fragments of tat overlapping a C-terminal arginine-rich sequence also appear to bind to TAR specifically (Cordingley et al., 1990; Roy et al., 1990b; Weeks et al., 1990). Second, experiments using hybrid proteins containing tat and an exogenous RNA-binding domain have demonstrated that tat can stimulate transcription from viral LTRs in which the TAR sequence is replaced by unrelated binding sites (Southgate et al., 1990; Selby & Peterlin, 1990). In the experiments of Southgate et al. (1990) a tat-rev fusion protein was constructed which was able to stimulate transcription when a rev-responsive element was placed immediately downstream of the HIV-LTR. Using a similar experimental strategy Selby & Peterlin (1990), showed that a tat-MS2 coat protein fusion protein could trans-activate an LTR in which TAR was replaced by the stem-loop structure recognized by the MS2 coat protein.

Here we define the binding site for tat on TAR more precisely and demonstrate a direct correlation between tat binding to TAR and trans-activation. Mutants that show significantly reduced tat binding in vitro show correspondingly reduced levels of tat-stimulated transcription in vivo.

Preparation of biologically active tat protein

In our original experiments HIV-1 tat was prepared from a β-galactosidase-tat fusion protein. This material was capable of binding to TAR RNA. The protein was also active in vivo. Transcription of an LTR-CAT reporter plasmid was activated by chloroquine-stimulated uptake of the purified tat protein into COSts cells (Dingwall et al., 1989).

More recently, we have prepared tat from a human growth hormone fusion protein (Ikehara et al., 1984) using an improved method (Figure 1). A key feature of the new method is that the cysteine residues in tat were protected by disulphide-exchange with 20mM 2-hydroxyethyldisulphide (Smithies, 1965) prior to cleavage of the fusion protein by cyanogen bromide (Dingwall et al., 1990). After purification the tat protein was refolded by stepwise dialysis against buffers containing progressively reduced concentrations of urea and 100µM Zn^{+2}. The purified, refolded protein migrates as a single band on polyacrylamide gels and remains soluble in low ionic strength buffers.

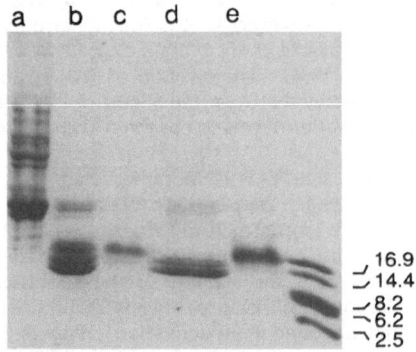

Figure 1. **Purification of HIV-1 tat.**
Tat protein was prepared from human growth hormone-tat fusion proteins (Ikehara et al., 1984). In brief, inclusion body pellets containing the fusion protein were solubilized in 7M GdHCl, 50mM Tris-HCl pH8.0, 2mM EDTA, 2mM DTT and the fusion protein was partially purified by ion-exchange chromatography on Q-Sepharose and S-Sepharose (Pharmacia) in buffers containing 6M Urea, 50mM Tris-HCl pH8.0, 2mM EDTA, 2mM DTT (lane a). The cysteine residues were protected by disulphide-exchange with 20mM 2-hydroxyethyldisulphide (Smithies, 1965), and the protein was cleaved by cyanogen bromide (lane b) as described previously (Dingwall et al., 1989). The basic tat protein (lane c) was then separated from the acidic human growth hormone peptides (lane d) by ion-exchange chromatography on Q-Sepharose and S-Sepharose in 6M Urea. The protecting groups were then removed from the protein by addition of 10mM DTT and the protein refolded by stepwise dialysis against 6M Urea, 5M Urea, 3M Urea, 1M Urea and 0M Urea in a buffer containing 50mM Tris-HCl pH8.0, 100mM DTT, 10mM ZnCl$_2$. Tat protein refolded in this manner elutes as a monomer from Superose 12 columns (Pharmacia) and migrates as a single band on polyacrylamide gels (lane e). For binding assays, protein concentrations were determined by amino acid analyses of the purified proteins.

Our preparations of refolded *tat* protein appear to be monomeric. Metal-linked dimers of *tat* are readily formed when either Zn^{+2} or Cd^{+2} ions are added to *tat* protein stored under acidic conditions (Frankel *et al.*, 1988). The Cd^{+2} linked dimers can be easily detected because they migrate with reduced mobility on SDS gels. Our preparations of *tat* readily form dimers when Cd^{+2} is added as described by Frankel *et al.* (1988). However, we have found that after stepwise dialysis against buffers containing $100\mu M$ Cd^{+2} the protein continues to migrate as a monomer on SDS gels. Furthermore, *tat* protein refolded by stepwise dialysis in the presence of Zn^{+2} or Cd^{+2} elutes from Superose 12 columns at the expected position for the monomeric protein.

Optimal conditions for HIV-1 *tat* binding to TAR RNA

A wide variety of parameters affecting the binding of *tat* protein to TAR RNA were studied and binding conditions were optimized (Figure 2). Binding of *tat* to TAR has a distinct pH optimum at pH 8.0, but good binding is observed between pH 7.5 and pH 8.5 (Figure 2a). Complexes between *tat* and TAR are very sensitive to ionic strength. Complex formation is strongly inhibited by greater than 100mM NaCl or 100mM KCl, the optimum overall ionic strength (μ) is 0.08M (Figure 2b). Addition of Zn^{+2}, Cd^{+2}, Ca^{+2}, Mg^{+2}, Fe^{+2} or Fe^{+3} ions to the binding reactions does not affect complex formation. However, the reaction is extremely sensitive to EDTA and addition of greater than 0.1mM EDTA abolishes binding (Figure 2d). Complex formation is also somewhat temperature sensitive. Optimal binding is observed at 0°C. At 15°C and 30°C there is a 30% reduction in complex formation (Figure 2c).

Stoichiometry

In our original experiments to demonstrate complex formation with TAR RNA (Dingwall *et al.*, 1989), a large molar excess of *tat* protein (1,000 to 10,000-fold) was used. This suggested to us that a significant fraction of the *tat* in our preparations might not have been properly folded. The improvements in the preparation of *tat* protein described above, and the use of high concentrations of RNA in the binding assays, have now allowed us to demonstrate complex formation using equimolar concentrations of protein and RNA, and to study in detail the sequence requirements for *tat* binding to TAR RNA.

Figure 3(a) shows a saturation binding experiment in which a constant concentration of labelled RNA (10nM) was incubated with increasing concentrations of *tat* protein, and the fraction of RNA retained on nitrocellulose filters was measured. Anti-sense TAR RNA was used as a control, since anti-sense TAR is unable to form stable complexes with *tat* in gel-retardation assays (Dingwall *et al.*, 1989). Tat binds HIV-1 TAR RNA with an apparent dissociation constant of $K_d = 12nM$, assuming a 1:1 complex. Binding to anti-sense RNA is lower by a factor of at least 10 ($K_d > 140$ nM). Indistinguishable results were obtained using HIV-1 TAR sequences that have either the UUU or UCU bulge sequences found in infectious molecular clones (Table 1).

The stoichiometry of the complex formed between *tat* and TAR was determined by varying the TAR RNA concentrations over a wide range, while keeping the *tat* protein concentration constant at 40nM (a value near the K_d). A Scatchard plot of the data is shown in Figure 3(b). The intercept gives a value for the stoichiometry, ν, of approximately one, while the slope of the line, fitted to the data by the least squares method, indicates a K_d of 11.8nM, in excellent agreement with the value obtained from the saturation binding experiment.

These results demonstrate that *tat* forms a one-to-one complex with TAR and that essentially all the *tat* in our new preparations is capable of binding RNA. The experiments do not distinguish between the formation of complexes between monomeric *tat* and TAR RNA or the formation of larger complexes consisting of metal linked dimers of *tat* (Frankel *et al.*, 1988) and two molecules of TAR RNA. However, because the *tat* protein which has been refolded in the presence of Zn^{+2} is monomeric, we believe that the biologically active form of *tat* is likely to be a monomer.

The dissociation constant of *tat* for TAR (12nM) is typical for an RNA-protein interaction. A K_d of 3nM has been measured for both *rev* binding to RRE RNA (Heaphy *et al.*, 1990; Daly *et al.*, 1989) and factor binding to the iron-response element RNA stem-loop (Leibold *et al.*, 1990). The approximately 10-fold difference in affinity that we have observed between *tat* binding to wild-type TAR RNA (specific) and anti-sense TAR (non-specific) is also consistent with values obtained in studies of other RNA binding proteins. For example, *rev* shows a 20-fold difference in its affinity for sense and anti-sense RRE sequences (Daly *et al.*, 1989; Heaphy *et al.*, 1990), while the *E. coli* ribosomal protein S4 shows only a 6-fold difference in affinity between specific and non-specific binding (Vartikar & Draper, 1989).

135

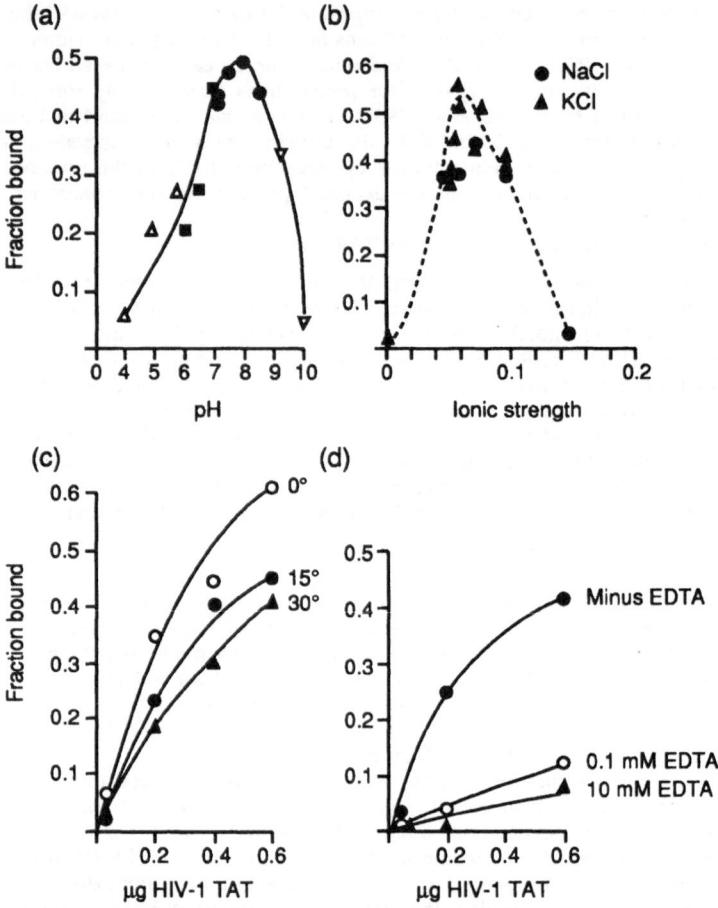

Figure 2.   Optimal conditions for HIV-1 *tat* binding to TAR RNA.
Panel a: pH dependency of *tat* binding to ^{32}P-labelled TAR RNA. Buffers of constant ionic strength (μ=0.07) but different pH were used. The Henderson-Hasselbach equation was used to determine buffer composition and the pH of the complete buffers was confirmed by direct measurements. Sodium acetate buffer ($\triangle$). Bis-Tris-HCl buffer ($\blacksquare$). Tris-HCl buffer ($\bullet$). Sodium borate buffer ($\triangledown$).   Panel b: Effect of increasing ionic strength on *tat* binding to ^{32}P-labelled TAR RNA. Binding reactions were carried out in Tris-HCl buffer pH 8.0 (μ=0.01) and the ionic strength was varied by the addition of NaCl ($\bullet$) or KCl ($\blacktriangle$). Panel c: Effect of temperature on *tat* binding to ^{32}P-labelled TAR RNA. Binding reactions were incubated at 0°C (O), 15°C ($\bullet$) and 30°C ($\blacktriangle$). The fraction of TAR RNA bound is plotted as a function of tat protein concentration. Panel d: Effect of EDTA on the binding of *tat* to ^{32}P-labelled TAR RNA. EDTA was added to standard binding reactions to give the final concentrations shown and the fraction of TAR RNA bound is plotted as a function of *tat* protein concentration.   Minus EDTA ($\bullet$). 1 mM EDTA (O). 10mM EDTA ($\blacktriangle$).

Tat recognizes a U-rich bulge in TAR RNA

Extensive mutagenesis studies, summarized in Figure 4, have defined sequences in TAR that are necessary for efficient *trans*-activation (Muesing *et al.*, 1987; Hauber & Cullen, 1988; Jakobovits *et al.*, 1988; Berkhout & Jeang, 1989; Selby *et al.*, 1989; Roy *et al.*, 1990a). Two notable features of the TAR RNA structure are the loop sequence (residues 30-35), and the "bulge" created by three non base-paired residues (residues 23-25) located near the apex of the double-helical stem. Both the bulge and the loop sequences in TAR are conserved between HIV-1 and the first stem-loop of HIV-2 TAR (Guyader *et al.*, 1987; Emerman *et al.*, 1987), and both regions are present in the minimal TAR sequence that is functionally active, located between nucleotides +19 and +42 (Muesing *et al.*, 1987; Hauber & Cullen, 1988; Jakobovits *et al.*, 1988; Berkhout & Jeang, 1989; Selby *et al.*, 1989; Roy *et al.*, 1990a).

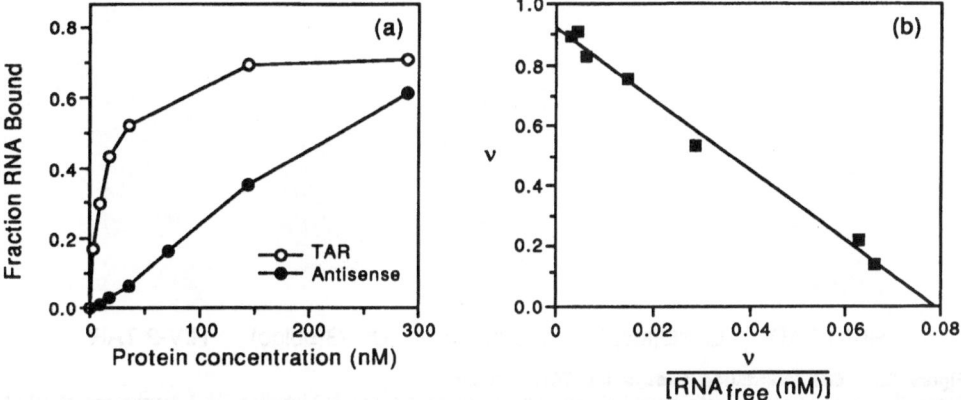

Figure 3. Saturation binding curve and Scatchard analysis of *tat* binding to TAR.
Reactions for saturation binding (Panel a) contained labelled TAR RNA or antisense TAR RNA at 10nM and between 0 to 300nM monomeric *tat* protein prepared by CNBr cleavage of the human growth hormone-*tat* fusion protein expressed in *E. coli*. For the Scatchard analysis (Panel b), the tat protein concentration was held constant at 40nM and the RNA concentration was varied between 7nM and 320nM. The total amount of labelled RNA was kept constant in the binding reactions, but the specific activity of the RNA was varied. The ordinate is the stoichiometry, v, the number of moles of RNA bound per mole of protein. The abscissa is the ratio of v to the free RNA concentration. After Dingwall *et al.*, 1990.

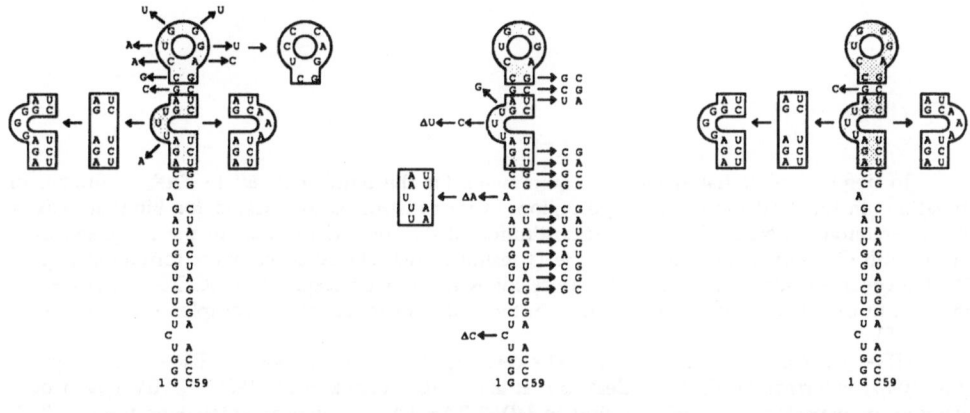

(a) Mutations in TAR that block *trans*-activation

(b) Mutations in TAR that do not affect *trans*-activation

(c) Mutations in TAR that block tat binding

Figure 4. Mutations in TAR known to affect *tat* binding and *trans*-activation.
Conserved sequence features present in both HIV-1 and HIV-2 TAR are boxed and shaded. Panel a: Mutations in TAR that block *trans*-activation. Point mutations are indicated by the arrows. More extensive changes involving multiple sites are indicated by the sequences in the open boxes. Panel b: Mutations in TAR that do not affect *trans*-activation. At almost every position in the TAR stem base substitutions that permit Watson-Crick base pairing do not affect *trans*-activation. The single unpaired bases in the stem are dispensable but a bulge containing at least two U residues is essential. Panel c: Mutations in TAR that block *tat* binding. Deletion of the U-rich bulge, replacement of the U-rich bulge by anti-sense sequences, replacement of the U residues in the bulge with G residues or the disruption of the structure by desta- bilisation of Watson-Crick base pairing at nearby residues in the TAR stem abolish *tat* binding to TAR. From Karn *et al.*, 1990.

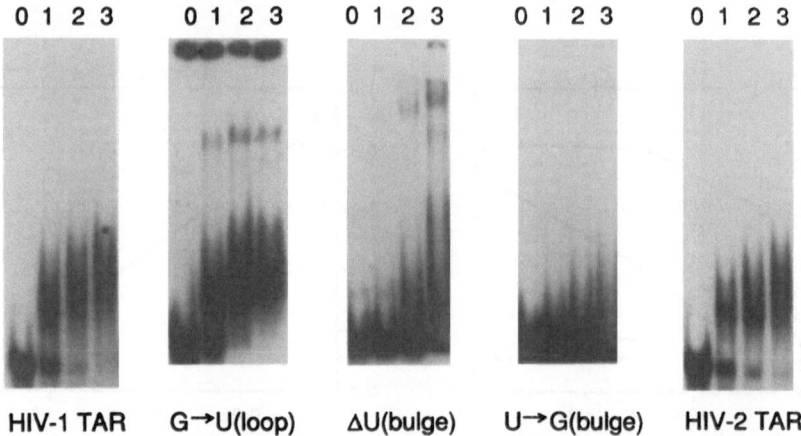

| 0 1 2 3 | 0 1 2 3 | 0 1 2 3 | 0 1 2 3 | 0 1 2 3 |

HIV-1 TAR   G→U(loop)   ΔU(bulge)   U→G(bulge)   HIV-2 TAR

Figure 5.   Gel retardation assays for TAR binding.
Autoradiographs show discrete complex formation between *tat* and ^{32}P-labelled RNA transcripts of HIV-1 TAR, HIV-2 TAR and the G→U (loop) mutation. No complexes were formed with the ΔU (bulge) or U→G (bulge), mutant sequences. Numbers above each gel lane refer to μl of HIV-1 *tat* protein at 200ng/μl added to each reaction mixture. Binding reaction mixtures (15μl) contained 1ng of uniformly labelled RNA probe, 0.5μg calf thymus DNA, 0.4μg yeast tRNA, 40units RNAsin (Promega), 50mM Tris-HCl pH7.9 and 20mM KCl and 0 to 600ng *tat* protein, prepared from *E. coli* β-galactosidase-*tat* fusion proteins, as described previously (Dingwall *et al.*, 1989,1990). After incubation at either 30°C or 4°C for 10-20 minutes, the reaction mixtures were applied to a 6% non-denaturing polyacrylamide gel (acrylamide : bis-acrylamide, 40:0.5) in 3.3mM sodium acetate, 6.7mM Tris (adjusted to pH7.9 with HCl). The gel buffer was recirculated during electrophoresis at 35mA for approximately 2 hours at room temperature. Gels were dried and exposed to X-ray film at -80°C using intensifying screens. After Dingwall *et al.*, 1990.

To define further the sequence requirements for the binding of *tat* to TAR, a number of mutations in the TAR loop and bulge regions were constructed and tested for binding activity in a gel-retardation assay (Figure 5). Substitution of the three U residues in the bulge (residues 23-25) with G residues, or deletion of the U residues (ΔU) abolished complex formation (Figure 5). However, substitution of the three G residues in the loop sequence CUGGGA (residues 30-35) with uridine had no discernible effect on the ability of *tat* to form a complex with TAR RNA (Figure 5).

HIV-1 *tat* is able to *trans*-activate LTRs carrying the first stem loop of HIV-2 TAR (Guyader *et al.*, 1987; Emerman *et al.*, 1987; Berkhout *et al.*, 1990a; Fenrick *et al.*, 1989). HIV-1 *tat* would therefore be expected to be able to bind to HIV-2 TAR RNA sequences. However, the two TAR sequences are not closely related, and only residues in the loop and the bulge regions are conserved between HIV-1 and HIV-2 TAR. In addition, the bulge sequence in HIV-2 TAR contains only two uridine residues. As shown in Figure 5 and Table I, binding of HIV-1 *tat* to the first stem-loop structure in HIV-2 TAR is observed in gel retardation and filter binding assays.

Although it is possible to calculate dissociation constants (K_d) for mutant TAR-RNAs from saturation binding experiments of the type shown in Figure 3, this method is unreliable when using RNA sequences with weak affinities for *tat* because it is not always possible to reach conditions of saturation. A more convenient and accurate measurement of the relative affinities of TAR and TAR mutants is $D_{1/2}$ (Table I and Figure 6), the concentration of competitor required to reduce binding by 50% (Lin & Riggs, 1972).

As a control experiment, unlabelled TAR RNA was competed against labelled TAR and a $D_{1/2}$ of 64nM was measured (Figure 6). Because this value is in close agreement with the calculated competitor RNA concentration at $D_{1/2}$ of 56nM (assuming K_d = 12nM), the experiment provides further evidence that the stoichiometry of the complex between *tat* and TAR is one to one.

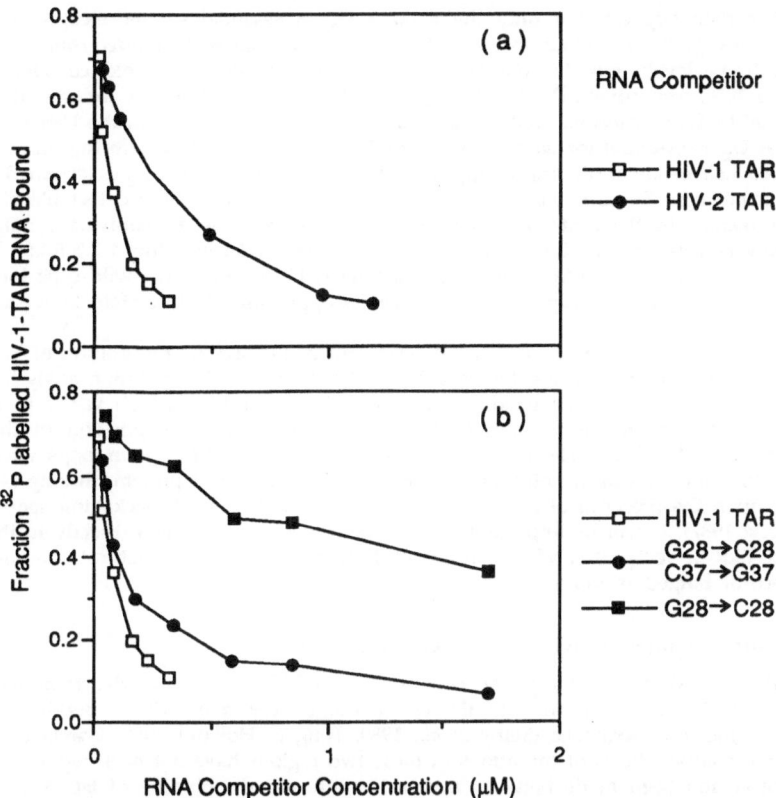

Figure 6. **Competition for *tat* binding between native HIV-1 TAR RNA and various mutant TAR sequences.**
The fraction of ^{32}P-labelled TAR RNA bound was plotted against the final concentration of competitor RNA. For competition experiments, binding reactions (500μl) were set up on ice and contained 80pM ^{32}P-labelled TAR RNA (17,600 dpm, 200μCi/nmol) and unlabelled TAR RNA to give a final concentration of 20nM, 0.5μg calf thymus DNA, 0.4μg yeast tRNA, 40 units RNasin (Promega), 50mM Tris-HCl pH 7.9 and 20mM KCl. After addition of 0 to 6μM competitor RNA, binding was initiated by addition of HIV-1 *tat* protein to a final concentration of 400ng/ml (40nM). The binding reaction mixtures were then filtered under reduced pressure through pre-washed nitrocellulose filters (Millipore 0.45μ pore size) and filters were dried and counted by liquid scintillation counting. The graphs do not include data obtained with competitor RNA at >2μM. After Dingwall *et al.*, 1990.

Figure 6 shows competition binding data for HIV-2 TAR and a number of loop and bulge mutations. The affinity of *tat* for HIV-2 TAR measured in direct binding assays, or in the competition binding assays, was approximately 2.5-fold lower than for HIV-1 TAR (Table I). Replacement of the three U residues in the bulge by G residues (U→G bulge) or deletion of the U-rich bulge (ΔU bulge) abolished *tat* binding to TAR in gel retardation assays (Figure 5) and produced, in both cases, an approximately seven-fold reduction in *tat* affinity (Table I). By contrast, replacement of the three G residues in the loop sequence by U residues (G→U loop) produced an RNA molecule which can bind *tat* with nearly the wild-type affinity and competes as effectively as the wild-type sequence for binding to ^{32}P-labelled HIV-1 TAR.

The data described here show that *tat* recognition of TAR requires the presence of a U-rich bulge in the context of an RNA double-helix. As long as Watson-Crick base pairing is maintained, the identity of most of the base pairs throughout the TAR stem appears not to be an important requirement for *trans*-activation (Figure 4 and Roy *et al.*, 1990a; Selby *et al.*, 1989; Feng & Holland, 1988) or for *tat* binding (Dingwall *et al.*, 1989, 1990). The bulge region need not contain only uridine residues in order to bind *tat* or to trans-activate the viral LTR. HIV-1 TAR sequences with either UCU or UUU bulge sequences found in infectious HIV-1 clones, can bind *tat* with equivalent affinities and both of these sequences are efficiently *trans*-activated.

In vivo data suggest that only the residue U_{23} is essential for *tat* recognition of TAR (Berkhout & Jeang, 1989). Mutation of U_{23} to A has been reported to reduce *trans*-activation to 18% of wild-type levels, but the mutations U_{24} to A and U_{25} to G only reduce *trans*-activation by 42% and 49%, respectively (Berkhout & Jeang, 1989). Our preliminary investigations of the binding of *tat* to TAR molecules containing a number of point mutations in the U-rich bulge also suggest that U_{23} is essential for *tat* recognition of TAR RNA. TAR RNA with the bulge sequence UCC showed half-maximal binding at approximately 20nM, close to the K_d determined for wild-type TAR sequences. By contrast, the bulge sequence CUC showed little detectable binding up to 20nM protein. In the first stem-loop of HIV-2 TAR there is an equivalent bulge which contains only two uridine residues, and the affinity for HIV-1 *tat* is reduced 2.5-fold. However, in the context of the HIV-1 TAR sequence, two bulged U residues gave wild-type binding. A TAR molecule with a single bulged U residue has an approximately three-fold lower affinity for *tat*.

Although U_{23} appears to be a major determinant in the specific recognition of TAR by *tat*, the conformation of TAR may also be critical for *tat* binding. Gel retardation analysis suggests that in an RNA duplex, bulged residues introduce a bend or kink which varies in curvature depending on the number and types of bases in the bulge and its position in the duplex (Bhattacharyya *et al.*, 1990). The presence of bulged residues in RNA stem-loops which act as protein binding sites has been noted before. For example, R17 coat protein binding is critically dependent upon the presence of a bulged A residue in the R17 RNA packaging sequence (Wu & Uhlenbeck, 1987). It will be important to determine whether *tat* binds directly to the bulged residues in TAR or recognizes other features of TAR created by the structural consequences of the presence of bulged residues.

Trans-activation requires a high affinity *tat*-binding site

Previous site-directed mutagenesis experiments on TAR have shown that *trans*-activation is indeed reduced by mutations in both the U-rich bulge (Roy *et al.*, 1990a; Berkhout & Jeang, 1989) and in the loop sequence (Selby *et al.*, 1989; Feng & Holland 1988; Garcia *et al.*, 1989). However, the relative effects of mutations in these two regions have not been reported, nor has a direct correlation been made between *trans*-activation and the binding of *tat* to TAR.

We have therefore tested a series of HIV LTR test plasmids containing the TAR mutations described above for *trans*-activation activity. Table I shows the results obtained when HeLa cells were co-transfected with a constant amount of each test plasmid and a variable amount of the *tat* expression plasmid pC63-4-1 (Dingwall *et al.*, 1990). In the presence of optimal levels of *tat* plasmids, viral LTRs carrying the HIV-1 TAR region showed increases over basal levels of CAT activity of between 40- and 60-fold. All plasmids with alterations in the bulge or loop regions were markedly inhibited. The quantitative effects of mutations in either the bulge region (ΔU, U→G) or the loop region (G→U) were similar, and resulted in 3- to 10-fold reductions in the maximal *tat*-stimulated CAT activity. The combined effects of mutations in both the bulge and loop are more severe, and there is no substantial increase in CAT activity over the basal levels even at high *tat* plasmid concentrations (Dingwall *et al.*, 1990).

The CAT assays show that there is a direct correlation between the effects of mutations in TAR RNA on *tat* binding and on *trans*-activation. Mutants which alter the U-rich bulge sequence, or which alter the structure of the U-rich bulge by disrupting base pairing in nearby residues of the TAR stem-loop structure, result in reduced *tat* binding and a corresponding loss of *trans*-activation efficiency. TAR RNAs derived from the first stem-loop of the HIV-2 LTR are partially *trans*-activated by HIV-1 *tat* (Guyader *et al.*, 1987; Emerman *et al.*, 1987; Berkhout *et al.*, 1990a; Fenrick *et al.*, 1989) and, as shown here, bind HIV-1 *tat* with reduced affinity.

What then is the role of the loop sequences in *trans*-activation? Sequences in the TAR RNA loop are not required for *tat* binding (Dingwall *et al.*, 1989, 1990; Roy *et al.*, 1990b), although these sequences are clearly required for maximal *trans*-activation (Feng & Holland, 1988; Berkhout & Jeang, 1989; Garcia *et al.*, 1989; Selby *et al.*, 1989; Dingwall *et al.*, 1990; Roy *et al.*, 1990a). One simple explanation for the behaviour of the loop mutations is that cellular RNA-binding proteins participate in *trans*-activation by binding to the loop sequences and a number of candidate proteins have been identified in HeLa cell extracts (Gaynor *et al.*, 1989; Gatignol *et al.*, 1989; Marciniak *et al.*, 1990). The genetic evidence does not indicate whether cellular proteins are used simply to help stabilize complexes between *tat* and TAR, or whether they play a more direct role in HIV transcription. However, because the *tat*-MS2 and *tat-rev* fusion proteins are functional in the absence of the loop sequences (Southgate *et al.*, 1990; Selby & Peterlin, 1990), it seems most likely that TAR-binding cellular proteins, if they exist, simply help to stabilize complexes betweeen *tat* and TAR. Both *rev* and MS2 show higher affinities for

Table I. Binding of HIV-1 *tat* to HIV-1 TAR and mutant TAR sequences and *tat*-stimulation of TAR mutants *in vivo*.

The dissociation constant for HIV-1 TAR was calculated from the Scatchard plot shown in Figure 2. Dissociation constants (K_d) for the TAR mutants were estimated by the half-maximal binding observed in filter binding assays and these are somewhat less reliable than for HIV-1 TAR since less data was used in the determination and saturation conditions were not always obtained. Replicate values in independent filter binding assays differ by less than 10%. A better measure of *tat* affinity for the mutant sequences is $D_{1/2}$, the concentration of competitor RNA that reduces the binding of HIV-1 *tat* to HIV-1 TAR RNA by 50%. The values for $D_{1/2}$ were obtained from the data shown in Figure 3. K_{Rel} is the ratio of the $D_{1/2}$ values for HIV-1 TAR and competitor RNA. Test plasmids were transfected into HeLa cells in the presence or absence of the *tat* expression plasmid pC63-4-1. CAT activity was determined and each assay used three different enzyme concentrations and was averaged and normalized. Values shown for the +*tat* determination show the CAT activity at the optimal C63-4-1 plasmid concentration. The relative CAT activity at optimal *tat* plasmid concentrations is expressed as a percentage of the value obtained for wild-type HIV-1 TAR (% wt). After Dingwall *et al,*. 1990.

TAR Sequence	K_d (nM)	$D_{1/2}$ (nM)	K_{Rel}	CAT activity -*tat*	CAT activity +*tat*	CAT activity (% wt)
HIV-1 (UUU)	12	64	1.00	2.2	89.0	100
HIV-1 (UCU)	12	—	—	—	—	—
HIV-2	30	160	0.40	—	—	—
ΔU (bulge)	>70	420	0.15	1.3	12.0	15
G→U (loop)	25	66	0.97	2.1	24.4	31
U→G (bulge)	>100	420	0.15	0.1	17.9	23
Anti-sense	>140	—	—	—	—	—

their target RNAs than *tat* does for TAR. It therefore seems likely that the fusion proteins would have similarly high affinities and this could have reduced the requirement for cellular factors artificially. Moreover, in the experiments of Southgate *et al.* (1990), unusually high levels of the *tat-rev* fusion protein may have been concentrated near the viral promoter because of the propensity of *rev* to aggregate (Heaphy *et al.*, 1990). Selby and Peterlin (1990) have also shown that mutations in the MS2 binding site that lead to decreased binding of the coat protein *in vitro* resulted in decreased *trans*-activation *in vivo*.

We have observed that TAR sequences with substitution of residues in either the bulge or loop domains are not completely inactive, suggesting that in the infected cell even a poor binding site can allow sufficient recruitment of tat to produce a measurable effect on transcription. Furthermore, double mutations in which both the U-rich bulge and loop are altered generally show lower levels of *trans*-activation than do the equivalent mutations in a single region (Dingwall *et al.*, 1990).

Trans-activation mechanism

The experiments described above strongly suggest that activation of HIV transcription by *tat* is a consequence of the direct binding of *tat* to nascent TAR RNA sequences. Peterlin and his colleagues (Kao *et al.*, 1987; Selby *et al.*, 1989; Selby & Peterlin, 1990) have proposed that *tat* acts as an anti-terminator and helps to overcome a block to elongation at or near the TAR site. Their proposal is based on observations that short, prematurely terminated RNA transcripts accumulate in the absence of *tat* (Kao *et al.*, 1987; Selby *et al.*, 1989; Toohey & Jones, 1989). Additional support for this hypothesis comes from nuclear run-on experiments which suggest that HIV-1 transcription initiation complexes are stablized in the presence of *tat* (Laspia *et al.*, 1989) and from evidence that TAR is functional on nascent chains (Berkhout *et al.*, 1989). However, a limitation of the Peterlin proposal is that TAR is unlikely to act directly as a site of anti-termination since mutations in TAR which abolish *trans*-activation, or deletions of TAR, do not result in constitutively high levels of LTR expression (Muesing *et al.*, 1987; Selby *et al.*, 1989; Cullen, 1986; Feng & Holland, 1988; Garcia *et al.*, 1989). Furthermore, it is difficult to "chase" the short RNA transcripts into full-length mRNAs. Equivalent levels of the short transcripts are seen in both the presence or absence of *tat*, whereas the levels of full-length transcripts are increased at least 10-fold in the presence of *tat* (Laspia *et al.*, 1989).

We favour a modified version of the Peterlin proposal, in which *tat* regulates elongation, but not initiation, by RNA polymerase II. We propose that *tat* and cellular co-factors assemble with the RNA polymerase II soon after transcription initiation in a reaction mediated by the protein binding sites present on the TAR-containing nascent transcripts. This modified transcription complex then stimulates viral mRNA production by overcoming additional blocks to elongation at a variety of distal sites. There is no need to postulate specific stop sites for RNA polymerase II downstream of TAR in the HIV genome since an unstable transcription complex in *tat*-minus cells could dissociate at any position that is thermodynamically most favourable (Platt, 1986; von Hippel *et al.*, 1984). For example, Laspia *et al.* (1989) have noted that transcription throughout the *E. coli* CAT gene is stablized by *tat*. It is not known whether there are specific sequences within the CAT gene which block elongation of polymerase in cells lacking *tat*, although it seems likely that pausing could occur at the many A:T-rich tracts present in the CAT sequence (Platt, 1986).

Other groups have proposed that *tat* interactions with TAR may indirectly influence the rate of transcription initiation (Laspia *et al.*, 1989; Sharp & Marciniak, 1989; Southgate *et al.*, 1990). However, there is only indirect evidence that *tat* plays a role in initiation, since the experimental measurements of *tat*-mediated transcription have relied exclusively on measurements of RNA transcript levels and these could be affected by changes in both the initiation and elongation rates of RNA polymerase. It should also be noted that there are no known examples of RNA binding proteins that affect transcription initiation, whereas there are many good examples in prokaryotic systems of RNA binding proteins, such as the bacteriophage λN protein (Barik *et al.*, 1987; Lazinski *et al.*, 1989), that control transcription elongation.

Finally, we note that because *tat* has a high affinity for a restricted binding site on TAR, compounds capable of interfering *in vitro* with this highly specific interaction may prove to be effective anti-HIV agents.

Acknowledgements

This work was supported by the MRC-AIDS Directed Programme. We are indebted to A. D. Lowe, S. M. Green and M. Singh for assistance with many of the experiments described here. J. K. is an Established Investigator of the American Heart Association.

References

Barik, S., Ghosh, B., Whalen, W., Lazinski, D., & Das, A. (1987) *Cell*, **50**, 885-899.
Berkhout, B. & Jeang, K.-T. (1989) *J. Virol.* **63**, 5501-5504.
Berkhout, B., Silverman, R. H. & Jeang, K.-T. (1989) *Cell*, **59**, 273-282.
Berkhout, B., Gatignol, A., Silver, J., & Jeang, K.-T. (1990a) *Nucl. Acids Res.* **18**, 1839-1846.
Berkhout, B., Gatignol, A., Rabson A. B. & Jeang, K.-T. (1990b) *Cell*, **62**, 757-767.
Bhattacharyya, A., Murchie, A. I. H. & Lilley, D. M. (1990) *Nature*, **343**, 484-487.
Cordingley, M. G., LaFemina, R. L., Callahan, P. L., Condra, J. H., Sardana, V. V., Graham, D. J., Nguyen, T. M., LeGrow, K., Gotlib, L., Schlabach, A. J. & Colonno. (1990) *Proc. Natl. Acad. Sci. U.S.A.* **87**, in press.
Cullen, B. R. (1986) *Cell*, **46**, 973-982.
Daly, T. J., Cook, K. S., Gary, G. S., Maione, T. E. & Rusche, J. R. (1989) *Nature*, **342**, 816-819.
Dayton, A. I., Sodroski, J. G., Rosen, C. A., Goh, W. C. & Haseltine, W. A. (1986) *Cell*, **44**, 941-947.
Dingwall, C. , Ernberg, I., Gait, M. J., Green, S. M., Heaphy, S., Karn, J., Lowe, A. D., Singh, M., Skinner, M. A., & Valerio, R. (1989) *Proc. Natl. Acad. Sci. U.S.A.* **86**, 6925-6929.
Dingwall, C. , Ernberg, I., Gait, M. J., Green, S. M., Heaphy, S., Karn, J., Lowe, A. D., Singh, M., & Skinner, M. A. (1990) *EMBO J.* **9**, in press
Emerman, M., Guyader, M., Montagnier, L., Baltimore, D. & Muesing, M. A. (1987) *EMBO J.* **6**, 3755-3760.
Fisher, A. G., Feinberg, M. B., Josephs, S. F., Harper, M. E., Marselle, L. M., Reyes, G., Gonda, M. A., Aldovini, A., Debouck, C., Gallo, R. C., & Wong-Staal, F. (1986) *Nature* , **320**, 367-371.
Fenrick, R., Malim, M. H., Hauber, J., Lee, S.-Y., Maizel, J. & Cullen, B. R. (1989) *J. Virol.* **63**, 5006-5012.
Feng, S. & Holland, E. C. (1988) *Nature*, **334**, 165-168.
Frankel, A. D., Bredt, D. S. & Pabo, C. O. (1988) *Science*, **240**, 70-73.

Garcia, J.A., Harrich, D., Soultanakis, E., Wu, F., Mitsuyasu, R. & Gaynor, R. B. (1989) *EMBO J.* **8**, 765-778.

Gatignol, A., Kumar, A., Rabson, A., & Jaeng K.-T. (1989) *Proc. Natl. Acad. Sci. U.S.A.* **86**, 7828-7832.

Gaynor, R., Soultanakis, E., Kuwabara, M., Garcia, J., & Sigman, D.S. (1989) *Proc. Natl. Acad. Sci. U.S.A.* **86**, 4858-4862.

Gorman, C. M., Moffat, L. F. & Howard, B. H. (1982) *Mol. Cell Biol.* **2**, 1044-1051.

Gorman, C. M., Padmanabliam, R. & Howard, B. H. (1983) *Science*, **222**, 551-553.

Guyader, M., Emerman, M., Sonigo, P., Clavel, F., Montagnier, L. & Alizon, M. (1987) *Nature*, **326**, 662-669.

Hauber, J., Perkins, A., Heimer, E. P., & Cullen, B. R. (1987) *Proc. Natl. Acad. Sci. USA*, **84**, 6364-6368.

Hauber, J. & Cullen, B. R. (1988) *J. Virol.* **62**, 673-679.

Heaphy, S., Dingwall, C., Ernberg, I., Gait, M. J., Green, S. M., Karn, J., Lowe, A. D., Singh, M. and Skinner, M. (1990) *Cell*, **60**, 685-693.

Ikehara, M., Ohtsuka, E., Tokunaga, T., Taniyama, Y., Iwai, S., Kitano, K., Miyamoto, T., Ohgi, T., Sakuragawa, Y., Fujiyama, K., Ikari, T., Kobayashi, M., Miyake, T., Shibahara, S., Ono, S., Ueda, T., Tanaka, T., Baba, H., Miki, T., Sakurai, A., Oishi, T., Chisaka, O. & Matsubara, K. (1984) *Proc. Natl. Acad. Sci. U.S.A.*, **81**, 5956-5960.

Jakobovits, A., Smith, D. H., Jakobovits, E. B. & Capon, D. J. (1988) *Mol. Cell Biol.* **8**, 2555-2561.

Jakobovits, A., Rosenthal, A. & Capon, D. J. (1990) *EMBO J.* **9**, 1165-1170.

Kao, S.-Y., Calman, A. F., Luciw, P. A. & Peterlin, B. M. (1987) *Nature*, **330**, 489-493.

Karn, J., Dingwall, C., Gait, M. J., Heaphy, S., & Skinner, M. A. in (F. Eckstein & D.M.J. Lilley, eds.) *Nucleic Acids & Molecular Biology*, V, in press.

Laspia, M. F., Rice, A. P. & Mathews, M. B. (1989) *Cell*, **59**, 283-292.

Lazinski, D., Grzadzielska, E. & Das, A. (1989) *Cell*, **59**, 207-218.

Leibold, E. A., Laundano, A., & Yu, Y. (1990) *Nucl. Acids Res.* **18**, 1819-1825.

Lin, S. -Y. & Riggs, A. D. (1972) *J. Mol. Biol.* **72**, 671-690.

Marciniak, R. A., Garcia-Glanco, M. A. & Sharp, P. A. (1990) *Proc. Natl. Acad. Sci. U.S.A.*, **87**, 3624-3628.

Muesing, M. A., Smith, D. H. & Capon, D. J. (1987) *Cell*, **48**, 691-701.

Peterlin B. M., Luciw, P. A., Barr, P. J., & Walker, M. D. (1986) *Proc. Natl. Acad. Sci. U.S.A.* **83**, 9734-9738.

Platt, T. (1986) *Ann. Rev. Biochem.* **55**, 339-372.

Rosen, C. A., Sodroski, J. G., Goh, W. C., Dayton, A. I., Loppke, J., & Haseltine, W. A. (1986) *Nature*, **319**, 555-559.

Roy, S., Parkin, N. T., Rosen, C. A., Itovitch, J. & Sonenberg, N. (1990a) *J. Virol.* **64**, 1402-1406.

Roy, S., Delling, U., Chen, C.-H., Rosen, C. A., & Sonenberg, N (1990b) *Genes & Dev.* **4**, 1365-1373

Selby, M. J., Bain, E. S., Luciw, P. & Peterlin, B. M. (1989) *Genes & Dev.* **3**, 547-55.

Selby, M. J., & Peterlin, B. M., (1990) *Cell*, **62**, 769-776.

Sharp P. A. & Marciniak, R. A.(1989) *Cell*, **59**, 229-230.

Smithies, O. (1965) *Science*, **150**, 1595-1598.

Southgate, C., Zapp, M. L., & Green, M. R. (1990) *Nature*, **345**, 640-642.

Toohey, M.G. & Jones, K.A. (1989) *Genes & Dev.* **3**, 265-283.

Weeks, K. M., Ampe, C., Schultz, S. C., Steitz, T. A. & Crothers, D. M. (1990) *Science*, **249**, 1281-1285.

Wu, H.-N. & Uhlenbeck, O. C. (1987) *Biochemistry*, **26**, 8221-8227.

Vartikar, J. V. & Draper E. D. (1989) *J. Mol. Biol.* **209**, 221-234.

von Hippel, P. H., Bear, D. G., Morgan, W. D., & McSwiggen, J. A. (1984) *Ann. Rev. Biochem.* **53**, 389-446.

THE HIV-1 TAT TRANSACTIVATOR CONTAINS AN ARGININE-GLYCINE-ASPARTYL (RGD)

CELL ADHESION SITE

David A. Brake and Christine Debouck

SmithKline Beecham Pharmaceuticals
Department of Molecular Genetics
King of Prussia, PA

Since the discovery of HIV-1 as the etiologic agent of acquired immunodeficiency syndrome (Barre-Sinoussi et al, 1983; Gallo et al, 1984), enormous progress has been made toward understanding this virus organization and its life cycle within target cells. HIV-1 has been shown to have the same 5' LTR-gag-pol-env-LTR 3' genomic structure as the classical avian and murine retroviruses (Figure 1; Ratner et al, 1985; Wain-Hobson et al,1985; Sanchez-Pescador et al, 1985; Muesing et al, 1985). It is therefore logical that early studies on HIV-1 focused on its gag, pol and env gene products for diagnostic, therapeutic and prophylactic purposes. Naturally, these initial studies greatly benefited from the knowledge accumulated over the years on the avian and murine systems (Weiss et al, 1982). In the therapeutic area, HIV-1 reverse transcriptase and later HIV-1 protease have been the object of intense investigations and inhibitor discovery efforts that have been reviewed recently (Goff, 1990; Debouck and Metcalf, 1990).

FUNCTION AND STRUCTURE OF THE HIV-1 TAT TRANSACTIVATOR

In addition to the gag, pol and env gene products, and in contrast with the avian and murine retroviruses, HIV-1 encodes a handful of smaller proteins that have been assigned regulatory or modulatory functions (see Cullen and Greene, 1990 for a review; Figure 1). This more complex genomic organization is also being uncovered in other lentiviruses such as visna (Davis et al, 1987) and equine infectious anemia virus (Dorn and Derse, 1988). The first of these auxiliary genes to be identified is tat which was discovered in 1985 (Arya et al, 1985; Sodroski et al, 1985), the same year that AZT was shown to efficiently and specifically inhibit HIV-1 reverse transcriptase (Mitsuya et al, 1985). The tat gene product is responsible for dramatically increasing the steady-state level of all HIV-1 messenger RNAs in the target cells (see Rosen and Pavlakis, 1990 for a review). Furthermore, tat is absolutely required for the replication of HIV-1 (Dayton et al, 1986; Fisher et al, 1986) and therefore constitutes an attractive therapeutic target for the treatment of HIV-1 infections.

Advances in Molecular Biology and Targeted Treatment for AIDS
Edited by A. Kumar, Plenum Press, New York, 1991

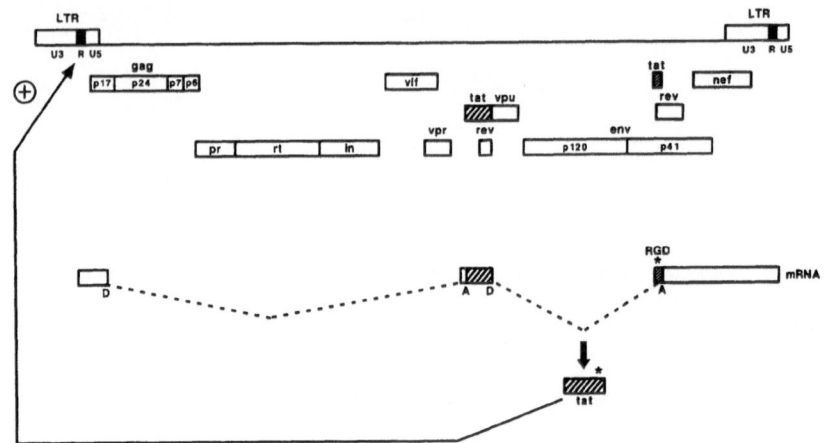

Fig.1. Genomic organization of HIV-1. The genes encoding HIV-1 structural proteins, enzymes and auxiliary factors are indicated on the map with their respective size, location and translation reading frame. The position and structure of the long terminal repeats (LTR) is also shown. The nomemclature for the proteins is as proposed by Leis et al, 1988. The structure of the tat mRNA is shown below the genetic map and indicates the position of the donor (D) and acceptor (A) splice sites (Arya et al, 1985; Sodroski et al, 1985). The tat coding region is marked by the hatched areas. The location of the RGD sequence (see text) is shown in the second tat exon. The arrow represents the LTR transactivation event mediated by tat.

Since its discovery five years ago, the tat gene has been dissected by site-specific and deletion mutagenesis and its function has been analyzed in a wide diversity of biochemical and biological assays (reviewed in Rosen and Pavlakis, 1990). Despite tremendous research efforts and the accumulation of novel and essential information on tat, its precise mechanism(s) of action remain unresolved. It is clear however that several cellular factors are taking part in this mechanism by interacting with tat itself (Nelbock et al, 1990) or with regulatory elements within the HIV-1 LTR (reviewed in Rosen and Pavlakis, 1990).

As shown in Figures 1 and 2, the tat gene is assembled through a double RNA splicing event and encodes an 86-amino acid-long protein. It was shown very early on that the first exon of tat encodes a 76-amino acid-long protein that is able to induce full transactivation of the HIV-1 LTR (Wright et al, 1986). Accordingly, most of the mutagenesis studies have focused on the three sequential domains, acidic, cysteine-rich and basic, that constitute the first exon of tat.

TAT CONTAINS AN RGD CELL ADHESION SEQUENCE

Interestingly, the second exon of tat (Figure 2) contains the tripeptide sequence Arginine-Glycine-Aspartyl (RGD) - highly conserved among HIV-1 isolates - which is a hallmark of a number of extracellular matrix proteins such as vitronectin, fibronectin, fibrinogen, and type I collagen, that interact with integrin cell adhesion receptors (Ruoslahti and Pierschbacher, 1987). We thus undertook to investigate if the RGD sequence in tat constitutes a cell attachment site.

To this end, purified recombinant _tat_ protein (Aldovini et al, 1986) was assayed for attachment to various cultured cell lines, including the human T lymphocytic cell lines HUT-78 and MOLT-4 and the human myelomonocytic cell line THP-1 which are replication-competent for HIV-1 (Levy et al, 1984; Kikukawa et al, 1986; Mikovits et al, 1990). An additional cell line, the L_8 rat skeletal muscle cell line, was also examined in this assay because of its known limited adhesion properties (Biesecker, 1990). Microtiter plate wells were coated with _tat_ protein overnight at 37°C, washed and then incubated with cells for one hour at 37°C. The amount of cells adhering to _tat_ was visualized and quantitated after fixing and staining the attached cells with 1% toluidine blue and then using a microplate reader. As shown in Figure 3, all four cell lines examined efficiently adhered to _tat_. Furthermore, this cell adherence was found to be dose-dependent with a half-maximal level of adhesion observed at _tat_ coating concentrations in the range of 0.1 to 0.5 µM (Brake et al, 1990a).

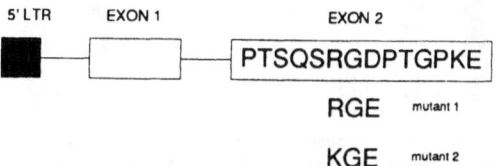

Fig.2.   Coding exons and partial amino acid sequence of HIV-1 _tat_. The amino acid sequence of the second exon (HTLV-IIIb isolate) is shown. The amino acid substitutions of two _tat_ mutants are indicated. (Reproduced from the Journal of Cell Biology, 111: 1275 (1990) by copyright permission of the Rockefeller University Press).

CELL ADHESION TO _TAT_ IS RGD-DEPENDENT

The cell adhesion to _tat_ was shown to require the RGD tripeptide sequence located in the second exon of _tat_ by using _tat_ proteins mutated in the RGD sequence. The single D80E and the double R78K,D80E substitutions were selected because they maintain the charge of _tat_ in this region and because an RGD to RGE substitution in fibronectin has been shown to prevent cell binding to this matrix protein (Obara et al, 1988). In all cases, the cell adherence to _tat_ RGE or KGE mutant proteins was dramatically reduced if not abrogated (Figure 3). These data suggest that _tat_ interacts directly with a cell surface receptor, presumably an integrin. This possibility was further substantiated in a series of experiments using reagents known to inhibit ligand-integrin interactions (Brake et al, 1990a). In addition, a peptide containing an RGD but not an RAD sequence as well as a monoclonal antibody directed toward the RGD epitope in _tat_ (Brake et al, 1990b) specifically inhibited the cell adherence to _tat_ (Brake et al, 1990a).

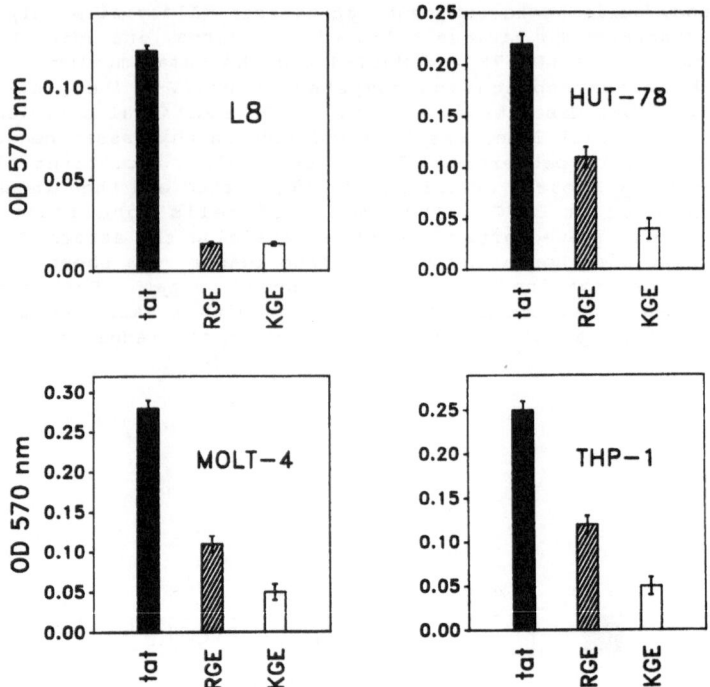

Fig.3.   Comparison of cell adhesion of cultured cell lines to wild-type,
          RGE and KGE mutant tat. The adherence of cells to protein-
          coated wells of a 96-well microtiter plate was measured by
          optical density reading (570 nm) after incubation and staining
          as described in the text. Shown is the adherence of rat
          skeletal L_8 myoblasts, human T lymphocytic HUT-78 cells, human T
          lymphocytic MOLT-4 cells, and human monocytic THP-1 cells to
          wells coated with purified recombinant wild-type, RGE, or KGE
          mutant tat protein (5 μg/ml sterile PBS). Note differences in
          the abscissa scales. (Reproduced from The Journal of Cell
          Biology, 111:1275 (1990) by copyright permission of the
          Rockefeller University Press).

CONCLUSIONS

    The identification of an RGD sequence in tat that efficiently and
specifically mediates cell adhesion was presented. It constitutes a property
not documented to date for a retrovirally encoded protein. It will be of
interest to examine if this property of HIV-1 tat mediates part or all of the
exogenous transactivation and cellular uptake that has been reported by other
laboratories (Frankel and Pabo, 1988; Frankel et al, 1989; Green and
Lowenstein, 1988). It is conceivable that this RGD-mediated cell binding to
tat could stimulate (with or without cellular uptake) HIV-1 gene expression in
latently infected cells. It could also stimulate the expression of certain
cellular genes leading to altered cell phenotypes such as the growth
stimulation of Kaposi's sarcoma-derived (Ensoli et al, 1990) or the induction
of skin lesions observed in tat-transgenic mice (Vogel et al, 1988). The
characterization of this cell adhesion site in tat and the elucidation of its
relevance to the HIV-1 infection process are currently in progress.

ACKNOWLEDGEMENT

This work was supported in part by NIH grant AI24845.

REFERENCES

Aldovini, A., Debouck, C., Feinberg, M.B., Rosenberg, M., Arya, S.K., and Wong-Staal, F., 1986, Synthesis of the complete trans-activation gene product of human T-lymphotropic virus type III in Escherichia coli, Proc. Natl. Acad. Sci. USA, 83:6672.

Arya, S.K., Guo, C., Josephs, S.F., and Wong-Staal, F., 1985, Transactivator gene of human T-lymphotropic virus type III (HTLV-III), Science, 229:69.

Barre-Sinoussi, F., Chermann, J.C., Rey, R., Nugeyre, M.T., Chamaret, S., Gruest, J., Dauguet, C., Axler-Blin, C., Brun-Vezinet, F., Rouzioux, C., Rozenbaum, W., and Montagnier, L., 1983, Isolation of a T-lymphotropic retrovirus from a patient at risk for acquired immunodeficiency syndrome (AIDS), Science, 220:868.

Biesecker, G., 1990, The complement SC5b-9 complex mediates cell adhesion through a vitronectin receptor, J. Immunol., 145:209.

Brake, D.A., Debouck, C., and Biesecker, G., 1990, Identification of an Arg-Gly-Asp (RGD) cell adhesion site in human immunodeficiency virus type 1 transactivation protein, tat, J. Cell Biol., 111:1275.

Brake, D.A., Goudsmit, J., Krone, W.J.A., Schammel, P., Appleby, A., Meloen, R.H., and Debouck, C., 1990, Characterization of murine monoclonal antibodies to the tat protein from human immunodeficiency virus type 1, J. Virol, 64:962.

Cullen, B.R., and Greene, W.C., 1990, Functions of the auxiliary gene products of the human immunodeficiency virus type 1, Virology, 178:1.

Davis, J.L., Molineaux, S., and Clements, J.E., 1987, Visna virus exhibits a complex transcriptional pattern: one aspect of gene expression shared with the acquired immunodeficiency syndrome retrovirus, J. Virol., 61:1325.

Dayton, A.I., Sodroski, J.G., Rosen, C.A., Goh, W.C., and Haseltine, W.A., 1986, The transactivator gene of the human T cell lymphotropic virus type III is required for replication, Cell, 44:941.

Debouck, C., and Metcalf, B.W., 1990, Human immunodeficiency virus protease: a target for AIDS therapy, Drug Dev. Res., 20: in press.

Dorn, P.L., and Derse, D., 1988, Cis- and trans-acting regulation of gene expression of equine infectious anemia virus, J. Virol., 62, 3522.

Ensoli, B., Barillari, G., Salahuddin, S.Z., Gallo, R.C., and Wong-Staal, F., 1990, Tat protein of HIV-1 stimulates growth of cells derived from Kaposi's sarcoma lesions of AIDS patients, Nature, 345:84.

Fisher, A.G., Feinberg, M.B., Josephs, S.F., Harper, M.E., Marselle, L.M., Reyes, G., Gonda, M.A., Aldovini, A., Debouck, C., Gallo, R.C., and Wong-Staal, F., 1986, The trans-activator gene of HTLV-III is essential for virus replication, Nature, 320:367.

Frankel, A.D. and Pabo, C.O., 1988, Cellular uptake of the tat protein from human immunodeficiency virus, Cell, 55:1189.

Frankel, A.D., Biancalana, S. and Hudson, D., 1989, Activity of synthetic peptides from the tat protein of human immunodeficiency virus type 1, Proc. Natl. Acad. Sci. USA, 86:7397.

Gallo, R.C., Salahuddin, S.Z., Popovic, M., Shearer, G.M., Kaplan, M., Haynes, B.F., Palker, T.J., Redfield, R., Oleske, J., Safai, B., White, G., Foster, P. amnd Markham, P., 1984, Frequent detection and isolation of cytopathic retroviruses (HTLV-III) from patients with AIDS and at risk for AIDS, Science, 224:500.

Green, M. and Lowenstein, P.M., 1988, Autonomous functional domains of chemically synthesized human immunodeficiency tat trans-activator protein, Cell, 55:1179.

Goff, S.P., 1990, Retroviral reverse transcriptase: synthesis, structure, and function, J. Acq. Imm. Def. Syn., 3:817.

Kikukawa, R., Koyanagi, Y., Harada, S., Kobayashi, N., Hatanaka, M., and Yamamoto, N., 1986, Differential susceptibility to the acquired immunodeficiency syndrome retrovirus in cloned cells of human leukemic T-cell line Molt-4, J.Virol., 57:1159.

Leis, J., Baltimore, D., Bishop, J.M., Coffin, J., Fleissner, E., Goff, S.P., Oroszlan, S., Robinson, H., Skalka, A.M., Temin, H.M., and Vogt, V., 1988, Standardized and simplified nomenclature for proteins common to all retroviruses, J. Virol., 62:1808.

Levy, J.A., Hoffman, A.D., Kramer, S.M., Landis, J.A., Shimabukuru, J.M., and Oshiro, L.S., 1984, Isolation of lymphocytopathic retroviruses from SanFrancisco patients with AIDS, Science, 235:840.

Mikovits, J.A., Raziuddin, Gonda, M., Ruta, M., Lohrey, N.C., Kung, H.-F., and Ruscetti, F.W., 1990, Negative regulation of human immune deficiency virus replication in monocytes. Distinction between restricted and latent expression in THP-1, J. Exp. Med., 171:1705.

Mitsuya, H., Weinhold, K.J., Furman, P.A., St. Clair, M.H., Lehrman, S.N., Gallo, R.C., Bolognesi, D., Barry, D.W., and Broder, S., 3'-Azido-3'-deoxythymidine (BW A509U): an antiviral agent that inhibits the infectivity and cytopathic effect of human T-lymphotropic virus type III/lymphadenopathy-associated virus in vitro, Proc. Natl. Acad. Sci. USA, 82:7096.

Muesing, M.A,., Smith, D.H., Cabradilla, C.D., Benton, C.V., Lasky, L.A., and Capon, D.J., 1985, Nucleic acid structure and expression of the human AIDS/lymphadenopathy retrovirus, Nature, 313:450.

Nelbock, P., Dillon, P.J., Perkins, A., and Rosen, C.A., 1990, A cDNA for a protein that interacts with the human immunodeficiency virus tat transactivator, Science, 248:1650.

Obara, M., Kang, M.S., Yamada, K.M., 1988, Site-directed mutagenesis of the cell-binding domain of human fibronectin: separable, synergistic sites mediate adhesive function, Cell, 53:649.

Ratner, L., Haseltine, W., Patarca, R., Livak, K.J., Starcich, B., Josephs, S.F., Doran, E.R., Rafalski, J.A., Whitehorn, E.A., Baumeister, K., Ivanoff, L., Petteway, S.R., Pearson, M.L., Lautenberger, J.A., Papas, T.S., Ghrayeb, J., Chang, N.T., Gallo, R.C., and Wong-Staal, F., 1985, Complete nucleotide sequence of the AIDS virus, HTLV-III, Nature, 313:277.

Rosen, C.A., and Pavlakis, G.N., 1990, Tat and Rev: positive regulators of HIV gene expression, AIDS, 4:499.

Ruoslahti, E., and Pierschbacher, M.D., 1987, New perspectives in cell adhesion: RGD and integrins, Science, 238:491.

Sanchez-Pescador, R., Power, M.D., Barr, P.J., Steimer, K.S., Stempien, M.M., Brown-Shimer, S.L., Gee, W.W., Renard, A., Randolph, A., Levy, J.A., Dina, D., and Luciw, P.A., 1985, Nucleotide sequence and expression of an AIDS-associated retrovirus (ARV-2), Science, 227:484.

Sodroski, J., Patarca, R., Rosen, C., Wong-Staal, F., and Haseltine, W., 1985, Location of the trans-activation region on the genome of human T-cell lymphotropic virus type III, Science, 229:74.

Vogel, J., Hinrich, S.H., Reynolds, R.K., Luciw, P.A., and Jay, G., 1988, The HIV tat gene induces dermal lesions resembling Kaposi's sarcoma in transgenic mice, Nature (Lond.), 335:606.

Wain-Hobson, S., Sonigo, P., Danos, O., Cole, S., and Alizon, M., 1985, Nucleotide sequence of the AIDS virus, LAV, Cell, 40:9.

Weiss, R., Teich, N., Varmus, H., and Coffin, J. (eds), 1982, "Molecular biology of tumor viruses, Cold Spring Harbor Laboratory", Cold Spring Harbor, NY.

INHIBITION OF HIV-1 REPLICATION BY MUTANT TAT PEPTIDES

Paul M. Loewenstein, Masaho Ishino, Na'eem Abdullah,
and Maurice Green

Institute for Molecular Virology
St. Louis University School of Medicine
St. Louis, Missouri 63110

SUMMARY

The HIV-1 Tat protein is a potent transactivator that is
essential for virus replication. We reported previously that
chemically synthesized Tat minimal domain peptides and full
length Tat (Tat86), in which alanine is substituted for Lys-
41, are taken up by cells and antagonize the expression of
microinjected HIV-LTR driven CAT gene. We find that several
Tat mutant peptides inhibit HIV-1 (HTLV-IIIB) replication.
Tat86,41Ala, the most active peptide, inhibits virus
replication 95-99% in acutely infected cells when added to the
culture medium at concentrations of 10^{-6} M to 10^{-7} M;
chronically infected cells are also inhibited but to a lesser
extent. Tat86,41Ala protects HIV-1 infected cells from virus-
induced cytopathogenicity and inhibits syncytia formation.
The mechanism of inhibition by Tat mutant peptides is unknown.
Unexpectedly, co-transfection of a Tat86 expressing plasmid
with plasmids expressing Tat86,41Ala does not inhibit
transactivation of endogenous HIV-LTRCAT in HeLa cells. Thus
the Tat86,41Ala peptide, when added to cells together with
Tat86 can block HIV-LTRCAT transactivation, but not when
expressed endogenously from a plasmid. These results imply a
novel mechanism of Tat antagonism.

INTRODUCTION

HIV-1 Tat is essential for viral gene expression and
virus replication (for reviews see Cullen and Green, 1989 and
Sharp and Marciniak, 1989). Tat is therefore an attractive
target for the development of antiviral therapies of AIDS. We
have shown that chemically synthesized Tat peptides, ranging
in size from 26 to 86 amino acids, are rapidly taken up by
cells and specifically transactivate the HIV-1 LTR promoter
(Green and Loewenstein, 1988). This unique ability of HIV-1
Tat peptides contrasts with the general inability of peptides

Advances in Molecular Biology and Targeted Treatment for AIDS
Edited by A. Kumar, Plenum Press, New York, 1991

and proteins to enter cells (Sternson, 1987). The ability of HIV-1 Tat peptides to "self-deliver" to cells provides a potential means to target antagonists to the site of Tat function. Our approach is to develop peptide inhibitors that encode both (i) a Tat "delivery" sequence as well as (ii) a Tat amino acid sequence that can antagonize Tat function. Since such peptides would be rapidly taken up by cells, the problem of drug delivery may be minimized, and Tat peptide antagonists might have direct use for AIDS therapy.

Here we describe a strategy for the development of Tat peptide antagonists and then briefly summarize studies on the identification of minimal domain Tat peptides and peptide antagonists of Tat function. Finally, we describe some current studies on the inhibition of HIV-1 replication by Tat peptides and preliminary studies on mechanism.

STRATEGY FOR THE DEVELOPMENT OF TAT PEPTIDE ANTAGONISTS OF HIV-1 TRANSACTIVATION

The molecular mechanism of Tat action is poorly understood. The functional structure of Tat is likely to be different from that of typical eukaryotic transactivators which contain two functional regions - a promoter DNA binding domain and an activation domain (for reviews see Ptashne, 1988 and Mitchell and Tjian, 1989). The biochemical target of Tat appears not to be DNA, but instead an RNA sequence termed TAR (transactivation response) element, located immediately downstream of the transcription start-site (for a review see Sharp and Marciniak, 1989) (Berkhout et al., 1989). Although Tat appears unrelated to typical transactivators, it is still reasonable to expect that Tat contains separate functional domains that (i) bind specifically within an intranuclear complex, presumably to TAR RNA, and (ii) that perform additional functions in transactivation. In support of this hypothesis, recent studies indicate that recombinant Tat binds specifically to TAR RNA (Dingwall et al., 1989). Further, the binding site is specified by the 6 amino acid portion of the basic region containing 5 arginines and a minimum of 8 adjacent non-sequence specific residues (Kamine et al., submitted for publication).

The strategy to identify Tat peptide antagonists involves first the biochemical mapping of Tat minimal functional domains that encode both the "delivery function" (cellular uptake and intranuclear targetting) and the "transactivation function". This is followed by synthesis of mutant peptides that retain the delivery function but are defective in the transactivation function. Mutant Tat peptides that are taken up by cells and bind to the Tat-specific target, but are defective in subsequent function(s), could interfere with wild type HIV-1 Tat function, and thus are potential antagonists of Tat. The overall strategy, shown in Figure 1, involves the following steps:(1) development of a rapid and sensitive assay for transactivation suitable for small Tat peptides that possess a short half life and lower activity; (2) identification of minimal domain deletion peptides that possess measurable transactivation activity; (3) identification of Tat deletion and amino acid substitution

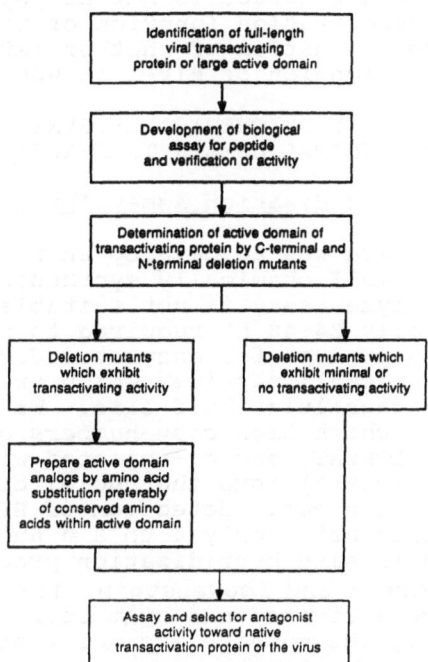

Fig. 1. Flowsheet of the Steps Employed in the Development of Peptide Antagonists of a Viral Transactivating Protein.

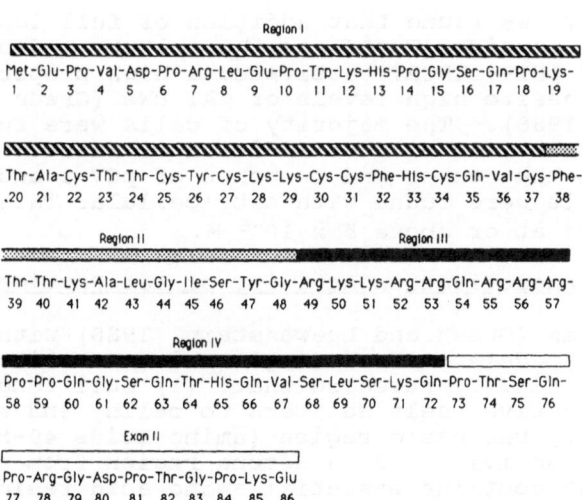

Region I

Met-Glu-Pro-Val-Asp-Pro-Arg-Leu-Glu-Pro-Trp-Lys-His-Pro-Gly-Ser-Gln-Pro-Lys-
1   2   3   4   5   6   7   8   9   10 11 12 13 14 15 16 17 18 19

Thr-Ala-Cys-Thr-Thr-Cys-Tyr-Cys-Lys-Lys-Cys-Cys-Phe-His-Cys-Gln-Val-Cys-Phe-
.20 21 22 23 24 25 26 27 28 29 30 31 32 33 34 35 36 37 38

Region II   Region III

Thr-Thr-Lys-Ala-Leu-Gly-Ile-Ser-Tyr-Gly-Arg-Lys-Lys-Arg-Arg-Gln-Arg-Arg-Arg-
39 40 41 42 43 44 45 46 47 48 49 50 51 52 53 54 55 56 57

Region IV

Pro-Pro-Gln-Gly-Ser-Gln-Thr-His-Gln-Val-Ser-Leu-Ser-Lys-Gln-Pro-Thr-Ser-Gln-
58 59 60 61 62 63 64 65 66 67 68 69 70 71 72 73 74 75 76

Exon II

Pro-Arg-Gly-Asp-Pro-Thr-Gly-Pro-Lys-Glu
77 78 79 80 81 82 83 84 85 86

Fig. 2. Amino Acid Sequence of HIV-1 Tat Protein. Superimposed on the sequence are tentative functional regions I to IV (from Green and Loewenstein, 1988). The absolute boundaries of these regions are not known.

peptides that are defective in transactivation; (4) determination of whether defective mutant peptides can antagonize the transactivation function of wild type Tat peptide; and (5) determination of whether Tat antagonist peptides inhibit replication of HIV-1 in human T lymphocytes.

TRANSACTIVATION ACTIVITY OF CHEMICALLY SYNTHESIZED TAT PEPTIDES - DEFINING MINIMAL DOMAIN TRANSACTIVATING PEPTIDES

Short-Term in situ Hybridization Assay for HIV-LTR Driven RNA

Our first objective was to develop an assay to measure transactivation by small chemically synthesized Tat peptides. The standard CAT enzyme assay is not suitable because of the long period (typically 24-48 h) required to accumulate sufficient CAT gene product for enzymatic detection; this would select against small peptides which are expected to possess a short intracellular half life. We developed a short-term assay in which *high copy numbers* of the chimeric reporter gene, HIV-LTRCAT, are co-injected with a Tat expression plasmid (pCV-1) into the nuclei of HeLa cells in order to facilitate the early detection of HIV-LTR driven CAT RNA. Cells are fixed after only 4 to 8 h and analyzed for CAT RNA by a sensitive in situ hybridization procedure using CAT ^{35}S-DNA as probe (Green and Loewenstein, 1988). We found that 60-90% of cells are activated and synthesize high levels of CAT RNA, as shown by the massive accumulation of dark photographic grains (^{35}S-DNA-RNA hybrids) above the nucleus; cells do not produce appreciable CAT RNA in the absence of co-injected Tat expression vector.

Rapid Uptake and Transactivation of HIV-LTRCAT by Added Tat86

Remarkably, we found that addition of full length synthetic Tat peptide (Tat86) to the culture medium of cells injected with pHIV-LTRCAT, activated a high proportion of cells to synthesize high levels of CAT RNA (Green and Loewenstein, 1988). The majority of cells were fully activated within 30 min after addition of Tat86. Little difference in the number and response of pCV-1 and Tat86 activated cells were found with extracellular Tat86 concentrations at or above 5 x 10^{-6} M.

Summary of Transactivation Activity of Tat Peptides

Our studies (Green and Loewenstein, 1988) with chemically synthesized Tat deletion and single amino substitution mutant peptides (see Fig. 2) revealed that: (1) full length Tat86 is biologically active, self delivers to cells, and activates HIV-LTRCAT; (2) the basic region (amino acids 49-57) is essential for activity; (3) a second region from about amino acids 37 to 48 contains essential amino acid residues; (4) sequences from about amino acids 58 to 86 are not essential for activity but a glutamine-rich region from about residues 58 to 72 enhances activity; and (5) deletion peptides lacking the cysteine-rich region possess minimal but clear activity when assayed by the sensitive, rapid, short-term RNA assay, but not readily by the long-term enzymatic assay. The report that a 26 amino acid peptide (Tat37-62) did not exhibit transactivation activity Frankel et al., 1989) does not disagree with our findings: we reported that this peptide was active by the RNA assay and not by the CAT enzyme assay. Frankel et al. (1989) did not attempt to measure activity by the short-term CAT RNA assay. As described above, the lack of activity in the long-term assay is probably due to lower inherent activity and short half life.

SEARCH FOR DEFECTIVE TAT MUTANT PEPTIDES - POTENTIAL ANTAGONISTS OF HIV LTR TRANSACTIVATION

Tat mutant peptides that can be taken up by cells and compete with wild type Tat could function as antagonists of HIV-LTR transactivation and of virus replication. We focused on Tat deletion peptides with significant transactivation activity, because they would be expected to retain putative "delivery" and "activation" functions. Tat deletion peptides with substitutions in critical conserved amino acids that inactivate the activation function but not the delivery function are potential antagonists. In addition, small deletion peptides are more readily synthesized and are more suited for use in AIDS therapy.

Transactivation by Minimal Domain Tat Peptides and Inactivation by Amino Acid Substitution

The purpose of our studies was not to map domains that are important for maximal transactivation activity but to identify minimally active backbones in which substitution of critical amino acids could lead to the generation of Tat antagonists. To identify amino acid residues important for transactivation, we utilized two minimally active Tat deletion peptides, Tat37-72 and Tat37-62, as backbones for amino acid substitution. Transactivation by unsubstituted Tat37-72 and Tat37-62 required much higher levels than Tat86 for demonstrable activity (Fig. 3A and 3D). These Tat deletion peptides exhibited the same target specificity as Tat86, as shown by their inability to transactivate a Tar negative mutant [pCD23 (117/+42), Fig. 3B and 3E]. Single amino acid substitutions in minimal domain peptides of alanine for Tyr-47 (Fig. 3C and 3F), Ser-46, or Lys-41 (not shown) (Green et al., 1989) inactivated transactivation. Finally, no activity was observed when Tat37-72 or Tat37-62 was added to cells microinjected with CAT driven by other promoters (data not shown). These experiments support the conclusion that the minimal domain peptides Tat37-62 and Tat37-72 transactivate HIV-LTRCAT by the same mechanism as does full length Tat86 and that transactivation is specific for the Tat amino acid sequence.

Identification of Tat Peptide Antagonists by Competition with Tat86 Added to pHIV-LTRCAT Microinjected Cells

To test whether defective Tat peptides are antagonists of HIV-LTR transactivation, they were added together with Tat86 to the media of cells injected with pHIV-LTRCAT; transactivation was measured 6 h later by in situ hybridization (see scheme in Fig. 4). Tat37-72,46Ala (Fig. 5, panel B) and Tat37-72,47Ala (Fig. 5, panel C) greatly reduced

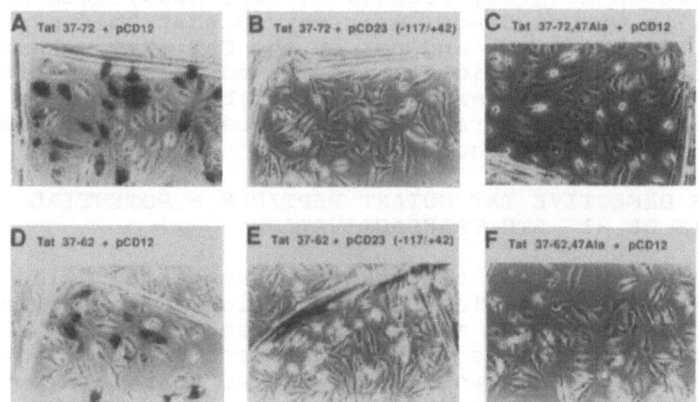

Fig. 3. Trans-Activation of HIV-LTRCAT Microinjected into HeLa
Cell Nuclei by Exogenous Treatment with Tat37-62 or Tat37-
72.   Cells on cover slips were microinjected with pCD12 (A,
C, D, F) which contains a functional HIV- LTR or with pCD23
(-117/+42) (B, E), a defective TAR deletion mutant and
treated for 6 h with 10^{-5} M Tat37-72 (A, B), Tat37-62 (D,
E), Tat37-72,47Ala (C) or Tat37-62,47Ala (F).   CAT RNA was
analyzed by in situ hybridization(from Green et al., 1989).

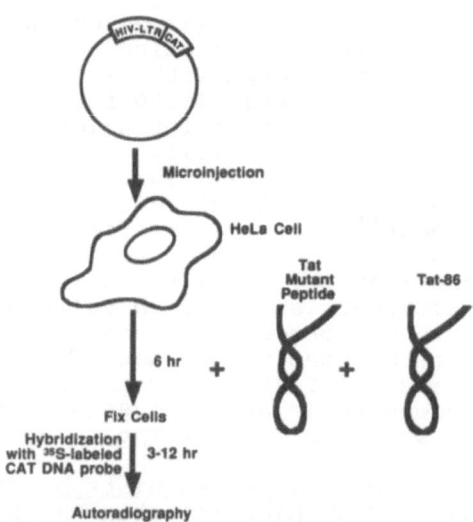

Fig. 4.   Assay for Measuring Inhibition of Tat86 Activity by
Tat Mutant Peptides Added to cells.   HeLa cells (about 100
per experiment) were microinjected with 100 ug/ml of pCD12
and treated with Tat86 and Tat mutant peptide added to the
culture medium.   Cells were fixed 6 h later and analyzed
for CAT RNA by in situ hybridization.

the synthesis of CAT RNA as compared to control cells treated with Tat86 alone (Fig. 5, panel A). Similar inhibition was obtained with Tat37-72,41Ala; and with the Tat37-62 backbone containing alanine substituted for Ser-46, Tyr-47, or Lys-41 (Green et al., 1989).

Activity of Tat Peptide Antagonists Added to Cells Containing Integrated HIV-LTRCAT

To determine whether Tat peptides would inhibit Tat induced transactivation of endogeneous HIV-LTRCAT (a model analogous to infection of cells infected by HIV-1), we utilized the HeLa HL3T1 cell line which contains integrated HIV-LTRCAT (Felber and Pavlakis, 1988). Frankel and Pabo (1988) reported that treatment of HL3T1 cells for 60 h with recombinant Tat in the presence of chloroquine and trypsin inhibitor, activates expression of endogenous HIV-LTRCAT as measured by the CAT enzyme assay. Under these conditions, sufficient CAT gene product must accumulate for detection by the CAT enzyme assay (Gormon et al., 1982). When biologically active synthetic Tat86 was added to HL3T1 cells together with Tat37-62 or Tat37-72 containing Ala substituted for Ser-46, Tyr-47, or Lys-41, reproducible antagonist activity was not obtained by single treatments with Tat peptide antagonist (see Fig. 6 for assay scheme); low inhibitory activity could sometimes be demonstrated by repeated 6 h peptide treatment. Thus the short Tat peptide antagonists identified in the short-term RNA assay exhibit less inibitory activity in the long-term CAT assay. We believe that antagonists built into these shorter peptide backbones are both inherently less active (Green et al., 1989) and are more rapidly degraded by serum and intracellular proteases. In support of this notion, Tat86 is stable in medium containing serum for up to 40 h whereas Tat37-72 is degraded (J. Symington, unpublished data).

Chemical Synthesis and Antagonist Activity of Tat86,41Ala

To test whether full length and presumably more stable Tat mutant peptides would antagonize Tat transactivation of endogenous HIV-LTRCAT, we synthesized Tat86,41Ala, Tat86,46Ala, and Tat86,47Ala by manual t-BOC solid phase chemistry (Green et al., 1989). We found that that Tat86,46Ala and Tat86,47Ala possessed substantial transactivation activity in the full length molecule whereas Tat86,41Ala did not (Green et al, 1989). We speculate that the presence of Tat residues 1 to 36 may increase the effective concentration or overall efficacy of the Tat peptide within the nucleus, thus overriding effects of point mutatins in residues 46 and 47. Alternativelty, the N-terminal region could encode an indendendent transactivation function although no evidence has been found for this.

Tat86,41Ala preparations were relatively homogeneous by several criteria. SDS-polyacrylamide gel electrophoresis revealed a major band at about 14 kDa by Coomassie blue staining (Fig. 7A, lane 2). Analysis of a second preparation by silver staining also showed predominantly one band (Fig.7B, lane 2) that co-migrated with recombinant Tat86 (Fig. 7B, lane 1). HPLC reverse phase chromatography revealed a major peak

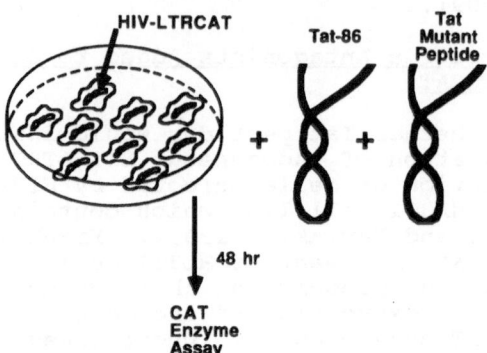

Tat Antagonist Assay with
Endogenous HIV-LTRCAT

HIV-LTRCAT   Tat-86   Tat Mutant Peptide

48 hr

CAT Enzyme Assay

Fig. 5. Microphotographs Showing Inhibition of Tat86
Induced Transactivation by Co-Treatment with Tat37-72,46Ala
or Tat37-72,47Ala. HeLa cells were injected with pCD12 and
treated with 1 x 10^{-6} M Tat86 alone (A) or together with 10^{-5}
M Tat37-72,46Ala (B) or 10^{-5} M Tat37-72,47Ala (C). Cells
were fixed after 6 h and CAT RNA analyzed by in situ
hybridization (from Green et al., 1989).

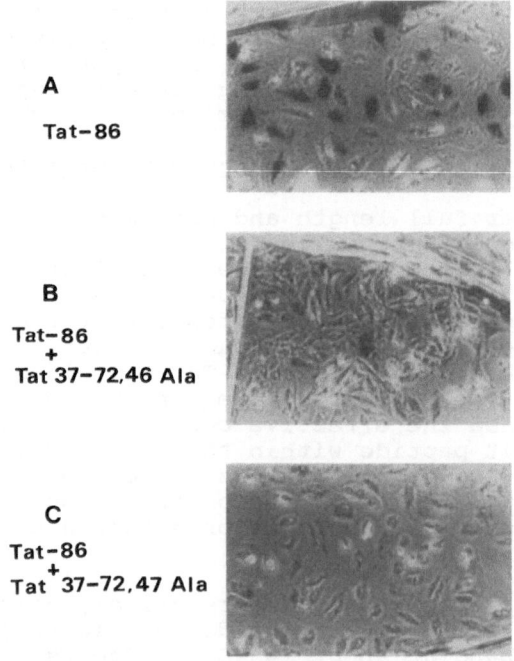

A

Tat-86

B

Tat-86
+
Tat 37-72,46 Ala

C

Tat-86
+
Tat 37-72,47 Ala

Fig. 6. Assay for Inhibition by Tat Mutant Peptides of Tat86
Transactivity of Endogenous HIV-LTRCAT. HL3T1 cells
containing integrated HIV-LTRCAT are treated with Tat86 and
Tat mutant peptide and analyzed for CAT enzyme activity 48 h
later (from Green et al., 1989).

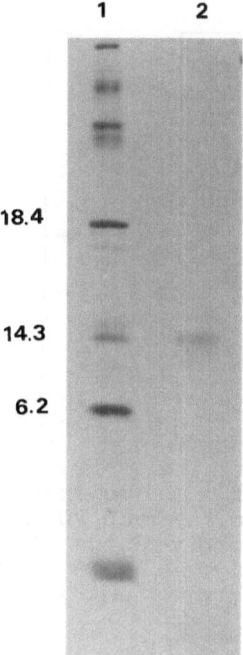

Fig. 7A. Polyacrylamide Gel Electrophoresis of Synthetic
Tat86,41Ala Peptide.   Lane 2.   Tat86,41Ala peptide, purified
by exclusion chromatography, was analyzed by electrophoresis
on a polyacrylamide-sodium dodecyl sulfate gel and stained
with Coomassie blue.   Lane 1, molecular size markers.

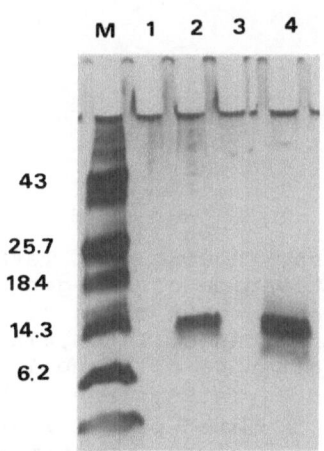

Fig. 7B.   Polyacrylamide Gel Electrophoresis of Chemically
Synthesized Tat86,41Ala and Recombinant Tat86 peptides.
Peptide was analyzed by electrophoresis on a sodium dodecyl
sulfate-15% polyacrylamide gel and treated with silver
stain.   Lane M, molecular size markers; lane 2, recombinant
Tat86; lane 4, Tat86,41Ala.

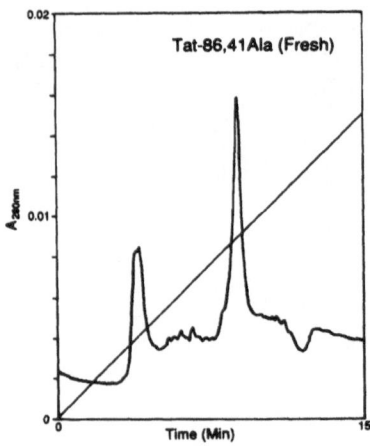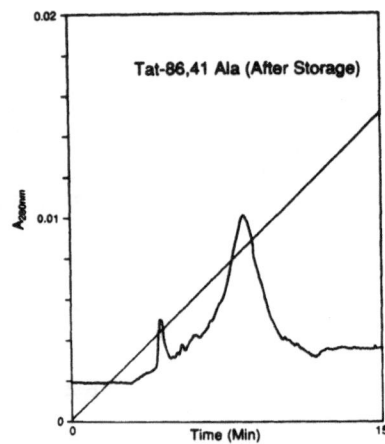

Fig. 7C. HPLC Profile of a Crude Preparation of Chemically
Synthesized Tat86,41Ala. About 20 ug of freshly dissolved
peptide (upper panel) and peptide after 24 h at -20 deg
(lower panel) were analyzed by C8 reverse phase HPLC using a
0 to 100% acrylonitrile gradient in 0.1% TFA.

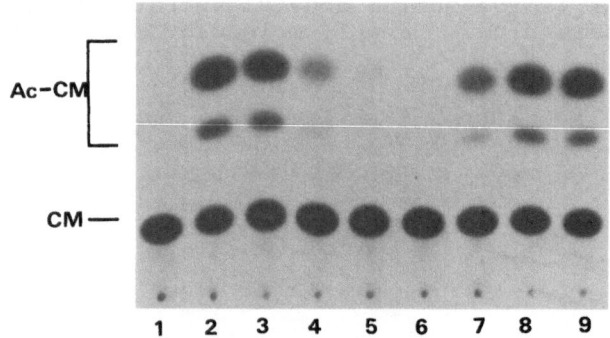

Fig. 8. CAT Enzyme Assay Showing That Tat86,41Ala Strongly
Blocks Tat86 Transactivation of Endogenous HIV-LTRCAT.
HL3T3 cells were treated with Tat86 (1×10^{-7} M) or Tat86
plus various concentrations of Tat86,41Ala and analyzed for
CAT enzyme activity after 48 h. Lane 1, no peptide
treatment; lane 2 and 3, Tat86; lane 4, Tat86 plus 5×10^{-6}
M Tat86, 41Ala; lane 5, Tat86 plus 1×10^{-6} M Tat86,41Ala;
lane 6, Tat86 plus 5×10^{-7} M Tat86,41Ala; lane 7, Tat86
plus 2×10^{-7} M Tat86,41Ala; lane 8, Tat86 plus 1×10^{-7} M
Tat86,41Ala; lane 9, Tat86 plus 5×10^{-8} M Tat86,41Ala (from
Green et al., 1989).

160

in freshly dissolved preparations (Fig. 7 C, upper panel) that broadened upon storage in solution (Fig. 7C, lane 2). This behavior is typical of chemically synthesized peptides of greater than 50 residues and is thought to reflect the formation of relatively stable, slowly exchanging folded structures of the peptide chain (for a review see Kent, 1988).

Tat86,41Ala exhibited no significant transactivation activity in both the RNA assay and the HL3T1 CAT enzyme assay. The effect of Tat86,41Ala on Tat86 transactivation in the HL3T1 antagonist assay (Fig. 6) was determined using different ratios of Tat86,41Ala to Tat86 (Fig. 8; Green et al., 1989). Strong inhibition was obtained. For example, 1×10^{-6} M and 5×10^{-7} M Tat-86,41Ala inhibited transactivation by 1×10^{-7} M Tat-86 by 99% (Fig. 8, lanes 5 and 6).

SUMMARY OF ANTAGONIST ACTIVITY OF HIV-1 TAT PEPTIDES ON HIVLTR-DRIVEN CAT EXPRESSION IN HELA CELLS

Our studies (Green et al., 1989; unpublished data) have revealed that: **first**, Tat minimal domain peptides (Tat37-62, Tat37-72, and Tat37-86) in which Lys-41, Ser-46, or Tyr-47 are replaced with alanine, are moderate antagonists of the expression of HIVLTR-CAT, as measured by the short-term assay for CAT RNA but not by the long-term CAT enzymne assay; and **second** Tat86,41Ala, but not Tat86,46Ala or Tat86,47Ala, is a strong antagonist of the expression of HIVLTR-CAT as measured both by the short-term RNA assay and by the long-term CAT enzyme assay.

INHIBITION OF HIV-1 REPLICATION IN HUMAN MOLT4 T LYMPHOCYTES BY TAT86,41ALA PEPTIDES

Acute HIV-1 Assay for Tat Peptide Antagonists

Tat86,41Ala, the most effective of the Tat antagonists in the long-term CAT enzyme assay, was tested for ability to inhibit the replication of HIV-1 in acutely infected Molt4 cells. Peptide was added in most experiments at 2 h postinfection and medium replaced with fresh medium containing peptide every day or every other day (Fig. 9). Virus replication was quantified by reverse transcriptase activity of released HIV-1 particles at 4 or 6 days postinfection; poly rA/oligo dT served as template/primer and ^{32}P-thymidine triphosphate as labeled precursor.

The effect on HIV replication of Tat86,41Ala added 6 h prior to HIV infection and every other day. In initial experiments, Tat86,41Ala was added to the culture medium 6 h prior to virus infection (Table 1). Over 99% inhibition of virus replication was observed with 2×10^{-6} M peptide added every other day. We note that there was marked protection of infected cells from destruction in the presence of peptide. Infected Molt4 cells increased in number to 10^6 cells/ml in peptide treated cultures as compared to 3×10^5 cells/ml in infected culture not treated with peptide. Tat86,41Ala, however, did not affect the growth or metabolism of uninfected Molt4 cells at the concentrations used for these studies

(manuscript in preparation). Thus Tat86,41Ala protected cells
from HIV-1 induced cytopathicity and facilitated cell
proliferation, presumably by inhibiting virus replication

The effect on HIV-1 replication of Tat86,41Ala and Tat86
added at 1 h postinfection and every other day. We next asked
whether addition of peptide after virus infection would
inhibit virus replication (Table 2). At 2 x 10^{-6} M
Tat86,41Ala, virus replication was inhibited by 99%, about the
same level as observed with peptide added before virus
infection (see Table 1). Tat86 at low levels appeared to
stimulate virus formation, and at high levels, 2 x 10^{-6} M,
appeared to slightly inhibit virus replication. These results
suggest a possible Tat feedback mechanism.

The effect on HIV-1 replication of Tat86,41Ala and of
Tat86,41Ala,47Ala added every day. We next asked whether more
frequent addition of peptide at lower concentrations would
result in more effective inhibition (Table 3). At 2 x 10^{-7} M
Tat86,41Ala, an 80% inhibition was observed by daily addition
of peptide (Table 3) as compared to about 25% inhibition with
peptide added every other day (Table 2). Further, the
antagonist activity of the chemically synthesized double
mutant peptide, Tat86,41Ala,47Ala, was examined; inhibitory
activity equivalent to that of Tat86,41Ala was observed (Table
2).

The Effect of Tat86,41Ala on HIV-1 Replication in Chronically
Infected Cells

Few agents are known that inhibit HIV-1 replication in
chronically infected cells. We tested different levels of
Tat86,41Ala or Tat86 in the chronic HIV-1 assay (see assay
scheme in Fig. 10). A maximum of about 60% inhibition was
found at the highest level tested, 1 x 10^{-6} M, as compared
with 99% inhibition in the acute assay (Fig. 11).

Inhibition by Tat86,41Ala Peptide of Syncytia Formation upon
Co-Cultivation of Molt4 and Molt4/HTLV-IIIB Lymphocytes

HIV-1 infected cells expressing viral envelope
glycoprotein gp120 will fuse with cells bearing the CD4
receptor (Sodroski et al., 1986). The loss of single cells to
form large syncytia has been exploited to provide a
quantitative assay thought to reflect virus spread by cell to
cell contact (Tochikura et al., 1988). Equal numbers of Molt4
cells and chronically infected Molt4 cells are co-cultivated
for 20 h (see scheme in Fig. 12). The decrease in single cell
number in cultures treated with drug is used to calculate the
fusion index and the % inhibition. Treatment with Tat86,41Ala
was found to inhibit syncytia formation by 82% at 2 x 10^{-6} M
peptide (Table 4). These results suggest that Tat peptide may
prevent cell to cell spread of virus.

Mechanism of Tat86,41Ala Inhibition of Transactivation

As described above, the chemically synthesized peptide,
Tat86,41Ala, inhibits transactivation by Tat86 when both

ACUTE HIV-1 ASSAY FOR PEPTIDE ANTAGONISTS

2×10^5 Molt4 lymphocyte cells/ml

↓

Infect with HIV-1 (strain HTLV-IIIB)
(1-10 infectious units/cell)

↓

Add peptide at 2 h postinfection

↓

Add fresh peptide every day (or every other day)

↓

Harvest cell supernatant at 4 and 6 days postinfection

↓

Isolate virus particles and analyze viral reverse
transcriptase activity

Fig. 9. Flow Sheet Showing the Acute Assay for Inhibition of HIV-1 Replication by Tat Peptide Antagonists.

CHRONIC HIV-1 ASSAY FOR PEPTIDE ANTAGONISTS

2×10^5 HTLV-IIIB chronically infected
Molt4 lymphocyte cells/ml

↓

Add peptide to cells

↓

Add fresh peptide every day (or every other day)

↓

Harvest cell supernatant at 4 and 6 days postinfection

↓

Isolate virus particles and analyze viral reverse
transcriptase activity

Fig. 10. Flow Sheet Showing the Chronic Assay for Inhibition of HIV-1 Replication by Tat Peptide Antagonists.

Table 1. Inhibition by Tat86,41Ala Peptide of HIV-1
Replication in Human Lymphocytes - Acute Assay (Pretreatment
and Peptide Addition Every Other Day)

Reverse Transcriptase Activity

Peptide Concentration	4 days		6 days		Cell Count x10^5/ml
	Cpm x10^3	Inhibition (%)	Cpm x10^3	Inhibition (%)	
None	28	--	172	--	2.96
None	28	--	145	--	2.98
3 x 10^{-8} M	25	11	129	19	3.01
3 x 10^{-8} M	25	11	132	17	3.48
1.5 x 10^{-7} M	18	36	59	63	3.94
1.5 x 10^{-7} M	14	50	66	58	4.56
1.5 x 10^{-6} M	2	92	4	98	9.16
1.5 x 10^{-6} M	3	91	2	99	10.00

Molt4 cells in microwells (1.0 ml, 2 x 10^5 cells) were treated
with peptide for 6 hr and infected with HTLV-IIIB. Cell
medium was replaced after 2 and 4 days with medium (RPMI-1640,
10% FBS) containing fresh peptide. Cell supernatant fluids
were harvested at 4 and 6 days and analyzed for virus-
containing reverse transcriptase as follows. Virus particles
were precipitated by addition of 0.5 ml of 30% PEG-6000/0.4 M
NaCl and incubation at 4 deg overnight. Virus was centrifuged
at 4500 x G for 30 min and solubilized in 100 ul of 0.8 M
NaCl/50 mM Tris.HCl, pH 7.9/0.5% Triton-X100/0.5 mM PMSF/1 mM
DTT/20% glycerol. Reverse transcriptase was determined on 20
ul aliquots by incorporation of ^{32}P-dTTP using poly rA.dT$_{12-18}$
as template-primer.

Table 2. <u>Inhibition by Tat86,41Ala Peptide of HIV-1</u>
<u>Replication in Human Lymphocytes - Acute Assay (Peptide Added</u>
<u>Every Other Day)</u>

		Reverse Transcriptase (6 Days After Infection)	
Peptide	Concentration	Cpm x10^3	Inhibition (%)
None		145	-
Tat86,41Ala	1 x 10^{-7} M	105	28
Tat86,41Ala	2.5 x 10^{-7} M	108	26
Tat86,41Ala	5 x 10^{-7} M	90	38
Tat86,41Ala	1 x 10^{-6} M	40	72
Tat86,41Ala	2 x 10^{-6} M	0	99.9
Tat86	1 x 10^{-7} M	210	+45
Tat86	2.5 x 10^{-7} M	210	+45
Tat86	5 x 10^{-7} M	163	+12
Tat86	1 x 10^{-6} M	119	18
Tat86	2 x 10^{-6} M	118	19

Molt4 cells (1.36 x 10^7 cells) were centrifuged at 200 x G for
5 min at room temperature and resuspended in 1 ml of RPMI/10%
fetal bovine serum. HTLV-IIIB (200 ul of a 40 x virus
concentrate) was added and virus adsorbed at 37 degrees for 1
h with agitation. Cells were centrifuged, washed with medium,
and resuspended at 400,000 cell per ml in RPMI/10% fetal
bovine serum. One ml portions were seeded into wells of 24
well plates and the indicated amount of peptide was added.
Cell culture media were replaced after 2 and 4 days with fresh
media containing peptide. Cell supernatant fluids were
harvested at 6 days and analyzed for virus particles
containing reverse transcriptase as described in Fig. 1. The
data are the average results of duplicate cultures.

Table 3. Inhibition by Tat Mutant Peptides of HIV-1 Replication in Human Lymphocytes – Acute Assay (Peptide Added Daily)

Peptide (1.5 x 10^{-7} M)	Reverse Transcriptase Activity 4 Days Postinfection	
	Cpm x10^3	Inhibition (%)
None	281	--
None	213	--
Tat86,41Ala	41	83
Tat86,41Ala	50	80
Tat86,41Ala,47Ala	31	87
Tat86,41Ala,47Ala	18	93

Molt4 cells in microwells (1.0 ml, 3 x 10^5 cells) were infected with HTLV-IIIB and treatment with peptides initiated at 2 hr postinfection. Fresh peptide was added at days 1, 2, and 3. Cell supernatant fluids were harvested at day 4 for analysis of reverse transcriptase in virus particles. No evidence of toxicity of peptides was noted and viable cell counts ranged from 4 x 10^5 to 7 x 10^5 cells per ml. Reverse transcriptase content was determined on 20 ul aliquots of solubilized virus (100 ul total) by incorporation of ^{32}P–dTTP as described in Fig. 1.

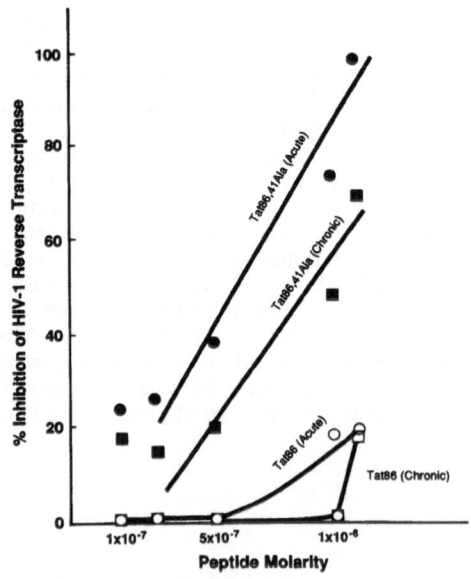

Fig. 11. Dose Response Measurements of the Inhibition of HIV-1 Replication by Tat86,41Ala and Tat86. Results of both the acute and chronic assays are plotted.

Table 4 **Inhibition by Tat86,41Ala Peptide of HIV-1 Induced Syncytia Formation Upon Cocultivation of Molt4 and Molt4/HTLV-IIIB Lymphocyte Cultures**

Peptide Concentration	Viable Cells per ml x 10^5	Fusion Index	Syncytia Formation Inhibition (%)
None (0 time)	3.92	0	-
None (20 h)	0.84	3.61	0
1 x 10^{-7} M	1.45	1.70	53
2.5 x 10^{-7} M	1.69	1.34	63
5 x 10^{-7} M	1.70	1.36	62
1 x 10^{-6} M	2.01	0.96	73
2 x 10^{-6} M	2.38	0.65	82

Molt4 cells and Molt4/HTLV-IIIB cells were centrifuged at 200 x g for 5 min and resuspended in RPMI/10% FBS at 3.92 x 10^5 cells per ml. Each cell suspension was treated with the indicated amount of Tat86,41Ala for 1 h at 37 deg. Mixtures of 0.5 ml of Molt4 and Molt/HTLV-IIIB cells were seeded in wells of a 24 well cell culture plate and incubated at 37 deg in a 5% CO_2 incubator. After 20 h, single viable cells were counted with a hemocytometer. The fusion index and % inhibition of syncytia formation was calculated as shown in Fig. 12.

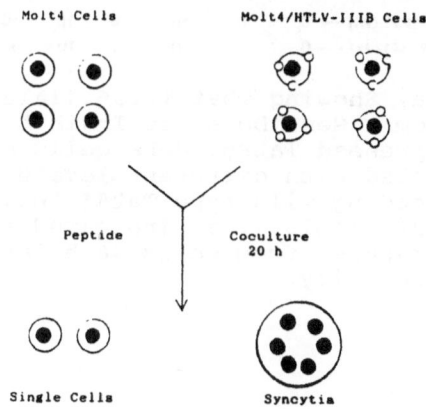

$$\text{Fusion Index (FI)} = \frac{\text{Cell number in control Molt4 well}}{\text{Cell number in peptide-treated well}} - 1$$

$$\% \text{ Inhibition} = 100 - \frac{\text{FI for peptide-treated cells}}{\text{FI for untreated cells}} \times 100$$

Fig. 12. Schematic of the Cell Fusion Assay for Inhibition of Syncytia Formation by Tat Peptide Antagonists.

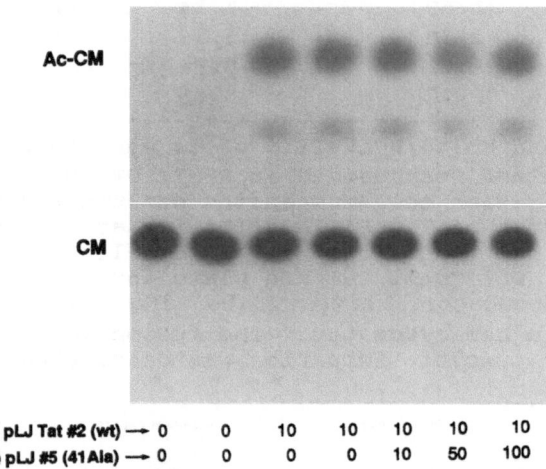

| (ng) pLJ Tat #2 (wt) → 0 | 0 | 10 | 10 | 10 | 10 | 10 |
| (ng) pLJ #5 (41Ala) → 0 | 0 | 0 | 0 | 10 | 50 | 100 |

Fig. 13. CAT Assay Showing That Tat86,41Ala Expressed
  Endogenously from a Gene Does Not Inhibit Transactivation by
  Endogenously expressed Tat86. HeLa cells on 60 mm plates
  were co-transfected with different levels of the expression
  plasmid pLJ, encoding wild type Tat86 [pLJ Tat#2(wt)] and/or
  Tat86,41Ala [pLJ#5(41Ala)], as indicated in the figure.
  Extracts were prepared from cells 48 h later and analyzed
  for CAT enzyme activity.

peptides are added together to cells containing microinjected or integrated HIV-LTRCAT. The implication is that inhibition reflects the transdominant negative behavior of Tat86,41Ala at its nuclear target site. To test this expectation, we constructed plasmids containing the mutated Tat41Ala gene and analyzed their ability to inhibit the wild type Tat gene by co-transfection analysis. HL3T1 cells were co-tranfected with wild type Tat in expression vector pLJ (pLJ Tat clone #2) and increasing levels of the Tat86,41Ala construct (pLJ,41Ala, clone #5) (Fig. 13). Surprisingly, high levels of the Tat86,41Ala plasmid did not inhibit transactivation of integrated HIV-LTRCAT by the Tat plasmid (Fig. 13). Similar results were obtained with several different isolates of the Tat86,41Ala plasmid (N. Abdullah, unpublised). We conclude that the endogenously expressed Tat86,41Ala gene product is not transdominant, although it inhibits transactivation by exogenous Tat86 and antagonize HIV-1 replicaation in infected Molt4 cells.

DISCUSSION

Mechanism of Tat antagonist activity

Here we report studies whose goal is the development of Tat peptide antagonists of HIV-1 replication. In the course of these studies, we identified several antagonists of wild type Tat, most notably Tat86,41Ala. Based on several considerations, we anticipated that antagonist Tat peptides would represent dominant negative mutants in essential protein domains. First, dominant negative mutant genes of regulatory proteins have been described that antagonize wild type function (for a review see Herskowitz, 1987). Second, it is reasonable to expect that Tat functions in an intracellular complex that involves specific binding of Tat domains to its target. Third, the Tat protein, although different from the typical transactivator, would be expected to possess a modular structure, i.e. to encode more than one functional domain.

As reported here, the chemically synthesized peptide Tat86,41Ala inhibits Tat86 transactivation when both peptides are added together to cells containing microinjected or integrated HIV-LTRCAT. Unexpectedly, the mutant gene did not exhibit transdominant behavior by the conventional transfection assay. Similar lack of transdominant activity by the Tat86,41Ala gene has been observed by B. Cullen (personal communication; Tiley et al, in press). The mechanism of transactivation by Tat is poorly understood. Therefore one can only speculate on the reasons for the differences in antagonist activity between Tat86,41Ala added to cells and the same Tat mutant peptide generated in vivo by DNA transfection. One possibility is that the true antagonist may be a minor component with high transdominant activity that is generated during peptide synthesis. For example, standard cleavage procedures may not completely remove from large peptides,the side chain blocking groups of cysteine and arginine, 4-methylbenzyl and tosyl, respectively. A Tat peptide molecule with a blocked residue in an essential amino acid(s) could exhibit a transdominant phenotype. We have not yet found evidence for this possibility.

Another possibility is that, contrary to expectation, the Tat86,41Ala antagonist acts at an extracellular level, perhaps by binding to a cellular receptor. The binding of exogenously added Tat86,41Ala to the cell surface could provide an explanation for its ability to inhibit HIV-1 replication in both acutely and chronically infected Molt4 lymphocytes. The assay for virus replication measures virus particles released into the supernatant fluid by budding from the cell surface, a process that could be inhibited by Tat peptides at the cell surface. Further, the reduction in virus-induced cytopathogenicity by Tat86,41Ala, could also reflect its cell surface binding property, i.e. reduced levels of virus budding would diminish membrane damage. Finally, this same hypothetical phenomenon could explain inhibition of cell fusion. For example, the binding of Tat86,41Ala on the surface of Molt4 and Molt4/HTLV-IIIB cells might reduce cell to cell attachment. Additionally, reduced levels of cell surface gp120 due to decreased virus budding could diminish syncytia formation. There is also the possibility of multiple modes of action. Clearly, further studies are needed to elucidate the mechanism of Tat antagonism.

Tat Antagonists of HIV-1 Replication. Despite an uncertain mechanism of inhibition, Tat86,41Ala peptide at 10^{-6} to 10^{-7} M is a strong antagonist of HIV-1 replication in Molt4 cells, exhibiting 95-99% inhibition depending upon treatment schedule. Furthermore, infected cells are protected from cytopathic destruction by virus at these levels. No peptide toxicity was observed at 10^{-5} M peptide concentration by several criteria, including cell viability, protein synthesis, and DNA synthesis (manuscript in preparation). Could Tat86,41Ala be useful for AIDS therapy? The expense of chemical synthesis and the difficulty of assuring purity of large peptides made by solid phase chemistry may preclude its practical use as a drug. Our current efforts are therefore aimed at identifying small Tat peptide antagonists and exploring the use of peptide mimetics to stabilize them against degradation by intracellular proteases. We believe this effort is worthwile in view of the ability of Tat peptides to self deliver to infected cells, to inhibit the replication of virus, and to counteract virus-induced pathogenicity. Our current findings indicate that cysteine residues that are essential for transactivation activity of full length Tat are not required for Tat antagonist activity; and that small peptides lacking the cysteine rich region can inhibit HIV-1 replication (manuscript in preparation).

ACKNOWLEDGEMENTS

We thank L. Ratner and W. Wold for evaluation of the manuscript, and C. Mulhall for editorial assistance. The work was supported by National Institutes of Health grant AI28201. M.G. holds Research Career Award 5 K06 AI04739 from the National Institute of Allergy and Infectious Diseases.

REFERENCES

Berkhout, B., Silverman, R.H., and Jeang, K.-T., 1989, Tat trans-activates the human immunodeficiency virus through a nascent RNA target, <u>Cell</u> 59:273.

Cullen, B.R., and Green, W.C., 1989, Regulatory pathways governing HIV-1 replication, <u>Cell</u> 58:423.

Dingwall, C., Ernberg, I., Gait, M.J., Green, S.M., Heaphy, S., Karn, J., Lowe, A.D., Singh, M., Skinner, M.A., and Valerio, R., 1989, <u>Proc. Natl. Acad. Sci. U.S.A</u>. 86:6925.

Felber, B.K., and Pavlakis, G.N., 1988, A quantitative bioassay for HIV-1 based on trans-activation, <u>Science</u> 239:184.

Frankel, A.D.., and Pabo, C.O., 1988, Cellular uptake of the Tat protein from human immunodeficiency virus, <u>Cell</u> 55:1189.

Frankel , A.D., Biancalana, S., and Hudson, D., 1989, Activity of synthetic peptides from the Tat protein of human immunodeficiency virus, <u>Proc. Natl. Acad. Sci. U.S.A.</u> 86:7397.

Gormon, C.M., Moffatt, L,.F., and Howard, B.H., 1982, Recombinant genomes which express chloramphenicol acetyltransferase in mammalian cells, <u>Mol. Cell. Biol.</u> 2:1044.

Green, M., and Loewenstein, P.M., 1988, Autonomous functional domains of chemically synthesized human immunodeficiency virus tat trans-activator protein, <u>Cell</u> 55:1179.

Green, M., Ishino, M., and Loewenstein, P.M., 1989, Mutational analysis of HIV-1 Tat minimal domain peptides: Identification of trans-dominant mutants that suppress HIV-LTR-driven gene expression, <u>Cell</u> 58:215.

Hauber, J., and Cullen, B.R.,1988, Mutational analysis of the transactivation-responsive region of the human immunodeficiency virus type 1 long terminal repeat, <u>J. Virol</u>. 62:673.

Herskowitz,I., 1987, Functional inactivation of genes by dominant negataive mutations, <u>Nature</u> 329:219.

Kamine, J., Loewenstein, P., and Green, M., 1990, Mapping of HIV-1 Tat protein sequences required for binding to Tar RNA, submitted.

Kent, S.B.H., 1988, Chemical synthesis of peptidesi and proteins, <u>in</u>: "Ann Rev. of Biochem," C. C. Richardson, ed., Ann. Reviews, Palo Alto.

Mitchell, P.J., and Tjian, R., 1989, Transcriptional regulation in mammalian cells by sequence-specific DNA binding proteins, <u>Science</u> 245:371.

Ptashne, M., 1988, How eukaryotic transcriptional activators work, <u>Nature</u> 335:683.

Sharp, P.A., and Marciniak, R.A., 1990, HIV TAR: An RNA enhancer?, <u>Cell</u> 59:229.

Sodroski, J., Goh, W.C., Rosen, C., Campbell, K., and Haseltine, W.A., 1986, Role of the HTLV-III/LAV envelope in syncytium formation and cytopathicity, <u>Nature (London)</u> 322:470.

Sternson ,L.A., 1987, Obstacles to polypeptide delivery, <u>Ann. NY Acad. Sci</u>. 57:19.

Tiley , L.S., Brown, P.H., and Cullen, B.R., 1990, Does the human immunodeficiency virus Tat trans-activator contain a discrete activation domain? <u>Virology</u>, in press .

Tochikura, T.S., Nakashima, H., Tanabe,A., and Yamamoto, N.,

1988, Human immunodeficiency virus (HIV)-induced cell fusion: Quantification and its application for the simple and rapid screening of anti-HIV substances in vitro, <u>Virology</u> 164:542.

RNA - PROTEIN INTERACTIONS REQUIRED FOR REV MEDIATED REGULATION OF HIV GENE EXPRESSION

Craig A. Rosen, Alan W. Cochrane, Patrick J. Dillon, Henrik S. Olsen

Department of Molecular Oncology & Virology
Roche Institute of Molecular Biology
Roche Research Center
340 Kingsland Street
Nutley, New Jersey 07110-1199

INTRODUCTION

Regulation of HIV gene expression is controlled by the interaction of cellular and viral transacting factors with cis-acting elements dispersed within the viral DNA and RNA. As a member of the lentivirus family of retroviruses, HIV encodes several non-structural regulatory proteins in addition to the *gag*, *pol*, and *env* genes common to all retroviruses. Two of these proteins, referred to as Tat[1,2] and Rev[3,4], are essential positive regulators of gene expression. The Rev protein is a 19 kD nuclear[5,6] phosphoprotein[7,8] required for expression of structural gene products[3,4]. Regulation is achieved via a mechanism involving a productive export of nuclear entrapped structural mRNAs to the cytoplasm[6,9-11]. Although the precise mechanism for this process remains unknown, interaction of the Rev protein with a structured RNA target sequence (RRE)[12] present in the *env* gene is required.

Rev forms a stable complex with RRE RNA

The minimal region required for Rev response, (termed RRE[12]), has been mapped to a region encompassing nucleotides 7300-7539 in the HIV genome[12,13]. We examined whether purified Rev protein could form a stable complex with RNA containing the RRE. Optimal expression of Rev in *E. coli* was achieved by creating a synthetic Rev gene that contains codons most frequently used in *E. coli*[14]. Rev was purified using an anti-Rev monoclonal immunoaffinity column (Fig. 1A). To examine Rev-RNA interactions an *env* gene fragment spanning the RRE was cloned downstream of the T7 promoter and ^{32}P labeled *in vitro* transcribed RNA was incubated with purified Rev protein in the presence of 5-10 μg of tRNA, digested with RNAse T[1], then analyzed on non-denaturing polyacrylamide gels[15,16]. As shown in Fig. 1B lane 1, incubation of HIV-1 RRE RNA with Rev resulted in the formation of an RNAse resistant RNA-protein complex. The smaller products present after RNAse digestion represent regions of secondary structure that are resistant to mild RNAse treatment. The specificity of complex formation was assessed by preincubating the reaction with an excess of specific RNA competitor (nt7204-7562) or nonspecific competitor (nt7498-7562) prior to addition of labeled RNA. While unlabeled specific competitor RNA inhibited complex formation, similar quantities of the nonspecific competitor did not (Fig. 1B). The purity of the preparation of Rev used in the assays indicates that Rev is responsible for the RNA protein complexes observed. To confirm this we have shown that inclusion of an anti-Rev antibody either blocked or altered Rev-RNA interaction[15].

Deletion analysis of the RRE RNA indicated that deletions in the major stem structure (Fig. 2A) impaired the ability of Rev to interact. To examine the importance of base pairing interaction within this structure, the stem was recreated by annealing two independently synthesized RNAs that comprised the 5' and 3' portions of the stem. For this experiment, RNA corresponding to nucleotide 7204-7494 was synthesized in the

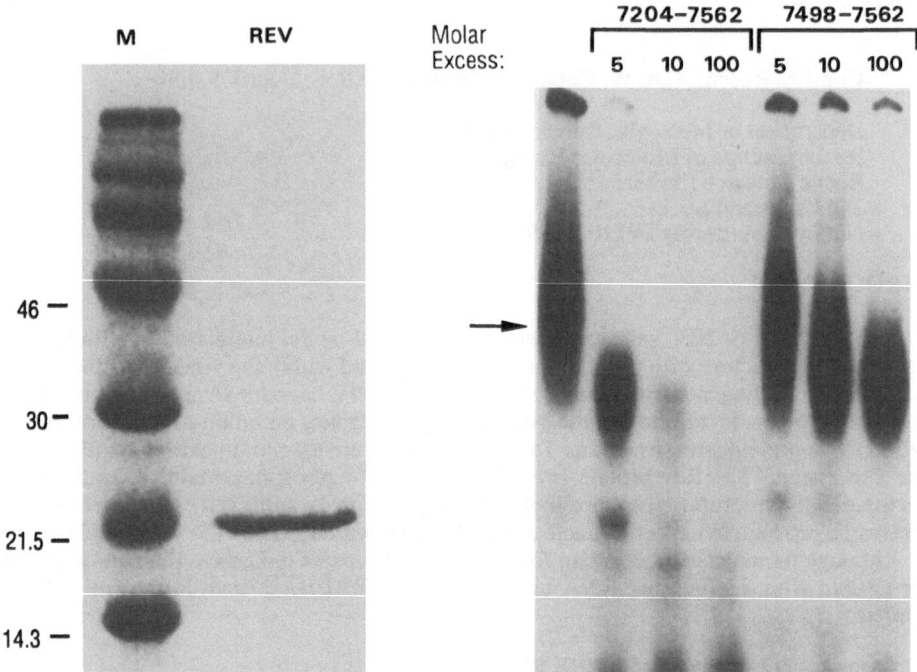

Figure 1. Purification of Rev from E. coli and complex formation with RRE RNA[15].

(a) An aliquot of purified Rev protein from *E. coli* was analyzed on a 14% polyacrylamide gel. Protein was visualized by staining with coommassie blue. Lanes: m, molecular weight markers; Rev, Rev protein. B, Complex formation was carried out using purified Rev protein and *in vitro* transcribed RNA corresponding to the HIV-1 RRE. The specificity of complex formation was assessed by preincubation of Rev with the indicated amount of specific (7204-7562) or nonspecific (7598-7562) competitor RNA. The molar excess of competitor RNA used is indicated.

absence of [^{32}P]UTP and the semi-complementary piece of RNA spanning nucleotides 7498-7562 was labeled with ^{32}P Although neither of these RNAs could independently form a complex with Rev (not shown), annealing (Fig. 3A) followed by incubation with Rev protein (Fig. 3B) restored complex formation[15]. Thus, the predicted stem structure is likely to form and its formation appears to be critical for Rev interaction.

Secondary structure is the major determinant for HIV-1 Rev protein interaction with RNA

The finding that the intact major stem structure is required for Rev-RRE interaction could be interpreted in several ways: (i) base pairing interactions in this region are required for the appropriate formation of the other hairpin structures, one or a combination of which, may comprise a site for Rev interaction; (ii) Rev interacts directly with the stem structure itself; or (iii) Rev interacts with the large stem structure and other elements of secondary structure simultaneously. To distinguish between these possibilities, nucleotide changes designed to alter the primary nucleotide sequence of the major stem were made. Individual changes produced small bulges, removed bulges, or maintained a different secondary structure. Each of these mutants bound to Rev, albeit some apparently to a lesser extent than others[16]. These results suggested that Rev-RRE interaction is not conferred by primary nucleotide sequence present in the stem structure.

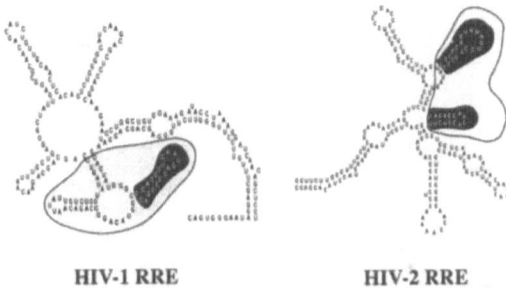

HIV-1 RRE   HIV-2 RRE

Fig. 2. Predicted secondary structure of the HIV-1 and HIV-2 RRE elements. The predicted RNA secondary structures generated within the HIV-1[12] and HIV-2 RRE[17] elements are shown. The putative Rev binding sites in the RREs are shown in black[16,18,19].

We reasoned that the absence of Rev binding obtained with the RREs which had major deletions in the stem probably reflected the formation of alternate secondary structures which lack the structural elements required for Rev binding. This possibility was examined further by creating mutations designed to delete individual hairpin loop structures within the stem. We observed that essentially all regions could be deleted with the exception of the stem structures depicted by shading in Fig. 2[16]. The importance of this structure was examined further by altering the nucleotide sequence in the loop. This mutation, however, had no obvious effect on Rev-RRE interaction. In contrast, a second mutation designed to disrupt formation of the hairpin prevented Rev binding. However, a compensatory mutation for the one above which was predicted to reconstitute secondary structure of this region restored Rev-RRE interaction even though the primary nucleotide sequence was dissimilar from that of the original hairpin loop. In subsequent studies we have been unable to define a single region of the RRE where mutation dramatically alters Rev binding while maintaining secondary structure of the region. Similar findings have been reported by several groups[18-20].

The data suggests that complex base pairing interactions within the HIV-1 RRE element generate hairpin loop structures that are essential for Rev-RRE interaction. Structure, rather than simply primary sequence, appears to be the major determinant for this interaction. If a specific primary sequence is required, the nature of the mutations made suggest the determinant is small in size and that its recognition is dependent upon context within the secondary structure of this region. Interaction of Rev with RRE RNA contrasts significantly with many DNA protein interactions where primary nucleotide sequence as opposed to structure, is a primary determinant for protein recognition. The interaction of Rev with RRE RNA also differs from other well characterized protein-RNA interactions in which the nucleotide sequence and the secondary structure are equally important for protein recognition.

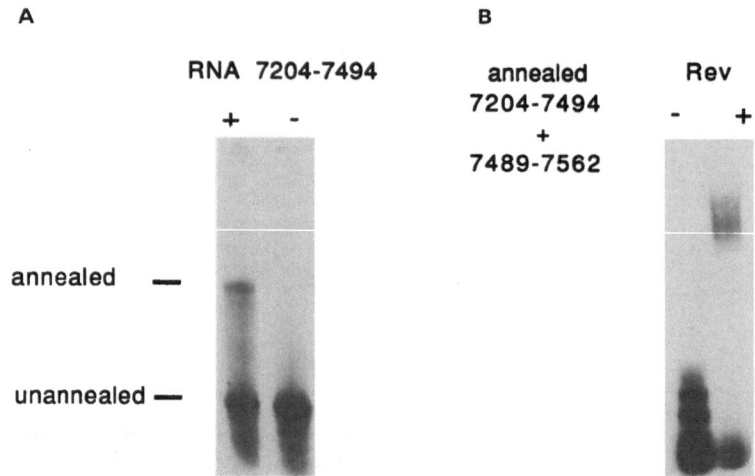

Fig. 3. Restoration of complex formation with reannealed RNA[15]. (a) The RRE structure was recreated by annealing *in vitro* transcribed RNA spanning nucleotides 7204-7494 (synthesized unlabeled) with RNA spanning nucleotides 7498-7562 (synthesized with [^{32}P]UTP). (b) The annealed RNAs shown were incubated in the presence (+) or absence (-) of Rev and complex formation was assessed using the RNA gel shift assay depicted in Fig. 1B.

Interaction of HIV-2 Rev with HIV-2 and HIV-1 RRE RNA

The evolutionarily and structurally related HIV-2 virus displays many similarities with HIV-1 including the presence of tat and rev genes. Although the HIV-1 and HIV-2 Tat proteins function in a reciprocal manner through each other's transacting responsive regions (TAR)[21], the function of Rev appears to be more restrictive. While HIV-1 Rev can function with the HIV-2 RRE, the reciprocal case does not hold true[17,22]. To understand the basis for this non-reciprocity the nature of the HIV-2 Rev interaction with different RRE elements was examined. For these studies, the HIV-2 Rev protein was purified from *E. coli*. For examination of Rev-RRE interactions, the HIV-2 *env* gene region containing the putative HIV-2 RRE (nucleotides 7608-7973) from HIV-2$_{ROD}$ was amplified by PCR and cloned downstream of the T7 promoter present in the Bluescript transcription vector. Protein-RNA interactions were assessed using the qualitative gel shift assay. It was found that the HIV-2 Rev protein bound to both the HIV-2 RRE RNA, and RNA corresponding to nucleotides present in the putative SIV RRE[22]. Very weak, if any binding, was observed with HIV-1 RRE RNA. However, the HIV-1 Rev protein was able to interact with all three of the RRE elements. This observation is consistent with the inability of HIV-2 Rev to regulate HIV-1 gene expression *in vivo*.

Using a strategy similar to that employed for delineation for the HIV-1 RRE binding site, mutations and deletions in hairpin loop structures were introduced into the HIV-2 RRE. While major deletions in the stem abolished Rev interaction, small changes in base pairing within the stem structure had little effect on RNA interaction. This indicates that stem formation may be required for the overall generation of the smaller secondary structures within the HIV-2 (Fig. 2B) RRE of which one, or more, may be important for Rev interaction. It was also observed that significant alterations could be made in regions of the RRE other than those areas depicted by shading in Fig. 2B. As observed in HIV-1 RRE RNA, compensatory mutations which regenerate the structures by base pairing interaction were able to interact with HIV-1 and HIV-2 Rev and were functional *in vivo*. Significantly, the region of the HIV-2 RRE RNA identified as being important for interaction with HIV-1, bears no resemblance with primary nucleotide sequence or secondary structure to the element required for binding of HIV-1 Rev within its own RRE. These observations offer compelling evidence that RNA secondary structure is a major determinant for both HIV-1 and HIV-2 Rev binding and further suggests that different secondary structures can be recognized by the HIV-1 Rev protein. The results may also provide insight relating to the ability of the HTLV-1 Rex protein to complement HIV gene expression, while HIV-1 Rev cannot complement HTLV-1[23,24]. The prediction would be that HTLV-1 Rex recognizes an alternate secondary structure in the HIV-1 RRE and that a suitable binding site for HIV-1 Rev is absent in the HTLV-1 RRE. Our findings would also be consistent with a group of RNA structures forming a binding domain in which Rev or Rex would have several contact points. Thus, it is likely that the human immunodeficiency virus Rev, and the human T-cell leukemia Rex proteins belong to a class of regulatory molecules that function through recognition of RNA secondary structure.

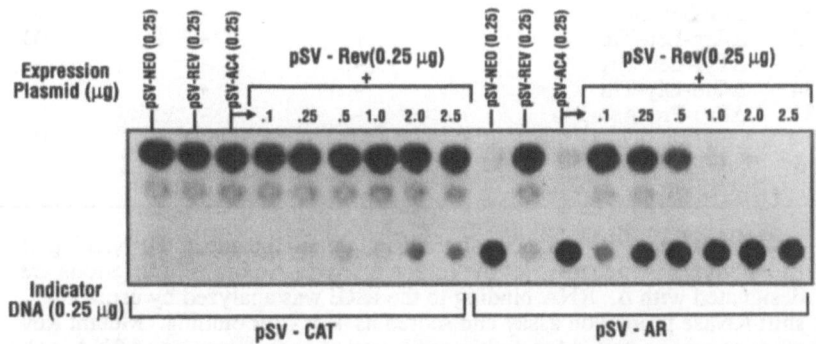

Fig. 4. *Trans*-dominant suppression of Rev function in a Rev-dependent heterologous gene expression assay[25]. COS cells were cotransfected with 250 ng of plasmid pSVCAT or pSVAR and the indicated amount of Rev expression plasmid or nonspecific competitor pSVNEO. The CAT assays shown were from cell lysates prepared 48 h after transfection. *Trans*-dominant suppression was assessed by increasing the amount of pSV-AC4 plasmid DNA in the presence of a constant amount (250 ng) of pSVRev in the cotransfection assays.

Identification of the HIV-1 Rev RNA binding domain

Earlier studies have delineated several functional domains within HIV-1 Rev[26-29]. In a recent study, mutational analysis of Rev was undertaken to identify the domain required for interaction with RRE RNA. Site directed mutations shown in Table 1 were constructed by oligonucleotide mutagenesis. Mutations were made within a modified synthetic Rev cDNA containing a 5' sequence encoding a stretch of histidine residues which facilitates purification[14]. Function of this modified Rev was comparable to that of authentic protein and its localization in the nucleus and nucleolus was also consistent with that of authentic Rev protein. Each of the mutations shown in Table 1 were made within the Rev insert present in a bacterial expression vector. Bacterial lysates prepared from exponentially growing cultures were incubated with *in vitro* transcribed RRE RNA transcripts and binding was measured in the qualitative gel shift ribonuclease T$_1$ protection assay. We found that many of the mutations which destroyed Rev's ability to bind RRE RNA centered near or within the basic stretch of amino acids (Table 1)[25]. As anticipated, each of the Rev proteins which failed to interact with RRE RNA *in vitro* were non-functional *in vivo*[25]. The lack of function was not a reflection of lower levels of expression since Rev-immunoreactive products were detected in each of the transfections.

Table 1 Functional characterization of Rev protein mutants

Mutation and localization	Amino acid changes	RRE RNA binding	Rev-Rev association	Trans-activation
R 38-39	Arg-Arg to His-Glu	-	+	-
RΔ48-51	ΔArg-Gln-Arg-Gln Glu47 to Asp	-	+	-
RΔ45-51	ΔTrp-Arg-Glu-Arg-Gln-Arg-Gln	-	+	-
R85	Cys to Phe	+	+	+
R75	Leu to Arg	+	+	TD
RΔ70-72	ΔPro-Val-Pro	+	+	+
RΔ59-61	ΔIle-Leu-Gly	-	+	-
RΔ55-57	ΔIle-Ser-Glu	-	-	-
RΔ80-82	ΔArg-Leu-Thr	+	+	TD
RΔ63-65	ΔTyr-Leu-Gly	-	+	-
RΔ32-34	ΔGlu-Gly-Thr	-	+	-
RΔ28-31	ΔPro-Pro-Asn-Pro	-	-	-
RevAC-4	Leu78-Glu79 to Asp-Leu	+	ND	TD
Rev-916	Ser92 to stop	+	+	+

Functional characterization of Rev protein mutants [32]. Naming of Rev mutants is indicative of the region altered. Amino acid deletions are designated with Δ. RNA binding to the RRE was analyzed by using a gel-shift RNase protection assay and scored as + or - for binding. Mutant Rev proteins were assayed for multimer formation in the absence of RNA, (+), (-), or not determined (ND). Mutant Rev proteins were analyzed for function in the Rev-dependent transient gene expression system[13] and scored as positive (+), negative (-), or transdominant suppression (TD).

Many of the well characterized protein-DNA and protein-RNA associations are governed through the formation of protein complexes. To examine whether Rev interacts with RRE RNA as a monomer or multimeric structure, a truncated Rev protein (Rev-916) lacking the 25 carboxy terminal amino acids was used in the binding assay. If Rev-RRE interaction occurs through complex formation, the mixing of a full length and truncated Rev in the binding reactions should produce intermediate size complexes upon gel analysis. We found that intermediate complexes were indeed formed when the truncated Rev-916 was combined with full length protein (Table 1)[25]. Lack of a single intermediate size band between the two species suggests that the Rev complex is composed of more than two subunits.

Does Rev function require cellular accessory factors?

We were interested in examining whether the interaction of Rev with RRE RNA was in itself sufficient to restore gene expression. It was previously shown that mutation of amino acids 78 and 79 produces a Rev protein that is a transdominant repressor of Rev function[27]. To examine the basis for the repressive effect exerted by the transdominant Rev, (designated here as Rev AC-4), the ability of the mutant to repress Rev-mediated regulation of heterologous gene expression was tested[25]. The results, depicted in Fig. 4, show that the Rev AC-4 is indeed a transdominant suppressor of Rev function, in accord with previous findings.

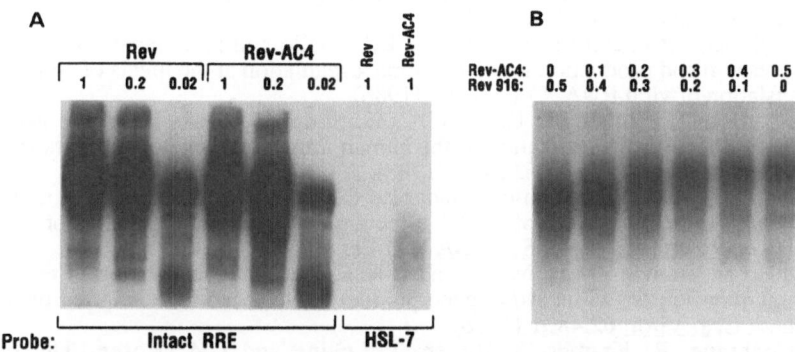

Fig. 5. Analysis of RevAC-4 interaction with RRE RNA[25]. (a) The amount shown in cleared *E. coli* lysate expressing Rev or RevAC-4 was incubated with intact RRE RNA or RRE RNA HSL-7, and Rev-RRE interaction was measured by use of the gel mobility-shift assay. (b) The ability of RevAC-4 to form multimers was examined by including the amounts shown of Rev-916 (truncated Rev) and RevAC-4 with RRE RNA in the normal binding reaction.

To elucidate the basis of the transdominant effects exerted by Rev AC-4, its ability to interact with HIV-1 RRE RNA was examined. As shown in Fig. 5, Rev AC-4 forms a stable interaction with RRE RNA comparable to that observed with authentic Rev protein. Further studies with additional mutants in this region as well as mutations within the same region of HTLV-1 Rex protein suggests this region in both proteins functions as an activation domain. The inability of the transdominant Rev to restore gene expression while still interacting with RRE RNA indicates that binding alone is insufficient for function. This suggests that binding of additional cellular factors to either Rev or the Rev RRE complex is necessary for function. Thus, the transdominant suppression observed for these mutants is most likely due to competition for functional Rev binding on RRE RNA. The accumulated data therefore suggests that at least two steps are required for Rev function; association of Rev with RRE RNA followed by interaction of cellular factors with the Rev RRE complex to mediate the productive export of structural mRNA from the nucleus to the cytoplasm. Identification of these putative host cell factors involved in the latter part of this pathway awaits further study.

ACKNOWLEDGEMENTS

We thank K. Eilenberg for preparation of the manuscript. This work was supported in part by a National Cooperative Drug Discovery grant (NCDDG). A. Cochrane is a recipient of an AmFAR/F.L.and E.L. Cummings Memorial Fund Scholarship.

REFERENCES

1. S.K. Arya, C. Guo, S.F. Josephs and F. Wong-Staal, Trans-activator gene of human T-lymphotrophic virus type III (HTLV-III), Science 229:69 (1985).
2. J. Sodroski, C.A. Rosen and W.A. Haseltine, Trans-acting transcriptional activation of the long terminal repeat of human T lymphotropic viruses in infected cells, Science 225:381 (1984).
3. J. Sodroski, W.C. Goh, C. Rosen, A. Dayton, E. Terwilliger and W.A. Haseltine, A second post-transcriptional transactivator gene required for the HTLV-III replication, Nature 321:412 (1986).
4. M.B. Feinberg, R.F. Jarrett, A. Aldovini, R.C. Gallo and F. Wong-Staal, HTLV-III expression and production involve complex regulation at the levels of splicing and translation of viral RNA, Cell 46:807 (1986).
5. B.R. Cullen, J. Hauber, K. Campbell, J.G. Sodroski, W.A. Haseltine and C.A. Rosen, Subcellular localization of the human immunodeficiency virus trans-acting art gene product, J. Virol. 62:2498 (1988).
6. B.K. Felber, M. Hadzopoulou-Cladaras, C. Cladaras, T. Copeland and G.N. Pavlakis, The rev protein of HIV-1 affects the stability and transport of the viral mRNA, Proc. Natl. Acad. Sci. USA 86:1495 (1989).
7. J. Hauber, M. Bouvier, M.H. Malim and B.R. Cullen, Phosphorylation of the human immunodeficiency virus art/trs gene product is enhanced by activation of protein kinase C, J. Virol. 62:4801 (1988).
8. A. Cochrane, R. Kramer, S. Ruben, J. Levine and C.A. Rosen, The human immunodeficiency virus rev protein is a nuclear phosphoprotein, Virology 171:264 (1989).
9. M. Hadzopoulou-Cladaras, B.K. Felber, C. Cladaras, A.A. Athanassopoulos, A. Tse and G.N. Pavlakis, The rev (trs/art) protein of the human immunodeficiency virus type 1 affects viral mRNA and protein expression via a cis-acting sequence in the env region, J. Virol. 63:1265 (1989).
10. M.L. Hammarskjöld, J. Heijmer, B. Hammarskjöld, I. Sangwan, L. Albert and D. Rekosh, Regulation of human immunodeficiency virus env expression by the rev gene product, J. Virol. 63:1959 (1989).
11. M. Emerman, R. Vazeux and K. Peden, The rev gene product of the human immunodeficiency virus affects envelope-specific RNA localization, Cell 57:1155 (1989).
12. M.H. Malim, J. Hauber, S.-Y. Le, J.V. Maizel and B.R. Cullen, The HIV-1 rev transactivator acts through a structured target sequence to activate nuclear export of unspliced viral mRNA, Nature 338:254 (1989).
13. C.A. Rosen, E. Terwilliger, A. Dayton, J.G. Sodroski and W.A. Haseltine, Intragenic cis-acting art gene-responsive sequences of the human immunodeficiency virus, Proc. Natl. Acad. Sci. USA 85:2071 (1988).
14. A.W. Cochrane, C.H. Chen, R. Kramer, L. Tomchak and C.A. Rosen, Purification of biologically active human immunodeficiency virus rev protein from E. coli, Virology 173:335 (1989).

15. A.W. Cochrane, C.-H. Chen and C. Rosen, Specific interaction of the HIV Rev transactivator protein with a structured region in the *env* mRNA, Proc. Natl. Acad. Sci. USA 87:1198 (1990).
16. H. Olsen, P. Nelbock, A. Cochrane and C. Rosen, Secondary structure is the major determinant for interaction of HIV Rev protein with RNA, Science 247:845 (1990).
17. M.H. Malim, S. Bohnlein, R. Fenrick, S.-Y. Lee, J.V. Maizel and B.R. Cullen, Functional comparison of the Rev transactivators encoded by different primate immunodeficiency virus species, Proc. Natl. Acad. Sci. USA 86:8222 (1989).
18. M.H. Malim, L.S. Tiley, D.F. McCarn, J.R. Rusche, J. Hauber and B.R. Cullen, HIV-1 structural gene expression requires binding of the Rev transactivator to its RNA target sequence, Cell 60:675 (1990).
19. S. Heaphy, C. Dingwall, I. Ernberg, M.J. Gait, S.M. Green, J. Karn, A.D. Lowe, et al, HIV-1 regulator of virion expression (Rev) protein binds to an RNA stem-loop structure located within the Rev response element, Cell 60:685 (1990).
20. E. Dayton, D. Powell and A. Dayton, Functional analysis of CAR, the target sequence for the rev protein of HIV-1, Science 246:1625 (1989).
21. M. Emerman, M. Guyader, L. Montagnier, D. Baltimore and M. Muesing, The specificity of the human immunodeficiency virus type 2 transactivator is different from that of human immunodeficiency virus type 1, EMBO J. 6:3755 (1987).
22. P.J. Dillon, P. Nelbock, A. Perkins and C.A. Rosen, Function of the human immunodeficiency virus types 1 and 2 Rev proteins is dependent upon their ability to interact with a structured region present in the *env* gene mRNA, J. Virol. (1990).(In Press)
23. L. Rimsky, J. Hauber, M. Dukovich, M.H. Malim, H. Langlois, B.R. Cullen and W.C. Greene, Functional replacement of the HIV-1 rev protein by the HTLV-1 rex protein, Nature 335:738 (1988).
24. S.M. Hanly, L.T. Rimsky, M.H. Malim, J.J. Kim, J. Hauber, M. DucDodon, S.-Y. Le, et al, Comparative analysis of the HTLV-1 rex and HIV-1 rev trans-regulatory proteins and their RNA response elements, Genes Dev 3:1534 (1989).
25. H.S. Olsen, A.W. Cochrane, P.J. Dillon, C.M. Nalin and C.A. Rosen, Interaction of the human immunodeficiency virus type 1 Rev protein with a structured region in *env* mRNA is dependent upon multimer formation mediated through a basic stretch of amino acids, Genes Dev (1990).
26. A.W. Cochrane, E. Golub, D. Volsky, S. Ruben and C.A. Rosen, Functional significance of phosphorylation to the human immunodeficiency virus rev protein, J. Virol. 63:4438 (1989).
27. M.H. Malim, S. Bohnlein, J. Hauber and B.R. Cullen, Functional dissection of the HIV-1 rev trans-activator - derivation of a trans-dominant repressor of rev function, Cell 58:205 (1989).
28. A. Perkins, A.W. Cochrane, S.M. Ruben and C.A. Rosen, Structural and functional characterization of the human immunodeficiency virus rev protein, J. AIDS 2:256 (1989).
29. A.W. Cochrane, A. Perkins and C.A. Rosen, Identification of sequences important in the nucleolar localization of human immunodeficiency virus Rev: relevance of nucleolar localization to function, J. Virol. 64:881 (1990).

A 5' SPLICE SITE IS ESSENTIAL FOR REV AND REX REGULATION OF HIV ENVELOPE PROTEIN mRNA EXPRESSION

Xiaobin Lu, Nancy Lewis, David Rekosh and
Marie-Louise Hammarskjöld

Departments of Microbiology, Biochemistry, Biological
Sciences and Oral Biology, State University of New York at
Buffalo, Buffalo, N.Y. 14214

INTRODUCTION

The organization of the HIV genome is much more complex than that of most other retroviruses (for a review see (1)). This is due to the presence of several regulatory genes in addition to the structural genes *gag*, *pol* and *env*. This complexity is reflected by the large number of different subgenomic mRNAs found in HIV infected cells (2). A proper balance between these mRNAs has to be achieved to ensure appropriate expression of the different viral proteins.

The generation of the various mRNAs is accomplished through differential splicing. For example, the full length primary transcript sometimes remains unspliced to become a substrate for packaging and the mRNA for the gag and pol proteins. In another instance, the primary transcript is spliced once to become the mRNA for the env proteins. In this case, it still contains a complete intron, which when removed generates the mRNAs for some of the regulatory proteins. Thus splicing must be suppressed to a large extent in HIV infected cells to allow for the maintenance of one or two introns in some of the viral mRNAs and this intron containing RNA has to exit from the nucleus.

The HIV rev protein is a small nuclear phosphoprotein (3), that has been shown to be required for the export of the non-spliced and incompletely spliced *gag/pol* and *env* mRNAs from the nucleus (4, 5, 6, 7). Rev acts on a specific segment of RNA present only in these mRNAs known as the rev responsive element (RRE) (4, 6, 8, 9, 10). It has recently been shown that purified rev protein binds directly and specifically to the RRE *in vitro* (11, 12) and thus it seems likely that rev action involves direct binding to the mRNA. It has been speculated that rev functions to divert the structural mRNAs away from the splicing machinery in infected cells (13). In an alternative model, it has been suggested that the function of the protein may be to specifically disrupt splicing complexes containing the RRE (14).

The rex proteins of HTLVI and II are nuclear phosphoproteins that seem to serve similar functions as the rev protein (15, 16). The rex proteins can substitute for rev in HIV infected cells and have been shown to function by acting on the HIV RRE (17, 18). In contrast, rev does not work in conjunction with the rex responsive element present in the HTLV genome (19, 20, 21).

Advances in Molecular Biology and Targeted Treatment for AIDS
Edited by A. Kumar, Plenum Press, New York, 1991

We have been analysing the requirements for expression of the HIV structural proteins using a SV40 based mammalian vector system (4, 17, 22, 23). Expression of HIV envelope as well as gag/pol proteins in this system requires a functional rev protein *in trans* and the presence of the RRE *in cis*. In addition we have shown that the HTLVI and II rex proteins can substitute for rev in this system.

In this report we show that env expression in transfected cells requires a 5' splice site to be present in the *env* mRNA and that this RNA has to form a complex with U1 snRNA for both rev or rex regulation to occur.

RESULTS AND DISCUSSION

The original construct that we used to express the HIV envelope proteins is shown in Figure 1 (24). This plasmid, pSVSX1, contains the entire coding regions for *tat, rev, vpu* and *env*. Transcription is directed by the SV40 late promoter that is positioned upstream of these genes and terminates in rabbit β-globin sequences downstream of the insert. The β-globin segment contains a complete intron and a polyadenylation signal. pSVSX1 expresses large amounts of the HIV env proteins from a mRNA that remains unspliced. The rev and tat proteins are expressed from a doubly spliced RNA in which the intron within the HIV sequences as well as the β-globin intron has been removed. In addition, small amounts of the vpu protein is expressed from the unspliced mRNA (Li et al., manuscript in preparation). Expression of the env and as well as the vpu proteins requires rev and the RRE . In the absence of rev, unspliced RNA accumulates in the nucleus, whereas spliced RNA is transported to the cytoplasm.

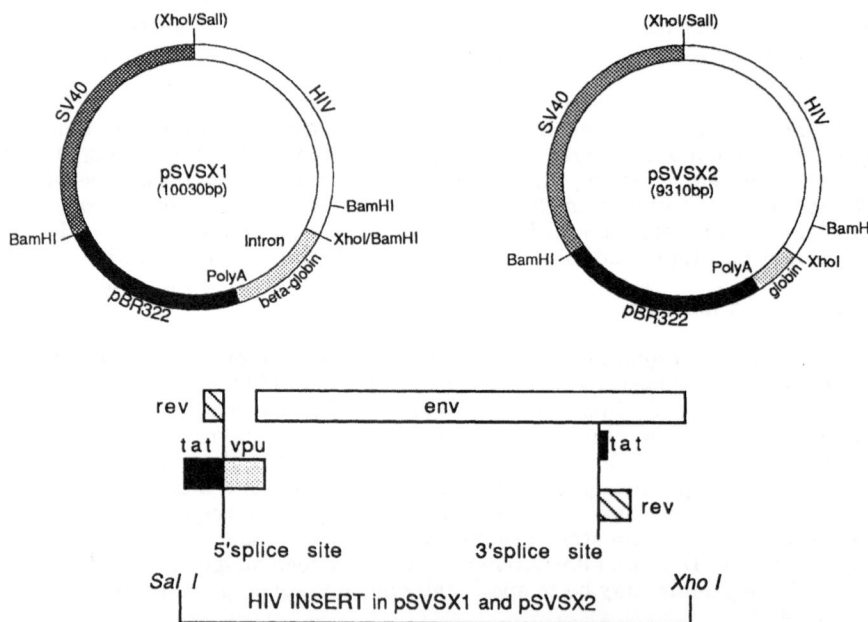

Fig. 1. **Expression plasmids pSVSX1 and pSVSX2 and the HIV insert they contain.** The plasmids are composed of sequences derived from HIV-1 (BH-10 clone), rabbit β-globin, pBR322 and SV40. The construction of pSVSX1 has been previously described (24). pSVSX2 is derived from pSVSX1 by deletion of a 720bp fragment which removes the rabbit β-globin intron.

To analyse the effect on rev regulation of the different splice sites in the RNA, we made several mutations in pSVSX1. In one of these (pSVSX2, see Figure 1), the β-globin intron was removed by deletion of a 720bp DNA fragment covering this intron. This deletion had no effect on env expression (data not shown). In contrast to this, deletions that removed the tat/rev 5' splice site upstream of the *env* gene completely abolished env expression. The same was true for point mutants of pSVSX1 in which the conserved G at the +1 position downstream of the splice site was changed to either a T or a C (data not shown). Insertion into the deletion mutants of a small synthetic oligonucleotide homologous to either the tat/rev 5' splice site or the rabbit β-globin 5' splice site restored env expression. Env expression could also be restored by insertion of a DNA fragment from the Epstein Barr Virus EBNA1 gene that contained a 5' splice site. Taken together, the results of the mutational analysis showed that the presence of a 5' splice site upstream of the *env* gene was required for rev regulated env expression in spite of the fact that the *env* mRNA remained unspliced (23).

An early step in a splicing event involves complex formation between an U1 snRNP and an intron containing RNA. This is mediated by specific base pairing between the U1 snRNA and the sequence surrounding the 5' splice site. This was directly demonstrated through the use of U1 suppressor mutations which restored base pairing and splicing at mutated non-functional 5' splice sites (25, 26, 27). To investigate whether the requirement for a 5' splice site upstream of env reflected a requirement for complex formation with U1 snRNA we performed co-transfection experiments with a construct expressing a mutated U1 snRNA and pSVSX1 containing a mutated 5' splice site. To do this we changed the G at the +5 position downstream of the tat/rev 5' splice site in pSVSX1 to a C (pSVSX1-5C) and used the corresponding U1 snRNA suppressor containing a G in position +4 (pUCBU1-4G). The results of this experiment are shown in figure 2. No env expression was obtained in cells transfected with the splice site mutant and a rev producing vector pRev1 (lane 7). However env expression was restored in cotransfections with both the U1 suppressor mutant and pRev1 (lanes 1-5). Env expression increased up to 10 µg of added suppressor. The fact that low levels of env were also seen the presence of the suppressor but in the absence of the pRev1 plasmid (lane 6) indicated that correct splicing of the suppressed env mRNA was obtained at low levels to restore some rev function.

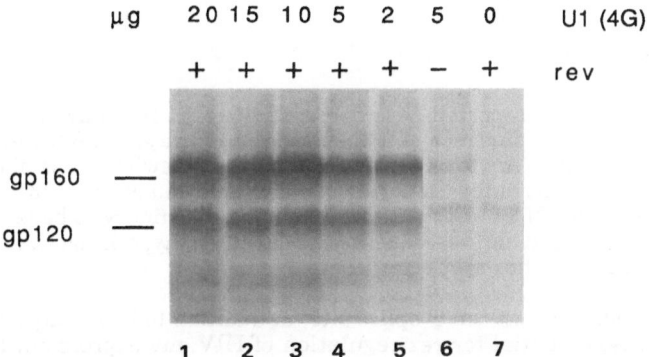

Fig.2. Suppression of a 5' splice site mutation. Lysates from cells cotransfected with pSVSX1(+5C),pRev1 and decreasing amounts of pUCBU1-4G (lanes 1-5) were western blotted with serum directed against HIV-1 envelope protein. The amounts of added pUCBU1-4G are indicated above the lanes. Also shown are controls in which cells were transfected with pSVSX1(+5C) and 5 ug of pUCBU1-4G in the absence of extra rev (lane 6) or with pSVSX1(+5C) and pRev1 in the absence of U1 snRNA suppressor (lane 7).

An analysis of the RNA in cells transfected with several different 5' splice site mutants showed only minute amounts of unspliced HIV specific RNA in both cytoplasm and nucleus. The results of this analysis have been presented in detail elsewhere (23). This and preliminary results from nuclear run-on experiments indicate that the stability of the HIV-specific mRNA is dramatically decreased in the absence of a functioning 5' splice site.

We have previously shown that the HTLV I and II rex proteins can substitute for the rev proteins in cotransfections with rev⁻ deletion mutants of pSVSX1 (17). To analyze whether rex complementation of env expression also required complex formation between the *env* mRNA and U1 snRNA we performed an experiment in which a rev- variant of pSVSX1-5C was cotransfected with plasmids expressing the rex proteins either alone or together with the U1 suppressor pUCBU1-4G. As controls, cells were also cotransfected with pRev1 in the presence or absence of pUCBU1-4G. The result of this analysis is shown in Figure 3. No env expression was detected when the mutant was transfected alone (lane 1) , together with pUCBU1-4G (lane 2) or together with plasmids expressing rev, rexI or rex II (lanes 3-5). In contrast, substantial amounts of the env proteins were expressed in cells which were cotransfected with the U1 suppressor construct and the rev, rexI or rexII expressing plasmids (lanes 6-8). The results of these experiments indicate that the 5' splice site and complex formation with U1 snRNA are essential also for rex regulated HIV1 env expression.

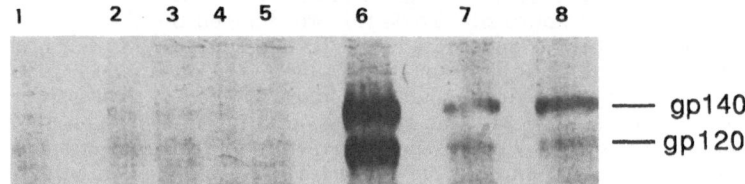

Fig. 3. Rev and rex regulation both require a 5' splice site.
Lysates of cells transfected with a rev⁻ derivative of pSVSX1(+5C) with or without pUCBU1-4G and rev or rex expressing plasmids were subjected to western blotting. The blot was developed with a serum directed against HIV-1 envelope protein.

CONCLUSIONS

The experiments reported here demonstrate that HIV env expression in our system requires the presence of a 5' splice site in the *env* mRNA and complex formation with U1 snRNA. RNA analysis, reported elsewhere, has shown that this interaction is essential for the formation of stable *env* mRNA (23). Thus it appears that the *env* RNA has to enter the splicing machinery to be protected from rapid breakdown. Based on these results, we favor the hypothesis that rev recognizes the RNA within a splicing complex.

We have shown that the 5' splice site, in addition to being important for rev regulation, is also essential for rex regulation of HIV env expression. In this context it should be noted that rex regulation of HTLV structural protein expression has been reported to also require a 5' splice site in the mRNA as well as the rex responsive element (16). It thus seems that distantly related viruses within the retrovirus family use very similar mechanisms for regulation of the expression of their structural proteins. This may be an indication that these viruses are tapping into the same previously unknown cellular regulatory pathway which is linked to the splicing machinery of the cell.

It seems likely, that future studies of the requirements for rev regulated transport and the factors that govern RNA processing in the HIV system, will yield important new information about the different nuclear pathways and the links that undoubtedly exist between them. Virus model systems have time and again proven invaluable in the quest for a better understanding of the molecular biology of the eukaryotic cell. Work on HIV was originally started because of its importance as a human pathogen. However it is becoming increasingly clear that this virus also provides us with novel and exciting tools in our exploration of the mechanisms that regulate fundamental cellular processes.

ACKNOWLEDGEMENT

We thank Drs. A. Weiner and Y. Zhuang (Yale University) for providing the U1snRNA expression vectors. Joy Van Lew provided expert cell culture assistance. This work was supported by a National Cooperative Drug Discovery Group/AIDS agreement with the National Institute of Allergy and Infectious Disease (AI25721). D.R. was the recipient of an RCDA from the National Cancer Institute (CA00905).

REFERENCES

1. Varmus, H. Retroviruses. Science. 240: 1427-1435, (1988).
2. Schwartz, S., B. K. Felber, D. M. Benko, E. M. Fenyo and G. N. Pavlakis. Cloning and functional analysis of multiply spliced mRNA species of human immunodeficiency virus type 1. J Virol. 64: 2519-2529, (1990).
3. Hauber, J., M. Bouvier, M. H. Malim and B. R. Cullen. Phosphorylation of the rev gene product of human immunodeficiency virus type 1. J Virol. 62: 4801-4804, (1988).
4. Hammarskjöld, M.-L., J. Heimer, B. Hammarskjöld, I. Sangwan, L. Albert and D. Rekosh. Regulation of human immunodeficiency virus env expression by the rev gene product. J Virol. 63: 1959- 1966, (1989).
5. Felber, B. K., M. Hadzopoulou-Cladaras, C. Cladaras, T. Copeland and G. N. Pavlakis. Rev protein of human immunodeficiency virus type 1 affects the stability and transport of the viral mRNA. Proc Natl Acad Sci U S A. 86: 1496-1499, (1989).
6. Malim, M. H., J. Hauber, S. V. Le, J. V. Maizel and B. R. Cullen. The HIV-1 rev trans-activator acts through a structured target sequence to activate nuclear export of unspliced viral mRNA. Nature. 338: 254-257, (1989).
7. Emerman, M., R. Vazeux and K. Peden. The rev gene product of the human immunodeficiency virus affects envelope-specific RNA localization. Cell. 57: 1155-1165, (1989).
8. Rosen, C. A., E. Terwilliger, A. Dayton, J. G. Sodroski and W. A. Haseltine. Intragenic cis-acting art gene-responsive sequences of the human immunodeficiency virus. Proc Natl Acad Sci USA. 85: 2071-2075, (1988).
9. Dayton, A. I., E. F. Terwilliger, J. Potz, M. Kowalski, J. G. Sodroski and W. A. Haseltine. Cis-acting sequences responsive to the rev gene product of the human immunodeficiency virus. J. Acquired Immune Deficiency Syndromes. 1: 441-452, (1988).
10. Hadzopoulou-Cladaras, M., B. K. Felber, C. Cladaras, A. Athanassopoulos, A. Tse and G. N. Pavlakis. The rev (trs/art) protein of human immunodeficiency virus type 1 affects viral mRNA and protein expression via a cis-acting sequence in the env region. J Virol. 63: 1265-1274, (1989).
11. Daly, T. J., K. S. Cook, G. S. Gray, T. E. Maione and J. R. Rusche. Specific binding of HIV-1 recombinant rev protein to the rev-responsive element in vitro. Nature. 342: 816-819, (1989).
12. Zapp, M. L. and M. R. Green. Sequence-specific RNA binding by the HIV-1 Rev protein. Nature. 342: 714-716, (1989).
13. Cullen, B. R. and W. C. Greene. Regulatory pathways governing HIV-1 replication. Cell. 58: 423-426, (1989).
14. Chang, D. D. and P. A. Sharp. Regulation by HIV rev depends upon recognition of splice sites. Cell. 59: 789-795, (1989).

15. Hidaka, M., J. Inoue, M. Yoshida and M. Seiki. Post-transcriptional regulator (rex) of HTLV-1 initiates expression of viral structural proteins but suppresses expression of regulatory proteins. Embo J. 7: 519-523, (1988).
16. Seiki, M., J. Inoue, M. Hidaka and M. Yoshida. Two cis-acting elements responsible for posttranscriptional trans-regulation of gene expression of human T-cell leukemia virus type I. Proc Natl Acad Sci U S A. 85: 7124-7128, (1988).
17. Lewis, N., J. Williams, D. Rekosh and M.-L. Hammarskjöld. Identification of a cis-Acting element in HIV-2 that Is responsive to the HIV-1 rev and HTLV-I and II rex proteins. J. Virol. 64: 1690-1697 (1990).
18. Rimsky, L., J. Hauber, M. Dukovich, M. H. Malim, A. Langlois, B. R. Cullen and W. C. Greene. Functional replacement of the HIV-1 rev protein by the HTLV-1 rex protein. Nature. 335: 738-740, (1988).
19. Ahmed, Y. F., S. M. Hanly, M. H. Malim, B. R. Cullen and W. C. Greene. Structure-function analyses of the HTLV-I rex and HIV-1 rev RNA response elements: insights into the mechanism of rex and rev action. Genes Dev. 4: 1014-1022, (1990).
20. Hanly, S. M., L. T. Rimsky, M. H. Malim, J. H. Kim, J. Hauber, D. M. Duc, S. Y. Le, J. V. Maizel, B. R. Cullen and W. C. Greene. Comparative analysis of the HTLV-I rex and HIV-1 rev trans-regulatory proteins and their RNA response elements. Genes Dev. 3: 1534-1544, (1989).
21. Felber, B.K., D. Derse, A Athanassopoulos, M. Campbell and G.N. Pavalakis. Cross-activation of the rex proteins of HTLV-I and BLV and of the rev protein of HIV-1 and nonreciprocal interactions with their RNA Responsive Elements. New Biologist 1: 318-330
22. Smith, A. J., M. I. Cho, M. L. Hammarskjöld and D. Rekosh. Human immunodeficiency virus type 1 Pr55gag and Pr160gag-pol expressed from a simian virus 40 late replacement vector are efficiently processed and assembled into viruslike particles. J Virol. 64: 2743-2750, (1990).
23. Lu, X., J. Heimer, D. Rekosh and M.-L. Hammarskjöld. U1 snRNA plays a direct role in the formation of a rev regulated HIV env mRNA that Remains Unspliced. Proc Natl Acad Sci U S A. 87: 7598-7602 (1990).
24. Rekosh, D., A. Nygren, P. Flodby, M. L. Hammarskjöld and H. Wigzell. Coexpression of human immunodeficiency virus envelope proteins and tat from a single simian virus 40 late replacement vector. . Proc Natl Acad Sci U S A. 85: 334-338, (1988).
25. Zhuang, Y. and A. M. Weiner. A compensatory base change in U1 snRNA suppresses a 5' splice site mutation. Cell. 46: 827-835, (1986).
26. Siliciano, P. G. and C. Guthrie. 5' splice site selection in yeast: genetic alterations in base-pairing with U1 reveal additional requirements. Genes Dev. 2: 1258-1267, (1988).
27. Seraphin, B., L. Kretzner and M. Rosbash. A U1 snRNA:pre-mRNA base pairing interaction is required early in yeast spliceosome assembly but does not uniquely define the 5' cleavage site. Embo J. 7: 2533-2538, (1988).

Molecular Functional Studies of HIV-1 REV and NEF Proteins

Sundararajan Venkatesan[1], Steven M. Holland[1], Nafees Ahmad[1], Paul Wingfield[3], Ratan K. Maitra[1], and H. Clifford Lane[2]

[1]Laboratory of Molecular Microbiology, [2]Laboratory of Immunoregulation, National Institute of Allergy and Infectious Diseases, and [2]Protein Expression Laboratory, National Institutes of Health, Bethesda, MD 20892.

INTRODUCTION

Human immunodeficiency virus type 1 (HIV-1), the etiological agent of AIDS, preferentially infects the CD4+ helper subset of human T lymphocytes. Clinically, HIV-1 infection is characterized by a chronic phase lasting several years with a paucity of infected lymphocytes in the circulation. Notwithstanding, HIV-1 infection of primary human T lymphocytes or CD4+ cell lines in vitro leads to massive acute infection and cell death (20). This dichotomy between the natural history of virus infection and its behavior in tissue culture implies the existence of viral and cellular determinants of viral latency and reactivation.

Early in the life-cycle of HIV-1, three small regulatory proteins, TAT, REV and NEF are expressed by a class of multiply spliced viral mRNAs (46, 56). Both TAT (5, 15, 22) and REV (13, 21, 59) are small nuclear proteins which are essential for virus replication and mediate their effects by interacting with viral mRNA targets. TAT specifically increases the steady state levels and the rates of initiation of HIV-1 LTR RNAs (12, 47, 53) containing a TAT responsive stem-loop sequence (TAR) between positions +19 and +43 of the mRNA (7, 12, 37). Although the optimal transcription from the HIV LTR is governed by the concerted interactions of cis elements in the HIV LTR with cellular transcriptional factors such as TATAA, SP1 and NFκB factors (24, 33, 37, 47, 49, 57, 58), TAT can activate a minimally configured LTR in concert with other unidentified cellular factors (8, 17, 60).

REV acts post-transcriptionally to modulate the extra-nuclear transport of unspliced or partially spliced viral mRNAs (18, 27, 41, 42). This occurs through interaction with a highly structured RNA sequence (14, 16, 31, 41, 52, 66) within the env ORF, referred to as the REV responsive element (RRE). Since RNA-protein interactions are fundamental to many biological processes (6, 38, 45, 54), study of the roles of primary and secondary structure of RNA in the REV/RRE interaction may serve as a paradigm for the mechanisms mRNA processing and transport, and in turn facilitate therapeutic strategies to interfere with HIV-1 replication.

In contrast to the TAT and REV proteins, NEF is a 27 kDal membrane-associated myristoylated protein (4, 23) which is dispensible for viral replication. Consistent with this observation, NEF proteins of HIV-1 isolates exhibit significant divergence at the amino acid level and in some cases, the nef ORF is prematurely terminated (48). Although there is a consensus for the positive regulatory functions of TAT and REV, the situation regarding NEF function is not as clear. In particular, there have been disagreements (28, 35) about whether NEF (negative factor) represses HIV-1 replication and LTR transcription, as its name implies (2, 3, 39, 51, 61). In view of the long period of latency before the onset of HIV-1 disease in humans (20, 32) and the restricted tropism of HIV (36), the discovery of a virus coded negative factor suggests that NEF may be one of the determinants of viral latency in vivo.

RRE-REV Interactions:

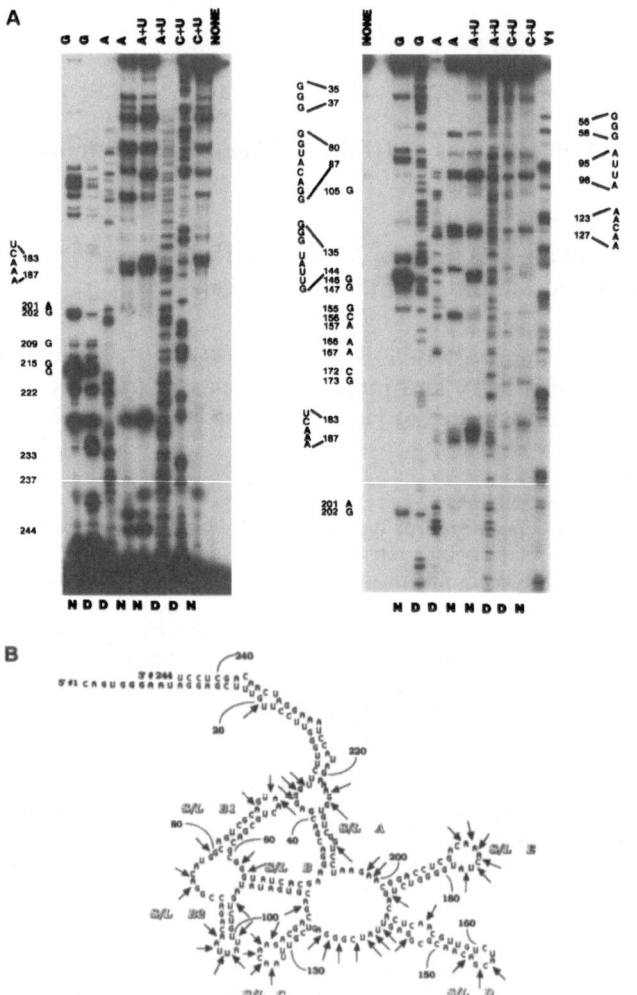

Figure 1. **1 A)** Partial ribonuclease digestions of RRE RNA under native and denatured conditions (31). The 244 bp RRE sequence in the proviral plasmid, pNL432 (positions 7749-7992) was amplified by the PCR and the amplified RRE fragment was cloned into the lone AspI site of the HIV-1 GAG reporter plasmid, pRMK4 (constructed by excising the sequence between the Asp I sites at positions 3823 and 9003 of the infectious provirus, pNL4-3; 1). The cloned RRE fragment was then amplified with primers tagged with the bacteriophage T7 RNA polymerase promoter. RRE RNAs were 3' end labeled with [^{32}P]pCP using T4 RNA ligase and purified by gel-filtration. Purity was checked by electrophoresis on urea/acrylamide (10%) gels. The sequence of the end-labeled RNAs was determined from partial digestions with base specific ribonucleases T1 (lanes G), U2 (lanes A), Phy M (A+U), B. cereus (C+U) under denaturing conditions. The secondary structure of the RRE RNA was determined by partial nuclease digestions using the above enzymes and the double-strand specific V1 RNAse at neutral pH in the presence of 0.2 M NaCl (lanes N) or under denatured conditions in 6M urea at pH 3.5 (lanes D). RNAse digestion products were resolved by urea/acrylamide (8 %) sequencing gels. **1 B)** The optimal computer predicted RRE structure is displayed (68). Arrows denote the observed cleavages in the native structure.

The RRE sequence is not highly conserved, but RREs from different primate immunodeficiency viruses can be functionally interchanged to a limited extent (43). The HIV-1 RRE can also be trans-activated in a non-reciprocal manner by the REX protein of the unrelated human retrovirus, HTLV-1 (19, 29). The Zuker RNA sub-optimal folding program (63, 67, 68) predicts a preferred secondary structure for RRE consisting of four distinct stem-loops (A, C, D, and E) and one branched stem-loop (B/B1/B2) surrounding a central single-stranded bubble (41, Fig. 1 B). To determine whether the RRE RNA assumes the computer predicted secondary structure in solution, 3' labeled in vitro synthesized RRE RNA was subjected to base-specific ribonuclease cleavages under native and denaturing conditions. The RNAse digests were resolved by urea-polyacrylamide gels. As illustrated by Fig. 1 A, the partial nuclease digestion profile of denatured RNA matched unambiguously with the primary structure of the RNA. Under native conditions of RNAse hydrolysis, the double stranded regions of the RRE RNA were resistant to nucleases. Except for minor discordances for single bases in stem loops A and B1, the computer predicted structure was confirmed (Fig. 1B).

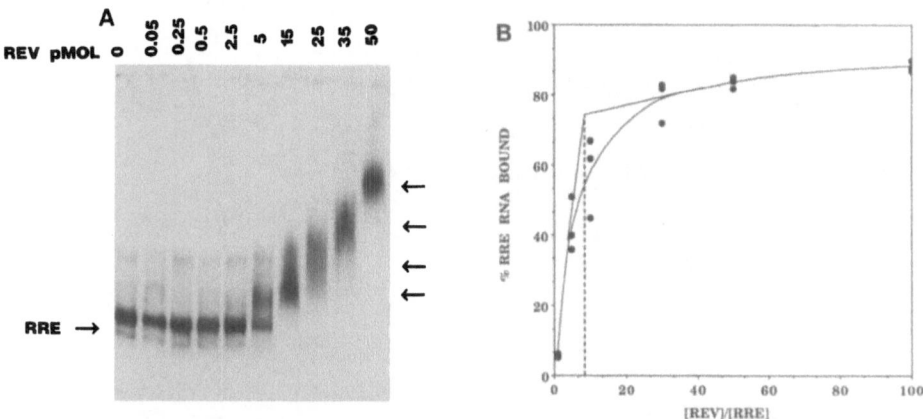

Figure 2. A) Titration of REV in RRE RNA gel retardation assay. Uniformly labeled RRE RNAs (0.5 pmol) were incubated for 10 min at 25°C with increasing amounts of purified HIV-1 REV protein in 10 µl of binding buffer (20 mM TRIS-HCl, pH 7.5, 0.05 M KCl, 1 mM DTT, 5 mM spermidine, 20 µg BSA with 40 µg carrier tRNA). Samples were electrophoresed through 4% polyacrylamide gels under native conditions, in 18.5% TBE buffer for 1-2 hr at 4°C at 350 V. The radioactive bands were visualized by dry-gel autoradiography. B) Filter-binding assays were done as described (31). Typically, 20-50 fmol of the labeled RRE was incubated at 25°C for 10 min in 10 µl of binding buffer with 1 µg of yeast tRNA and increasing concentrations of purified REV. The samples were filtered through pre-wetted 25 mm 0.45 µ nitrocellulose filters. Filters were rinsed twice with 0.5 ml of binding buffer (without BSA, spermidine or tRNA), air dried and the bound radioactivity determined. From the saturation curves of RRE/REV binding, the fractions of free and bound REV were computed and used to determine the average K_d of the wt RRE/REV interaction to be on the order of ~5X10^{-9} M. The RRE/REV ratio at the saturation break-point was used to determine the stoichiometry of the reaction.

RRE RNA has been shown to bind directly to purified REV expressed in E. coli (11, 14, 66). To determine the parameters of RRE/REV binding, we purified large quantities of REV from E. coli expressing REV from a thermoinducible bateriophage promoter system. REV was also expressed in eukaryotic cells infected with a vaccinia virus recombinant expressing REV from a bacteriophage T7 promoter. The E. coli expressed REV migrated as a single species of 19 kDal under denaturing conditions. At neutral pH, REV oligomerized to a stable tetramer at a concentration of ~1 mg/ml. When REV/RRE RNP formation was evaluated by RNA gel-retardation, a prominent gel-shift of RNA was observed at a ten-fold molar excess of REV over RRE (Fig. 2 A). With increasing amounts of REV, there was a stepwise increment in the size of the gel-retarded species with the maximally retarded species corresponding to an RNP species containing between four and eight molecules of REV (Fig. 2A). Nitro-cellulose filter binding was used to generate saturation curves of RRE/REV binding, and the fractional values for the amounts of free and bound REV at the saturation break-point were used to determine the average K_d of the wt RRE/REV interaction to be on the order of $\sim 5 \times 10^{-9}$ M (Fig. 2 B). Anti-sense RRE RNA

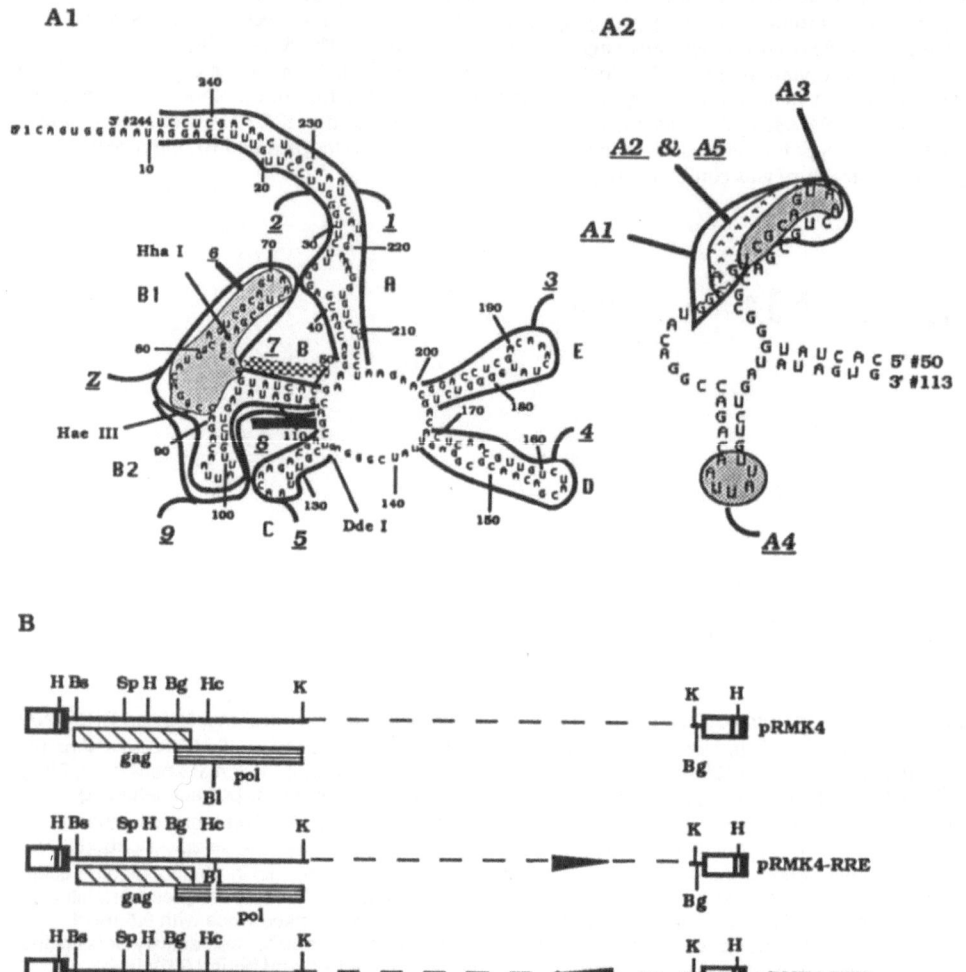

Figure 3. **A)** Diagram of the RRE deletions used in the study. The RRE mutants were generated by a two-step PCR amplification using mutant primer pairs. The mutant PCR fragments were inserted into the GAG reporter plasmid, pRMK4, and sequenced. The plasmids were used for transfection and to generate T7 promoter tagged RRE fragments. **A1:** Deletions designed to disrupt or excise individual stem loops. The numbered 244 nt RRE is in the predicted secondary structure. The structural dimensions of the individual deletions (_1_ through _7_) are denoted by the respective Bezèier curves wrapping the RRE structure. Mutants that removed more than one stem-loop (e.g. RRE34 lacking stem-loops D and E) or excised stem-loop A (RRE12) are not shown. The stem-loops are labeled **A-E** respectively. The HaeIII, HhaI and DdeI restriction sites are shown. **A2:** Map coordinates of the more circumscribed deletions in the branched stem-loop. Only the branched stem loop structure is shown here. **B)** Schematic diagram of the HIV-1 LTR linked GAG expression vectors, pRMK4 and pRMK4 containing RRE in the sense (pRMK4-RRE) and anti-sense (pRMK4-ERR) orientation (31).

(ERR) and HIV-1 TAR RNA were used as negative controls. Under the conditions used for filter-binding, less than 5% of these two RNAs was bound by REV even at a 200 fold molar excess. Kinetic analysis of binding yielded a stoichiometry of 8-10:1 for REV/RRE interaction.

To identify the minimal functional elements of RRE, we evaluated mutant RREs for REV binding in vitro and REV response in vivo in the context of a GAG expression plasmid. pRMK4 was constructed by excising the sequence between the Asp I sites at positions 3823 and 9003 of the infectious proviral plasmid, pNL4-3 (1). Mutations were engineered to specifically remove the individual stems and/or loops (designated A through E in Fig 3 A). Deletions RRE1 through RREZ, designed to excise or disrupt individual stem/loops, are illustrated in Fig. 3 A. Fig. 3 B shows the coordinates of deletions (RREA1 through RREA5) limited to the branched stem-loop, B/B1/B2. Every mutant RRE used in this study was analyzed using the sub-optimal RNA folding algorithm of Zuker (63, 67, 68). The structures of selected mutants were additionally verified by RNAse digestions. Mutant RRE RNAs were evaluated for RNP formation by gel mobility retardation and filter-binding experiments. Mutants were also transferred to a GAG reporter plasmid to measure their REV response in vivo. Individual GAG plasmids were co-electroporated with pHIV-CAT into HeLa cell lines expressing both TAT and REV, and transient GAG expression was quantitated by p24 ELISA of cell extracts and clarified culture fluids.

RRE deletion mutants which lacked one or more stem-loops (RRE1 through RREZ, Fig. 3 A) could be grouped into three sets according to REV binding results: those that bound REV almost as well as wt RRE, those that did not bind, and a group exhibiting intermediate levels of activity. Removal of stem-loops C, D, or E (RRE5, RRE4, RRE3) only resulted in marginal reduction in REV RNP formation by filter-binding or gel-retardation. These mutants also competed efficiently with wt RRE for REV binding (not shown). REV dependent GAG expression from this group of mutant RRE plasmids was slightly less than that of wt RRE (Table 1). Although the separate excision of the three 3' RRE loops was well tolerated, their combined excision (e.g. stem-loops D & E, RRE34) resulted in greater than 50% reduction in binding and in vivo response (Table 1). In contrast to the above mutants, deletions that eliminated the entire branched stem-loop B/B1/B2 (RREZ), deleted the B1 branch (RRE6), or which disrupted stem B (RRE7), were severely impaired for in vitro REV binding and were not responsive to REV in vivo (Table 1). These mutated RRE elements failed to compete with wt RRE for REV binding even at a 20 fold molar excess. A third set of deletions was intermediate in REV binding. Deletions which excised the 3' strand of stem B (RRE8) or stem-loop B2 and the 3' strand of stem B (RRE9) were ~20% as efficient as wt RRE in RNP formation and in vivo REV response (Table 1). REV recognition of mutants that had lost either the 3' strand of stem-loop A (RRE1) or both strands of stem-loop A (RRE12) was more severely impaired and could not be readily visualized by gel-retardation. Stem A mutants RRE1 and RRE12 were also negative for REV dependent GAG expression (Table 1). These gel retardation, filter binding and GAG expression results suggested that the minimal functional unit of RRE should include the stem loop A and the branched stem-loop B/B1/B2.

To identify the putative REV recognition elements in the B1 and B2 branches of stem-loop B, we studied the behavior of five discrete deletions in these regions (Fig. 3 B and Table 1). Excision of the small loop at the end of B2 (RREA4) reduced in vitro filter-binding only slightly. In vivo, the REV response of a plasmid containing the RREA4 mutation was only 20% compared to wt (Table 1). The REV binding of the B1 stem-loop mutants (RREA1, RREA2, RREA3 and RREA5) was even more impaired; they behaved like the "intermediate" group of RRE mutants in filter-binding assays. The two remaining B1 deletion mutants (RREA2 and RREA5), bound REV poorly at low protein concentrations; GAG expression from all four B1 stem-loop lesions was significantly reduced (Table 1).

Our studies show that REV binding to the RRE RNA is a prerequisite for REV function and that a minimal unit of the RRE composed of the branched stem-loop B/B1/B2 is sufficient for REV binding. Although stem-loops C, D and E could be singly excised from the full RRE molecule without severely affecting either binding or activation, 3' truncated RREs which retained the branched stem-loop B/B1/B2 yielded different results. The 5' 132 nt RRE fragment (RREDDE) bound REV more avidly than the wt RRE, but was not REV responsive in vivo. Except for a modified stem A, RREDDE folded into a structure similar to that of a biologically active mutant that had deleted the 3' stem-loops C, D and E (16, 44, 52). The 5' 58 nt of RRE (RREHHA) lacked the B/B1/B2 stem-loop and was not recognized by REV in vitro. When the modified stem A was excised from RREDDE, REV binding was significantly reduced. Therefore, the presence of stem A, even in a modified form as in RREDDE, appeared to enhance the REV binding of the branched stem-loop B/B1/B2.

TABLE 1

RRE Mutants [a]	Sequence Coordinates	Stem-loop Affected[b]	In vitro REV Binding [c]	p24 ELISA, (% of wt RRE [SEM]) [d]
MOCK	-	-	−	2 (0.4)
wt RRE	1-244	NA[e]	++++	100
ERR, Reverse RRE	244-1	NA	−	3 (1.8)
Major Deletions				
RRE1	Δ207-244	A	+/−	3 (1.2)
RRE2	Δ11-44	A	+	2 (0.8)
RRE12	Δ11-44+Δ207-244	A	+/−	1.5 (1.1)
RRE3	Δ174-199	E	++++	80 (18)
RRE4	Δ144-170	D	++++	72 (9)
RRE5	Δ117-134	C	++++	90 (8)
RRE34	Δ144-170+Δ174-199	D,E	++	35 (13)
RRE6	Δ57-89	B1	−	2 (1.8)
RRE7-ACG	Δ47-58	B	−	3 (1.2)
RRE8	Δ107-113	B	+++	22 (7)
RRE9	Δ90-114	B	++	15 (6)
RRE Z	Δ49-113	B/B1/B2	−	2 (1.4)
RRE-B/B1/B2	Δ1-49+Δ117-244	A,C,D,E	++	ND[f]
RRE-DDE	Δ132-244	A(3'),D,E	++++	1 (1.1)
RRE2-DDE	Δ11-44+Δ132-244	A,D,E	+/−	2 (0.3)
RRE-HAE	Δ87-244	A(3'),B2,C,D,E	+	2 (0.4)
RRE-HHA	Δ61-244	A(3'),B1,B2,C,D,E	−	1 (0.2)
B/B1/B2 Deletions				
RRE A1-ACG	Δ65-81	B1	++	13 (7)
RRE A2-ACG	Δ72-79	B1	++	15 (8)
RRE A3-ACG	Δ68-76	B1	++	15 (8)
RRE A5-ACG	Δ71-78	B1	++	14 (5)
RRE A4	Δ94-100	B2	+++	20 (10)

[a] All RRE sequences are 5'(62)AGC(64)3' except where noted as ACG.

[b] The stem-loops of the RRE structure (Fig. 1), disrupted by the deletions are identified.

[c] The results of RRE RNA filter-binding and gel mobility shifts are symbolically denoted.

[d] The normalized p24 GAG ELISA values represent averages from six or eight experiments with the standard error of the mean (SEM) shown in parentheses.

[e] NA: Not Applicable

[f] ND: Not done.

Deletion of the B2 stem-loop and the 3' strand of stem B (RRE9), did not completely eliminate REV binding or in vivo REV response. Excision of the loop at the end of B2 (RREA4) only marginally affected filter-binding to REV, although it reduced the in vivo response significantly. Therefore, the B2 stem-loop may not play a role in the initial steps of REV binding, but may stabilize the RNP once it is formed and be crucial for the in vivo REV response. In contrast, the B1 stem-loop appears to contain important recognition sequences. The RRE6 mutant which deleted stem-loop B1, resulted in the loss of REV response. However, base substitutions of stem-loops A, B, B1 & B2 have been reported to preserve the REV response as long as the structure of the individual stem-loops was maintained (16, 44, 52). Therefore, it has been suggested that this region is recognized on the basis of the secondary structure (52). Deletions which disrupted the predicted B1 stem (RREA1, RREA2, RREA3 & RREA5) markedly reduced but did not eliminate REV binding. The lack of a completely negative phenotype of the "structure-disrupting" B1 stem-loop mutants suggests that this region may not be recognized exclusively on the basis of secondary structure.

In contrast to the above mutants, a 12 nt deletion of the main stem B (RRE7, positions 47-58) gave a negative phenotype. The sequences absent in this mutant normally base-pair with the sequences at 107-113 in the wt RRE. In the mutant RRE8, REV binding as well as the in vivo response to REV were preserved despite the elimination of nt 107-113 and consequent disruption of secondary structure. Thus, the sequences corresponding to the RRE7 deletion may be recognized predominantly on the basis of their primary structure. Indeed, the 12 nt between nt 47-58 may not even have to be base-paired for REV binding. In support of this, the 89 nt RREHAE RNA bound REV, albeit inefficiently, despite lacking the downstream sequence that could have base-paired with the bases 49-55 (Fig. 3 A). Further, the Zuker program predicted preferred RNA structures for the partially responsive mutants RRE8 and RRE9 which showed the critical 12 nt between 47-58 to be unpaired. Recently, REV binding has been shown to specifically shield a 33 nt RRE RNA fragment, including the above 12 nt, from nuclease digestion (30). In this region of RRE, a 12 nt 5'..CACUAUGGGCGC..3' sequence is conserved in all HIV-1 isolates except for the substitution of G for U at position 4. Although this may reflect the strong codon conservation in this region of

the env ORF of HIV-1, a similar but somewhat degenerate 5'..^U_GGC^A_GAUGGG..3' sequence is also found in the RREs of HIV-2 and SIV isolates which use a different codon scheme in this region. Computer predicted RNA folding of all the HIV-1 RREs showed the 5'..CACUAUG..3' of the HIV-1 RRE sequence to be base-paired to equally conserved bases 3'..CGUGAUA..5' between positions 113-107. The latter sequence is also conserved in the HIV-2 and SIV RREs, which are responsive to HIV-1 REV. Unlike HIV-1, the above two 7 nt elements are not base-paired with each other in the HIV-2 and SIV RREs.

On the basis of the above findings, we propose that sequence-specific recognition of nucleotides 47-58 represents a critical step in RRE/REV interaction. The magnitude of this interaction is enhanced if this target sequence is base-paired. REV probably recognizes additional elements in the neighboring stem-loops. Sequences located in stem-loop A and B1 may be somewhat degenerate and not be readily identified by in vitro studies but could play important roles in vivo. Alternatively, stem-loops A, B1 and B2 may simply provide the secondary structural context for optimal REV recognition both in vitro and in vivo.

The inability of the 5' 132 bp of RRE, RREDDE, to function in vivo implies separate binding and activation domains within the RRE. Since REV function may require spliceosome formation (9), the missing component(s) of the REV/RRE functional unit may be snRNPs or other host nuclear factors. RREDDE RNA may lack the target sequences for these putative factors or they may not be correctly exposed after complexation with REV. It is possible that REV functions in vivo as a tetramer, and the oligomeric form of REV may not recognize RREDDE optimally in vivo. Two important functional domains of REV, namely, the nuclear localization domain and the activation domain have been tentatively identified. Certain of the mutants in the latter activation domain will competitively inhibit the function of wt REV protein (42). RREDDE may lack the contact points for the REV activation domain and the associated nuclear factors.

Functional Studies of HIV-1 NEF protein

The potential virological and cell-biological role(s) of HIV-1 NEF protein have become enmeshed in controversy. As mentioned before, the 27 kDal NEF protein is dispensable for replication. Early genetic studies showed that nef mutant viruses had better replication potential than NEF$^+$ viruses, suggesting that NEF may be negative factor (39, 61). We and others (2, 3, 51) have shown that NEF represses viral replication through a transcriptional inhibition of the HIV-1 LTR. A cis element at the 5' end of the LTR, referred to as the negative regulatory element (NRE, 55) appeared to be the target for these inhibitory effects (2). However, it has been suggested that artifacts of transfection may have contaminated previous studies and that NEF may have no effect on LTR transcription or viral replication (28, 35).

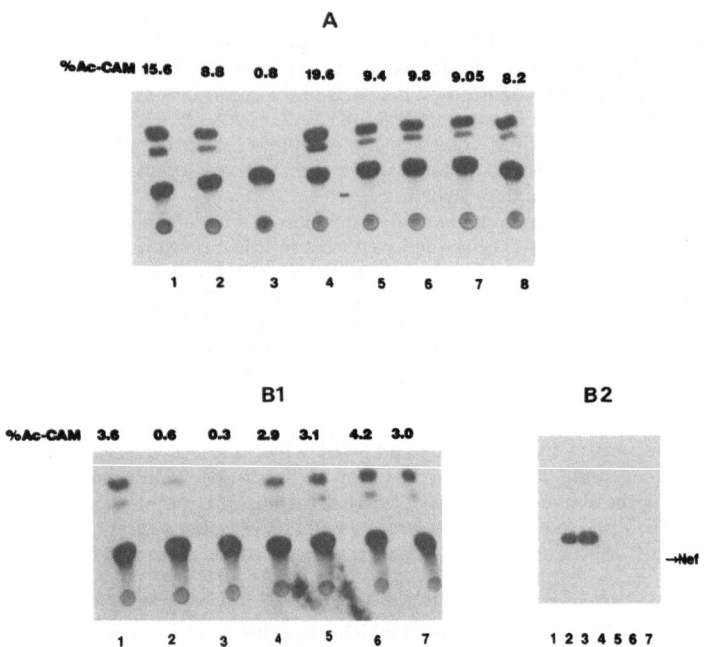

Figure 4. HIV-1 NEF mediated repression of basal HIV-1 LTR transcription. **A.** COS-1 cells were transfected with pHIV-CAT (lane 1); pHIV-CAT and 15 µg of pHIV-NEF in the sense orientation (lane 2); pHIV-CAT and 15 µg of pCMV-NEF in the sense orientation (lane 3); pHIV-CAT and 15 µg of pCMV-NEF in the anti-sense orientation (lane 4); pHIV-CAT with 10 or 15 µg of pAH8 (lanes 5 and 6); and pHIV-CAT with 10 or 15 µg of pAH19 (lanes 7 and 8). 5 µg of pHIV-CAT was used for each transfection.**B1.** Results from HeLa cells transfections with 5 µg of pHIV-CAT alone (lane 1); pHIV-CAT with 10 or 15 µg of pCMV-NEF (lanes 2-3); pHIV-CAT with 10 or 15 µg of pCMV-NEFXho (lanes 4-5); and pHIV-CAT with 10 or 15 µg of pHIV-NEFXho (lanes 6-7) are shown. Cell extracts were assayed for CAT activity. The CAT values are presented as %Ac-CAM and did not vary by more than 10%. 5 µg of pSV-βGAL was also cotransfected to monitor transfection efficiency in all the experiments. **B2.** NEF expression in the experiments of **B1** was examined by immunoblotting using a rabbit anti-NEF antiserum.

NEF is partly encoded by U3 sequences in the 3' HIV-1 LTR. It has been suggested that these U3 sequence elements in the nef ORF of the expression vectors, rather than NEF protein *per se*, may have competed for limiting transcriptional factors and thereby reduced LTR transcription (28, 35). Therefore, we used the same U3 sequences in a negative control plasmid in an anti-sense orientation with respect to the CMV promoter. In other experiments, COS-1 cells were co-transfected with an excess of LTR plasmids containing U3/R sequences (between nts. -158/+80 of HIV-1 LTR) in the sense (pAH8) or antisense orientation (pAH19) and pHIV-CAT. With pHIV-CAT and pCMV-NEF cotransfection, there was an eighteen-fold inhibition of basal LTR transcription (Fig. 4 A, lane 3). However, less than two-fold inhibition was obtained when the AvaI/HindIII fragment was co-transfected with pHIV-CAT (lanes 5-8). Co-transfection of pHIV-CAT with pHIV-NEF caused less than two fold repression of LTR transcription (Fig. 4A, lane 2). Since pHIV-NEF had complete copies of the LTR at both the 5' and 3' ends but did not express detectable NEF in the absence of TAT, some of the inhibition observed with the pHIV-NEF/pHIV-CAT co-transfection may have been due to DNA competition (40).

When HeLa cells were cotransfected with pHIV-CAT and increasing amounts of pCMV-NEF, there was a dose-dependent inhibition of LTR transcription correlating with an increase in the expression of NEF (Fig. 4 B1 and B2). To correlate NEF expression with the transcriptional repression of the HIV-1 LTR, we used the mutated NEF expression vectors, pCMV-NEFXho and pHIV-NEFXho. The nef ORF in these vectors has been ablated by a 4 bp insertion at the XhoI site upstream of the U3 domain of the 3' LTR, leading to a frame-shifted termination of NEF. With the pCMV-NEFXho (Fig. 4, B1 and B2, lanes 4-5), or pHIV-NEFXho (B1 and B2, lanes 6-7) co-transfections, there was neither NEF expression nor LTR transcriptional inhibition (40).

NEF expressing Jurkat T lymphoid cells have been used to examine the effects of NEF on HIV provirus expression and on the replication of natural and recombinant HIV-1 viruses (10). Infectivity studies with the NEF+ cells have demonstrated a gradient of repressive effects on different natural HIV-1 isolates. Strains isolated early in disease have been more susceptible to repression in the NEF+ cells than those later in disease. Long-term infection with various HIV-1 isolates was measured but the role of NEF during single cycle infection or in the context of LTR transcription alone was not addressed (10).

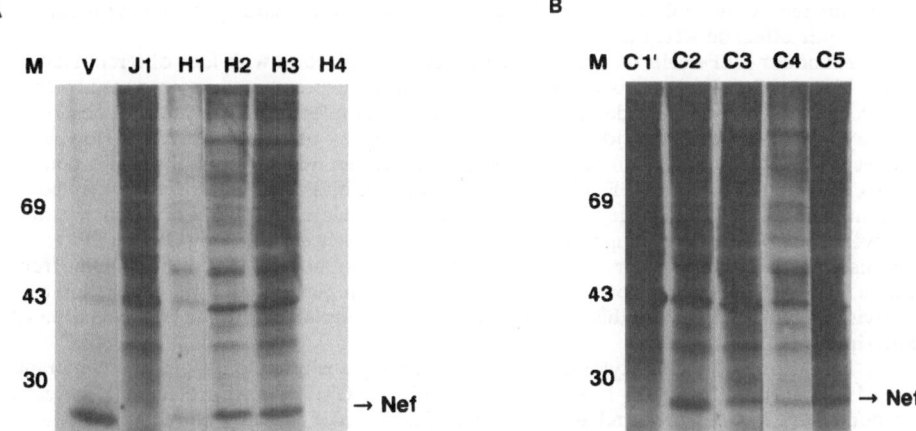

Figure 5.   Immune-detection of HIV-1 NEF in Jurkat T lymphocyte clones expressing NEF. Cells carrying a CMV promoter linked or an HIV-1 LTR linked nef were isolated by co-selection for neomycin resistance and expression of nef transcripts. Cells expressing nef related RNAs were labeled with [^{35}S] methionine and [^{3}H] leucine and cell extracts of these cells were immunoprecipitated with rabbit hyper-immune serum raised against purified NEF expressed in E. coli (2, 3) and resolved by SDS/PAGE. **A.** Jurkat cells expressing NEF from an HIV-1 LTR linked plasmid (2, 40). Lanes H1 through H4 are individual cell lines. J1 refers to the parental J1D1 Jurkat cells. Result in lane V is from Jurkat cells acutely infected with HIV-1. **B.** Lanes C2 through C5 are Jurkat cells expressing NEF from a CMV promoter (pCMV-NEF; 2). A plasmid containing NEF coding sequences in the anti-sense polarity (pCMV-NEF-anti) with respect to the CMV promoter is shown in lane C1'.

HeLa and Jurkat cell lines carrying a CMV promoter linked or an HIV-1 LTR linked nef were co-selected for neomycin resistance and NEF expression. With the exception of the anti-sense NEF integrants, all NEF cell lines expressed the 27 kDal NEF protein, detectable by immunoprecipitation. Four Jurkat pHIV-NEF clones (Fig. 5A, H1-H4) and four pCMV-NEF clones (Fig. 5B, C2-C5) and two HeLa NEF+ cell lines were selected in this manner. NEF expression has been presumed to modulate membrane signalling, the expression of receptors such as CD4 ,or some transcriptional factors which recognize sequences in the HIV-1 LTR (25, 26). However, in agreement with a recent report (10) we did not observe a correlation between CD4 downregulation and NEF expression (40).

We evaluated the magnitudes of LTR transcription under basal conditions or in the presence of TAT in the NEF+ and NEF- cell lines. Up to ten-fold repression of TAT activated HIV-1 LTR transcription was demonstrated in NEF+ cell lines. Basal LTR transcription was repressed to a lesser extent in both the the NEF+ HeLa and Jurkat cell lines. This may have reflected the generally poor levels of basal LTR transcription in the uninduced lymphocytes and HeLa cells. The magnitudes of LTR repression observed in the T lymphoid NEF+ and HeLa NEF+ cell lines were similar to those obtained with co-transfections using exogenous NEF plasmid.

The cloned NEF+ Jurkat T cells were moderately resistant to HIV-1 provirus expression under transient conditions and some of them were severely restricted for virus expression. Under transient expression conditions, peak virus production from NEF+ cells was delayed by 5-10 days, compared to NEF- cell lines. In CD4+ cells, the ultimate virus yields following provirus transfection probably reflect contributions from both the provirus expression and reinfection. Although we cannot completely exclude a block at the level of re-infection in NEF+, it is unlikely since the CD4 receptor was not down-modulated significantly in the NEF+ cells. It is possible that adverse physiological effects induced by NEF may have resulted in a pre-integration arrest following re-infection. Such a mechanism of retroviral arrest seems to be a property associated with resting T lymphocytes (65), but may not be operative in continuous cell lines. Within the kinetic window of our experiments, the cell doubling times and rates of DNA synthesis were similar in the NEF+ and the NEF- cells. Therefore, in NEF+ Jurkat cells, the reduced virus yields from provirus transfections can be correlated with repression of LTR transcription. With HeLa NEF+ cells, virus production from the transfected proviruses was also reduced to a modest extent (four-fold), but there was no delay in virus release. Since HeLa cells cannot be reinfected with the progeny virus (unlike NEF+ CD4+ cells), the reduction in virus production probably reflects a NEF effect on LTR transcription.

All of our NEF+ cells were somewhat resistant to infection with four different HIV-1 strains. None of the virus strains was totally arrested and the progeny virus produced by the NEF+ cells was somewhat retarded in subsequent rounds of infection. Over multiple observations, the delay in virus production and the decrease in the progeny titers were variable. However, in contrast to a previous report (10), we did not observe a differential effect on the replication of two isolates, SF2 and SF33 in our NEF+ cell lines. However, the NEF used in our studies may be different from that of Cheng-Meyer *et al* (10) and may regulate other viral or cellular pathways. However, it has been recently suggested that sequence differences in the env ORF of SF2 and SF33 proviruses may account for their different tropisms for T lymphocytes (64). NEF proteins from various HIV-1 isolates were shown recently to have pleiotropic effects on the replication of otherwise isogenic viruses, but this phenomenon was not correlated with the NEF effects on LTR transcription (62).

NEF is predominantly cytoplasmic and is probably anchored to the plasma membrane by its myristoyl group (23, 25, 26). It is unlikely that NEF is a GTP binding protein and, therefore may not behave like p21 ras or related G proteins (34, 50). NEF has a measurable auto-phosphorylation activity that uses purine ribo-nucleoside triphosphate substrates (25, 50). NEF proteins with a threonine at position 15 have a more potent auto-kinase activity than those with an alanine at this position such as was the case with the NEF used in these studies (25, 50). The threonine at position 15 of BRU NEF has also been presumed to be the site of protein kinase C mediated phosphorylation (26) and to influence the metabolic half-life of NEF proteins expressed in HIV-infected T4 lymphocytes (26). However, none of these putative properties of NEF allow straightforward deduction of NEF function or mechanism of action. The negative effect of NEF on HIV-1 LTR transcription is measurable under the controlled conditions we have developed and appears to be a reproducible assay for NEF function. The mechanisms of NEF mediated transcriptional repression are still unknown, but this approach should allow us to unravel the underlying biochemical mechanisms.

ACKNOWLEDGEMENTS

We thank Charles E. Buckler for help in computer analysis and Alicia Buckler-White for oligonucleotide synthesis. We also thank Angel Nebreda and Eugenio Santos of LMM, NIAID for sharing their unpublished results. This work was supported in part by the Intramural AIDS targeted Anti-Viral program administered by the Office of the Director, National Institutes of Health.

REFERENCES

1. Adachi, A., H. E. Gendelman, S. Koenig, T. Folks, R. Willey, A. Rabson, and M. A. Martin. (1986) Production of acquired immunodeficiency syndrome-associated retrovirus in human and nonhuman cells transfected with an infectious molecular clone. J. Virol. 59, 284-291.

2. Ahmad, N., and Venkatesan, S. (1988) Nef Protein of HIV-1 is a transcriptional repressor of HIV-1 LTR. Science 241, 1481-1485.

3. Ahmad, N., Maitra, R. K., and Venkatesan, S. (1989). HIV-1 Rev induced modulation of Nef protein underlies temporal regulation of viral replication. Proc. Natl. Acad. Sci. (USA) 86, 6111-6115.

4. Allan, J. S., Coligan, J. E., Lee, T. H. , McLane, M. F., Kanki, P. J., Groopman, J. E., and Essex, M. (1985) A new HTLV-III/LAV encoded antigen detected by antibodies from AIDS patients. Science 230, 810-813.

5. Arya, S.K., Guo, C., Josephs, S.F., and Wong-Staal, F. (1985). Trans-activator gene of human T-lymphotropic virus type III (HTLV-III). Science 229, 69-73.

6. Bandziulis, R. J., M. S. Swanson, and G. Dreyfuss. (1989) RNA binding proteins as developmental regulators. Genes & Develop. 3, 431-437.

7. Berkhout, B., Silverman, R., and Jeang, K-T. (1989) Tat trans-activates human immunodeficiency virus through a nascent RNA target. Cell 57, 273-282.

8. Berkhout, B. R., Gatignol, A., Rabson, A. R., and Jeang, K.-T. (1990) TAR-independent activation of the HIV-1 LTR: Evidence that Tat requires specific regions of the promoter. Cell (In press).

9. Chang, D. D., and P. A. Sharp. (1989) Regulation by HIV Rev depends upon recognition of splice sites. Cell 59, 789-795.

10. Cheng-Mayer, C., Iannello, P., Shaw, K., Luciw, P. A., and Levy, J. A. (1989) Differential effects of nef on HIV replication: Implications for viral pathogenesis in the host. Science 246, 1629-1632.

11. Cochrane, A. W., C.-H. Chen, and C. A. Rosen. (1990) Specific interaction of the human immunodeficiency virus Rev protein with a structured region in the env mRNA. Proc. Natl. Acad. Sci. USA 87, 1198-1202.

12. Cullen, B. R. (1986) Trans-activation of human immunodeficiency virus occurs via a bimodal mechanism. Cell 46, 973-81.

13. Cullen, B. R., J. Hauber, K. Campbell, J. G. Sodroski, W. A. Haseltine, and C. A. Rosen. (1988) Subcellular localization of the human immunodeficiency virus trans-acting rev gene product. J Virol. 62, 2498-2501.

14. Daly, T. J., Cook, K. S., Gray,G. S., Maione, T. E., and Rusche, J. R. (1989) Specific binding of HIV-1 recombinant Rev protein to the Rev-responsive element in vitro. Nature (London) 342, 816-819.

15. Dayton, A.I., Sodroski, J.G., Rosen, C.A., Goh, W.C., and Haseltine, W.A. (1986). The trans-activator gene of the human T-cell lymphotropic virus type III is required for replication. Cell 44, 941-947.

16. Dayton, E. T., Powell,D. M., and Dayton, A. I. (1989) Functional analysis of CAR, the target sequence for the Rev protein of HIV-1. Science 246, 1625-1629.

17. Duh, E. J., Maury, W. J., Folks, T. M., Fauci, A. S., and Rabson, A. B. (1989) Tumor necrosis factor a activates human immunodeficiency virus type 1 through induction of nuclear factor binding to the NF-kB sites in the long terminal repeat. Proc. Natl. Acad. Sci. (USA) 86, 5974-5978.

18. Emerman, M., R. Vazeur, and K. Peden. (1989) The rev gene product of the human immunodeficiency virus affects envelope-specific RNA localization. Cell 57, 1155-1165.

19. Felber, B. K. (1989) Cross-activation of the Rex proteins of HTLV-1 and BLV and of the REV protein of HIV-1 and non-reciprocal interactions with their RNA responsive elements. The New biologist 1, 318-330.

20. Fauci, A. S. (1988) The human immunodeficiency virus : infectivity and mechanisms of pathogenesis. Science 239, 617-622.

21. Feinberg, M.B., Jarrett, R.F., Aldovini, A., Gallo, R.C., and Wong-Staal, F. (1986). HTLV-III gene expression and production involve complex regulation at the levels of splicing and translation of RNA. Cell 46, 807-817.

22. Fisher, A.G., Feinberg, M.B., Josephs, S.F., Harper, M.E.,Marselle, L.M., Reyes, G., Gonda, M.A., Aldovini, A., Debouk, C., Gallo, R.C., and Wong-Staal, F. (1986). The transactivator gene of HTLV-III is essential for virus replication. Nature 320, 367-371.

23. Franchini, G., Robert-Guroff, M., Ghrayeb, J., Chang, N. T., and Wong-Staal, F. (1986) Cytoplasmic localization of the HTLV III 3' orf protein in cultured T cells. Virology 155, 593-599.

24. Garcia, J. A., Wu, F. K., Mitsuyasu, R., and Gaynor, R. B. (1987) Interactions of cellular proteins involved in the transcriptional regulation of the human immunodeficiency virus. The EMBO J. 6, 3761-3770.

25. Guy, B., Kieny, M.P., Riviere, Y., LePeuch, C., Dott, K., Girard,M., Montagnier, L., and Lecocq, J-P. (1987). HIV F/3' ORF encodes a phosphorylated GTP-binding protein resembling an oncogene product. Nature 330, 266-269.

26. Guy, B., Riviere, Y., Dott, K., Regnault, A., and Kieny, M. P. (1990). Mutational analysis of the HIV-1 nef protein. Virology 176, 413-425.

27. Hadzoupoulu-Cladaras, M., Felber,B. K., Cladaras, C., Athanasopoulos, A., Tse,A., and Pavlakis, G. N. (1989) The rev (trs/art) protein of human immunodeficiency virus type I affects viral mRNA and protein expression via cis acting sequences in the env region. J. Virol. 63, 1265-1274.

28. Hammes, S. R., Dixon, E. P., Malim, M. M., Cullen, B. R., and Greene, W. C. (1989). Nef protein of human immunodeficiency virus type 1: Evidence against its role as a transcriptional inhibitor. Proc. Natl. Acad. Sci. (USA) 86, 9549-9553.

29. Hanly, S. M., L. T. Rimsky,M. H., Malim, J. H. Kim, J. Hauber, M. Duc Dodon, S.-Y. Le, J. V. Maizel, B. R. Cullen, and W. C. Greene. (1989) Comparative analysis of the HTLV-I Rex and HIV-1 Rev trans-regulatory proteins and their RNA response elements. Genes & Develop. 3, 1534-1544.

30. Heaphy, S., C. Dingwall, I. Ernberg, M. J. Gait, S. M. Green, J. Karn, A. D. Lowe, M. Singh, and M. A. Skinner. (1990) HIV-1 regulator of virion expression (Rev) binds to an RNA stem-loop structure located in the Rev response element region. Cell 60, 685-693.

31. Holland, S. M., Ahmad, N., Maitra, R. K., Wingfield, P., and Venkatesan, S. (1990) Human immunodeficiency virus REV protein recognizes a target sequence in the RRE RNA within the context of RNA secondary structure. J. Virol., 64, 5966-5975.

32. Hoxie, J. A., Haggarty, B. S., Rackowski, J. L., Pillsbury, N., Levy, J. A. (1985) Persistent noncytopathic infection of normal human T lymphocytes with AIDS-associated retrovirus. Science 229, 1400-1402.

33. Jones, K. A., Kadonaga, J. T., Luciw, P. A., and Tjian, R. (1986) Activation of the AIDS retrovirus promoter by the cellular transcription factor, Sp1. Science 232, 755-759.

34. Kaminchik, J., Bashan, N., Pinchasi, D., Amit, B., Sarver, N., Johnston, M. I., Fischer, M., Yavin, Z., Gorecki, M., and Panet, A. (1990) Expression and biochemical characterization of human immunodeficiency virus type 1 nef gene product. J. Virol. 64: 3447-3454.

35. Kim, S., Ikeuchi, K., Byrn, R., Groopman, J., and Baltimore, D. (1989). Lack of a negative influence on viral growth by the nef gene of human immunodeficiency virus type 1. Proc. Natl. Acad. Sci. (USA) 86, 9544-9548.

36. Klatzmann, D., F. Barre-Sinoussi, M. T. Nugeryre, C. Daugnet, E. Vilmer, C. Griscelli, F. Brun-Vezinet, C. Rouzious, J. C. Gluckman, J. C. Chermann, and L. Montagnier. (1984) Selective trophism of lymphadenopathy associated virus (LAV) for helper-inducer T-lymphocytes. Science 225 59-63.

37. Laspia, M. F., Rice, A. P., and Mathews, M. (1989) HIV-1 Tat protein increases transcriptional initiation and stabilizes elongation. Cell 59, 283-292.

38. Lazinski, D., E. Grzadzielska, and A. Das. (1989) Sequence-specific recognition of RNA hairpins by bacteriophage antitermination requires a conserved arginine-rich motif. Cell 59, 207-218.

39. Luciw, P.A., Cheng-Mayer, C., and Levy, J.A. (1987). Mutational analysis of the human immunodeficiency virus, The orf-B region down-regulates virus replication. Proc. Natl. Acad. Sci. USA 84, 1434-1438.

40. Maitra, R. K., Ahmad, N. McCoy, S., Holland, S. M., Lane, H. C., and Venkatesan, S. (1990) Expression of human immunodeficiency virus, type 1 (HIV-1) NEF protein causes repression of HIV-1 provirus expression and LTR transcription in the context of NEF+ cell lines. Manuscript submitted.

41. Malim, M. H., Hauber, J., Le, S.-Y., Maizel,J. V., and Cullen,B. R. (1989) The HIV-1 rev trans-activator acts through a structured target sequence to activate nuclear export of unspliced viral mRNA. Nature (London) 338, 254-257.

42. Malim, M. H., S. Bohnlein, J. Hauber, and B. R. Cullen. (1989) Functional dissection of the HIV-1 Rev trans-activator; derivation of a trans-dominant repressor of Rev function. Cell 58, 205-214.

43. Malim, M. H., S. Bohnlein, R. Fenrick, S.-Y., Le, J. V. Maizel, and B. R. Cullen. (1989) Functional comparison of the Rev trans-activators encoded by different primate immunodeficiency virus species. Proc. Natl. Acad. Sci. (USA) 6, 8222-8226.

44. Malim, M. H., Tiley,L. S., McCarn, D. F., Rusche, J. R., Hauber,J., and Cullen, B. R. (1990) HIV-1 structural gene expression requires binding of the Rev trans-activator to its RNA target sequence. Cell 60, 675-683.

45. Mattaj, I. W. (1989) A binding consensus: RNA-protein interactions in splicing, snRNPs, and sex. Cell 57, 1-3.

46. Muesing, M. A., Smith, D. H., Cabradilla, C. D., Benton, C. V., Lasky, L. A., and Capon, D. J. (1985) Nucleic acid structure and expression of the human AIDS/lymphadenopathy retrovirus. Nature 313, 430-458.

47. Muesing, M. A., Smith, D. H., and Capon, D. J. (1987) Regulation of mRNA accumulation by a human immunodeficiency virus trans-acivator protein. Cell 48, 691-701.

48. Myers, G., Berzofsky, J. A., Rabson, A. B., and Smith. T. F. (1990) Human Retroviruses and AIDS. Los Alamos National Laboratory (Pub.), Los Alamos, New Mexico.

49. Nabel, G., and Baltimore, D. (1987) An inducible expression factor activates expression of human immunodeficiency virus in T cells. Nature 326, 711-713.

50. Nebreda, A. R., Bryan, T., Segade, F., Wingfield, P., Venkatesan, S. and Santos. E. (1990) Human immunodeficiency virus type 1 NEF is an unlikely candidate for GTP binding or oncogenic transformation. (Manuscript submitted).

51. Niederman, T. M. J., Thielan, B. J., and Ratner, L. (1989) Human immunodeficiency virus type 1 negative factor is a transcriptional silencer. Proc. Natl. Acad. Sci. (USA) 86, 1128-1132.

52. Olsen, H. S., Nelbrook, P., Cochrane, A. W., and Rosen, C. A. (1990) Secondary structure is the major determinant for interaction of HIV rev protein with RNA. Science 247, 845-848.

53. Peterlin, B. M., Luciw, P. A., Barr, P. J., and Walker, M. D. (1986) Elevated levels of mRNA can account for the transactivation of human immunodeficiency virus (HIV). Proc. Natl. Acad. Sci. (USA) 83, 9734-9738.

54. Query, C. C., R. C. Bentley, and J. D. Keene. (1989) An RNA recognition motif identified within a defined U1 RNA binding domain of the 70K U1 snRNP protein. Cell 57, 89-101.

55. Rosen, C. A., Sodroski, J. G., and Haseltine, W. A. (1985) Location of cis-acting regulatory sequences in the human T cell lymphotropic virus type III (HTLV-III/LAV) long terminal repeat. Cell 41, 813-823.

56. Schwartz, S., Felber, B. K., Benko, D. M., Fenyo, E.-M., and Pavlakis, G. N. (1990) Cloning and functional analysis of multiply spliced mRNA species of HIV-1. J. Virol. 64, 2519-2529.

57. Shaw, J-P., Utz, P. J., Durand, D. B., Toole, J. J., Emmel, E. A., and Crabtree, G. R. (1988) Identification of a putative regulator of early T cell activation genes. Science 241, 202-205

58. Siekevitz, M., Josephs, S.F., Dukovich, M., Peffer, N., Wong-Staal, F., Greene, W.C. (1988). Activation of the HIV-1 LTR by T Cell Mitogens and the Trans-Activator Protein of HTLV-1. Science 238, 1575-1578.

59. Sodroski, J., Goh, W.C., Rosen, C., Dayton, A., Terwilliger, E.,and Haseltine, W.A. (1986). A second post-transcriptional transactivator gene required for HTLV-III replication. Nature 321, 412-417.

60. Southgate, C., Zapp, M. L., and Green, M. R. (1990). Activation of transcription by HIV-1 Tat protein tethered to nascent RNA through another protein. Nature 345, 640-642.

61. Terwilliger, E., Sodroski, J.G., Rosen, C.A., and Haseltine, W.A. (1986). Effects of mutations within the 3' orf open reading frame region of human T-cell lymphotropic virus type III (HTLV-III/LAV) on replication and cytopathogenicity. J. Virol. 60, 754-760.

62. Terwilliger, E., Langhoff, E., Obaru, K., Haseltine, W. (1990) Pleiotropic effects upon HIV replication induced by the nef gene. VI International Conference on AIDS. vol. 2, p133.

63. Turner, D. H., N. Sugimoto, J. A. Jaeger, C. A. Longfellow, S. M. Freier, and R. Kierzek. (1987) Improved parameters for prediction of RNA structure. Cold Spring Harbor Symp. Quant. Biol. 52, 123-131.

64. York-Higgins, D., Cheng-Mayer, C., Bauer, D., Levy, J. A., and Dina, D. (1990) Human Immunodeficiency virus type 1 cellular host range, and cytopathicity are linked to the envelope region of the viral genome. J. Virol. 64, 4016-4020.

65. Zack, J. A., Arrigo, S. J., O'Brien, W. A., and Chen, I. S. Y. (1990) HIV-1 infection of quiescent primary lymphocytes: Molecular analysis reveals a labile, latent viral DNA. Cell 61, 21-222

66. Zapp, M. L., and M. R. Green. (1989) Sequence-specific RNA binding by the HIV-1 Rev protein. Nature (London) 342, 714-716.

67. Zuker, M. (1989) Computer prediction of RNA structure. Methods in Enzymol. 180, 262-288.

68. Zuker, M. (1989) On finding all suboptimal foldings of an RNA molecule. Science 244, 48-52.

STRUCTURE AND FUNCTION OF HIV AND SIV NEF PROTEINS

Lee Ratner, Thomas M.J. Niederman,
Homer Lozeron, and Martin Bryant

Departments of Medicine & Molecular Microbiology
Washington University School of Medicine
St. Louis, MO

INTRODUCTION

Unlike many avian and murine retroviruses, human retroviruses and some simian retroviruses are characterized by the presence of genes in addition to those which encode virion structural proteins (gag and env) and enzymatic proteins (pol) (Fig. 1) (reviewed in Bryant & Ratner, 1990). These additional genetic determinants bear no strong resemblance to normal cellular sequences, and are thus, distinct from onc genes found in acute transforming viruses. Some of these genes are essential for virus replication, and have been termed "regulatory" genes. Examples of such regulatory genes are the tax and tat genes of HTLV-I/II and HIV/SIV families, respectively, which enhance viral transcription, and the rex and rev genes of HTLV-I/II and HIV/SIV families, respectively, which promote the transport of unspliced and single spliced viral mRNAs from the nucleus to the cytoplasm.

There is also a wide array of "accessory" genes which are dispensable for virus replication, but may modulate some aspect of the virus life cycle. Two of these genes encode proteins found in the virus particle: the highly conserved viral protein R gene (vpr) found in all members of the HIV/SIV family with the exception of SIVagm, and the viral protein X gene (vpx) restricted to HIV-2 and SIV genomes with the exception of SIVcpz. In contrast, the viral protein U gene (vpu) is restricted to the genomes of HIV-1 and SIVcpz, and is not found in the virion, but promotes virus particle assembly. The virion infectivity factor gene (vif) is critical for the infectivity of free virus particles, but is not required for cell-cell transmission.

The negative factor gene (nef) stands out as the most controversial of these accessory genetic elements. It has been given multiple different names since its original discovery (3'orf, orf B, and orf E') (Ratner et al., 1985a; Gallo et al., 1988). It is f und in the genomes of all members of the HIV/SIV family of retroviruses (Fig. 1), yet the amino acid sequence of its predicted protein product is among the most divergent of all lentiviral proteins (Fig. 2) (Ratner et al., 1985b, Myers et al., 1990). The HIV-1 and SIVmac Nef proteins are identical at only 38% of residues (Chakrabarti et al., 1987). Nef proteins of different human and simian lentiviruses vary in size from 206 residues for HIV-1 Nef

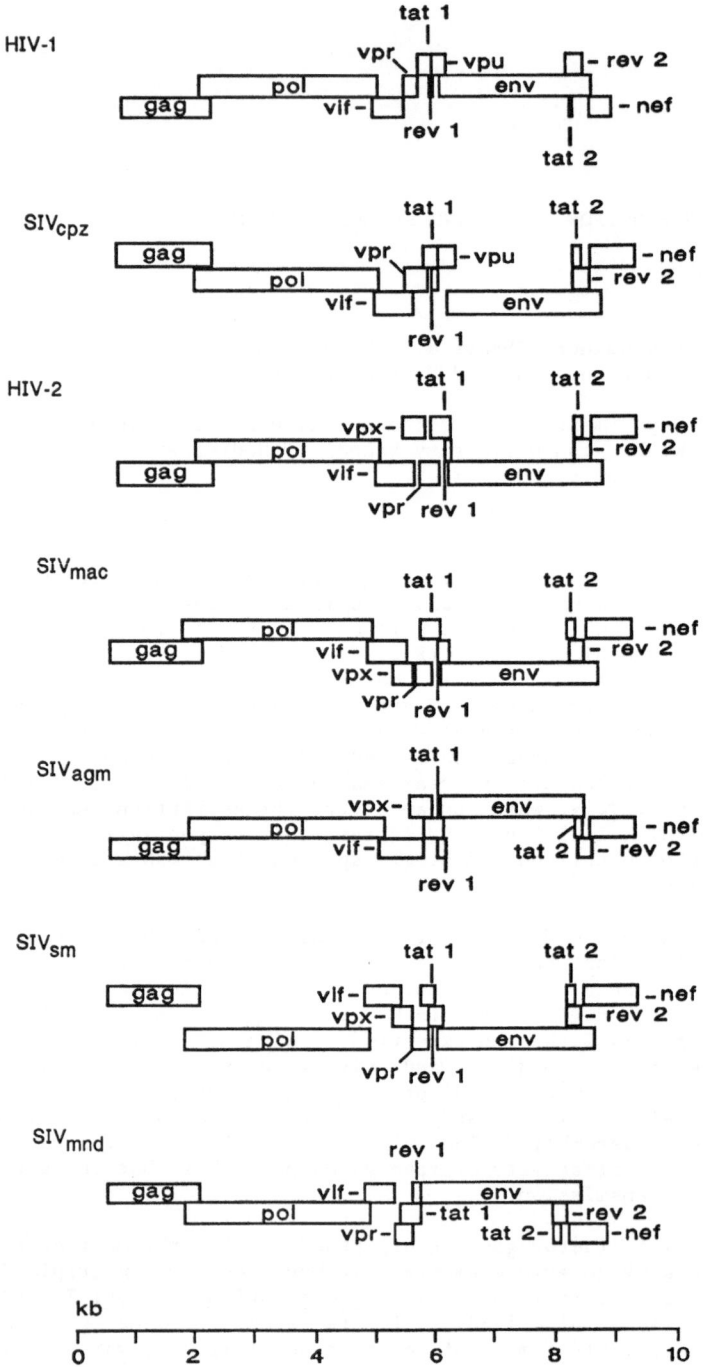

Fig. 1. Comparison of simian and human lentiviral genomes.

to 265 amino acids for HIV-2 NEF proteins (Myers et al., 1990). The myristoyl acceptor glycine residue at position 2 (prior to cleavage of the N-terminal methionine) is conserved in all Nef proteins as well as a core amino acid sequence between residues 72 and 148 of HIV-1 Nef which includes sequences with similarity to nucleotide binding proteins (Guy et al., 1986). However, marked variation is found in the N- and C-terminal domains. The protein kinase C phosphorylation acceptor site found in some HIV-1 Nef proteins is not found in a similar position in Nef proteins of HIV-2 and SIVs. In HIV-1 infected cells, the second methionine codon (residue 20) may also be utilized for translational initiation (Ahmad & Venketessan, 1988), but no corresponding methionine residue is present in HIV-2 or SIV Nef.

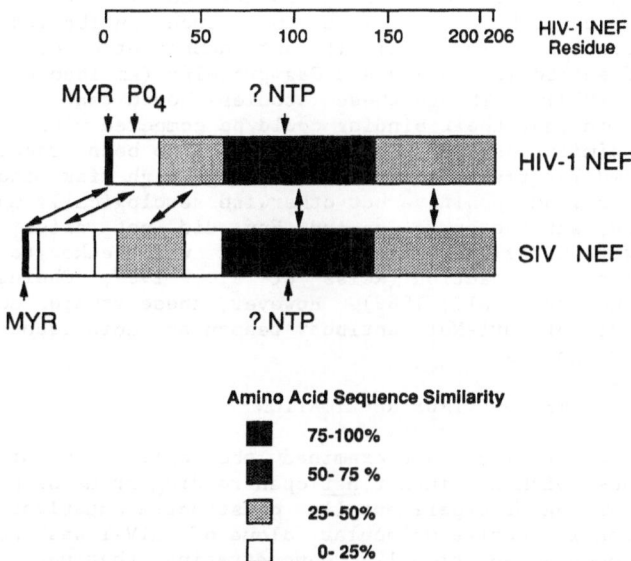

Fig. 2. Amino acid sequence homology between HIV-1 & SIVmac NEF proteins. The level of sequence similarity within each domain of the NEF proteins is indicated. Also indicated are the sites of myristoylation (MYR), possible nucleotide triphosphate binding (NTP), and phosphorylation by protein kinase C (PO$_4$).

Controversies have arisen over virtually all aspects of Nef studies including 1) the significance of antibodies to Nef, 2) the effects of Nef on virus replication, 3) the effects of Nef on CD4 expression, 4) the cellular localization of Nef, and 5) the nucleotide binding activities of Nef. This review will highlight a selected group of experimental findings bearing on these questions, offer explanations that may resolve experimental discrepancies, and suggest avenues of future investigation to clarify the role of Nef in the virus life cycle. This information should be critical not only for our understanding of viral pathogenesis, but also for the development of diagnostic reagents, vaccines, and therapeutic agents.

IMMUNE RESPONSES TO NEF

Nef has been shown to be immunogenic in humans and chimpanzees (Franchini et al., 1986; Sabatier et al., 1989; Wieland et al, 1990; Cheingsong-Popov et al., 1990; Bahraoui et al., 1990; Culmann et al., 1989). Humoral and cell mediated antibody responses have been identified, and responsible epitopes mapped in several regions of the protein (Arya, 1987; Gobert et al., 1990). One interesting finding is the difference in response to antibodies directed against N-terminal peptides depending upon whether there is an alanine or threonine at residue 15. Since this residue has been identified as a site for phosphorylation by protein kinase C, this modification may at least in part contribute to the alteration in immunological properties of the Nef protein (Guy et al., 1987).

Antibodies to Nef have been identified in 80% of HIV-1 infected humans. Two studies have suggested that such antibodies appear early after infection and in a significant number of cases may precede the appearance of antibodies to Env and Gag proteins (Ameisen et al, 1989a,b; Ranki et al, 1987). Though these studies showed that antibodies to Nef were specific in that their binding could be competed with recombinant Nef proteins, a similar antibody response has also been reported with sera from uninfected low risk individuals. Among high risk individuals, who are anti-Nef antibody positive but otherwise serologically unreactive with HIV-1 proteins, antibodies to Gag and Env did not arise at subsequent times, nor were HIV-1 sequences identified with leukocyte DNA by polymerase amplification reaction (Reiss et al., 1989; Cheingsong-Popov et al., 1990; Gluckman et al., 1989). However, these studies demonstrated the development of anti-Nef antibody responses coincident with anti-Gag and anti-Env responses.

EFFECTS OF HIV-1 NEF ON VIRUS REPLICATION

A number of studies have examined the replication of HIV-1 derived from proviruses with an intact nef open reading frame or proviruses with naturally occurring or experimentally constructed mutations in nef. The first functionally active molecular clone of HIV-1 was noted to have a termination codon at position 124, demonstrating that nef is dispensable for virus replication (Fisher et al., 1985). Other functionally active proviral clones have been constructed with or without an intact nef open reading frame (Ratner et al., 1987).

Terwilliger and colleagues (1986) first compared the replication of HIV-1 derived from a provirus with an intact nef gene to that of viruses derived from otherwise isogenic clones with deletions in nef. Their data using both C8166 and Jurkat-Tat lymphoid cells demonstrated a slightly enhanced rate of replication of the nef mutants compared to the parental virus. Similar findings were reported with primary lymphocytes and rhabdomyosarcoma cells by Luciw and colleagues (1987). Down-regulatory effects of nef on HIV-1 replication have been reported by Ahmad and Venketessan (1988), Niederman et al. (1989), Terwilieger et al. (1986), Luciw et al. (1987). and Cheng-Mayer et al. (1989). However, no significant effects of nef on virus replication were obtained by Kim et al. (1989).

To determine the step in replication affected by nef, Niederman and colleagues (1989) examined virus from two proviral clones identical except for the presence of a 45 nucleotide deletion in the 5'portion of the nef

gene of clone pHIV F-, compared to the parental clone, pHIV F+ (Fig. 3).
Virus derived from each clone infected a wide range of lymphoid cells
equally well, producing viral DNA at similar rates. These findings
demonstrated that _nef_ did not alter the early steps in virus replication
in uninfected lymphoid cells. In contrast, when clones pHIV F- and pHIV
F+ were transfected into CD4+ and CD4- lymphoid cells, 2-10-fold more
viral particles were produced in the absence than the presence of _nef_.

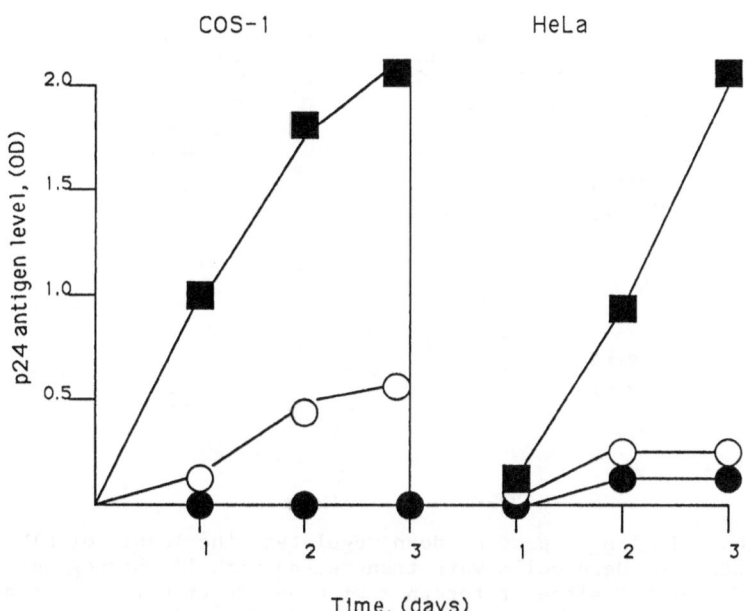

Fig. 3. NEF decreases the level of p24 GAG antigen. Ten micrograms of
pHIV F- (solid squares), pHIV F+ (open circles), or pSP (lacking HIV-1
sequences; closed circles) was transfected onto 50% confluent 100 mm
plates of COS-1 or HeLa cells. Samples of cell culture media were tested
for p24 antigen by an ELISA (DuPont). A sample of p24 (1 ng/ml) had an OD
values of 0.82, and OD values were linear with protein concentration with
the range of values reported. This experiment was repeated five times in
COS-1 cells, and twice each in HeLa, HOS, and SW480 cells with similar
results. Adapted with permission from Niederman et al. (1989).

In order to decipher the step in virus particle synthesis that was
suppressed by _nef_, Niederman and colleagues analyzed viral RNA levels by
Northern blot analysis, and noted a similar level of suppression to that
of virus particle production (Fig. 4). Nuclear run-on assays demonstrated that _nef_ suppressed the rate of RNA synthesis (Fig. 5). Similar
transcriptional suppressive effects were also reported by Ahmad and
Venketessan. However, no effects on the transcription of an indicator
gene, chloramphenicol acetyl transferase, were noted by Hammes and
colleagues (1989) and Bachelerie and colleagues (1989).

Ahmad and Venketessan (1988) suggested that the critical sequences for Nef response were in the negative regulatory element (NRE) of the LTR, between nucelotides -350 and -161 relative to the RNA initiation site. Guy and colleagues (1990a) have characterized a group of nuclear proteins binding to these LTR sequences, isolated from activated primary lymphocytes. DNA binding by one of these factors is downregulated by Nef.

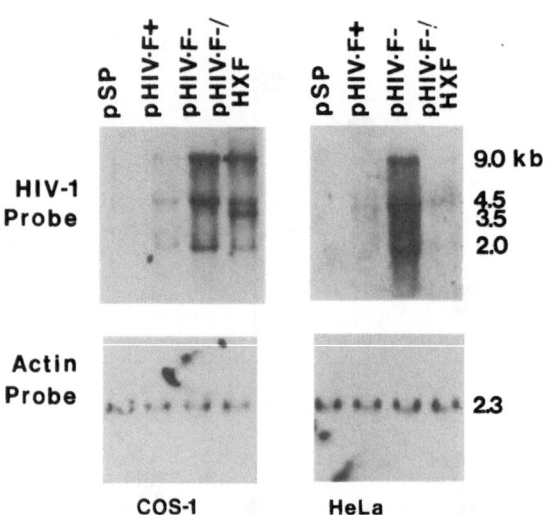

Fig. 4. The HIV-1 nef product down-regulates the level of HIV-1 RNAs. Plates of COS-1 or HeLa cells were transfected with 10 micrograms of pHIV F+, pHIV F-, of pSP alone or together with an additional 30 micrograms of nef expression clone pHXF. RNA was harvested after 3 days, and Northern blot analysis was performed with 35 micrograms of RNA using Optiblot filters. Filters were hybridized with a full-length HIV-1 genome probe, or an actin probe. Reprinted with permission from Niederman et al. (1989).

EFFECTS OF SIV NEF ON VIRUS REPLICATION AND PATHOGENICITY

Though the nef gene is present in all human and simian lentivirus genomes, its primary sequence exhibits a high level of sequence variation. The Nef protein expressed in SIVmac infected cells contains 250 amino acids, and has only 38% amino acid sequence similarity with the HIV-1 Nef protein. Nevertheless, its in vitro activity is remarkable similar to that of HIV-1 Nef (Niederman et al., 1990). SIVmac Nef was found to down-regulate virus production from transfected CD4- cells 3-5-fold. In addition, the steady state level of correctly initiated viral mRNAs and their rate of synthesis are depressed by Nef. SIVmac Nef had no effect on viral mRNA stability, as measured in actinomycin D treated cells. Thus, as in HIV-1 infection, SIVmac Nef appears to act primarily at the level of RNA synthesis to down-regulate virus production.

Recently, Kestler, Desrosiers, and their colleagues (1990) reported that <u>nef</u> encoded a pathogenicity factor in SIVmac infected animals. Only SIVmac strains with an intact <u>nef</u> gene or a <u>nef</u> gene capable of mutation to produce a full-length product were able to give rise to high levels of virus and produce simian AIDS. In contast, SIVmac with a deletion in the <u>nef</u> gene and unable to repair the defect, gave rise to significantly lower levels of virus expression <u>in vivo</u>, and caused no disease. No significant difference was noted in the titer or rate of generation of antibody responses to Env and Gag proteins with the divergent SIVmac strains used in this study.

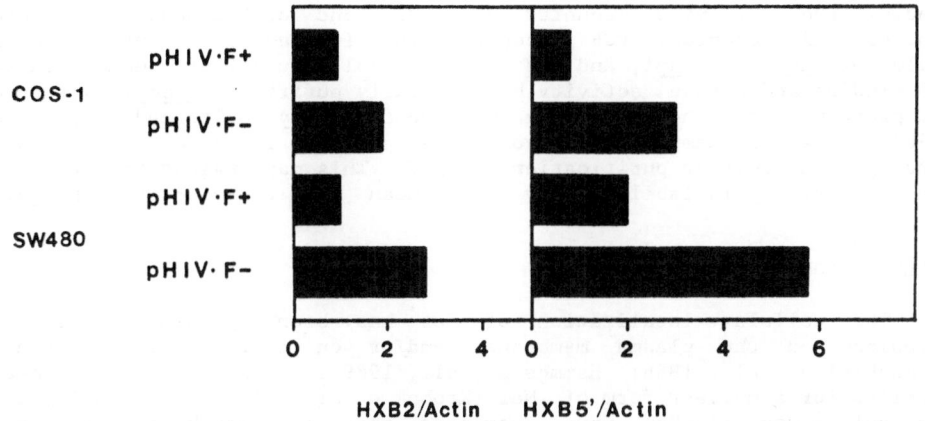

Hybridization Ratio

Fig. 5. The HIV-1 <u>nef</u> product down-regulates the rate of synthesis of HIV-1 RNAs. Plates were transfected as described in Fig. 4. Nuclear run-on assays were performed and the [32P]-labeled transcripts were hybridized to 0.5 microgram aliquots of pHXB2 (HIV-1 nucleotides 1-9213), pHXB5' (HIV-1 nucleotides 222-5580), or actin. The ratio of hybridization to HIV-1 sequences compared to actin is shown. Reprinted with permission from Niederman et al. (1989).

Can one reconcile these <u>in vitro</u> and <u>in vivo</u> findings? Perhaps down-regulation of SIVmac replication is essential for establishing a persistent infection <u>in vivo</u>. Rapid virus replication shortly after inoculation may lead to swift cell lysis and virus clearing by the immune system. Analysis of virus load in the plasma and cellular compartments within the first two months after infection may provide information relevant to this hypothesis. Alternatively, Nef may up-regulate virus infection or replication in another cell type that was not tested in these experiments, e.g. monocytes or lymphocytes. Another possibility is that immune responses directed against the Nef protein may be detrimental to the clearing of virus. Experiments to test the latter two hypotheses also may be very informative.

HETEROGENEITY OF NEF PROTEINS

Several different Nef proteins are detectable in HIV-1 infected cells. In HIV-IIIB infected cells, at least three distinct proteins are evident as bands on SDS polyacrylamide gel electrophoresis at 27, 25, and 17 kDa (Fig. 6). The 27 and 25 kDa proteins were capable of incorporating [3H]-myristic acid, though the 17 kDa form could not be myristoylated. Similar findings were obtained with Nef expression clones utilizing either an SV40 or CMV promoter (Fig. 6). It remains to be determined whether these different Nef proteins are different translation products, processed forms from a precursor, or proteins with differences in post-translational modifications.

Samuel and colleagues (1987) demonstrated similarity between amino acids 91 and 116 of Nef and those of nucleotide binding proteins, including the catalytic subunits of cAMP- and cGMP-dependent protein kinases, IL2 receptor, EGF receptor, insulin receptor, and oncogenes including ras, src, abl, and mos. Guy and colleagues (1987) demonstrated GTP binding and GTPase activity by partially purified E. coli expressed Nef proteins, and these findings were confirmed by Poulin and Levy (1989). However, Kaminchik and colleagues (1990) found a loss of GTP binding with further purification of Nef. This may suggest that the GTP binding activity is labile or may have been due to associated E. coli proteins.

CELLULAR LOCALIZATION OF NEF

The cellular localization of Nef has been reported to be in the cytoplasm, on the plasma membrane, and/or on intracellular membranes (Franchini et al., 1986; Hammes et al., 1989). Evidence has also been provided for a nuclear form of Nef (Krohn et al., 1990). Recently, Guy and colleagues (1990b) have suggested that Nef may be released from infected cells within vesicles. Differences in cell localization of the diverse molecular forms of Nef as noted above, as well as differencees in the primary amino acid sequence or post-translational modification may explain some of these differences.

SUMMARY OF PAST DISCOVERIES AND FUTURE GOALS

Our current understanding of Nef structure and function is summarized in Fig. 7. Nef transcripts are multiply spliced mRNAs expressed early after infection (Ahmad et al., 1989; Guatelli et al., 1990). Translation of the Nef protein is initiated from both the first and second AUG codons of the doubly spliced mRNAs, producing in the case of HIV-1 infected cells 206 and 187 amino acid long proteins, respectively. The larger form of Nef is myristoylated, phosphorylated, and glycosylated (Guy et al., 1990b). Myristoylation may mediate Nef binding to membranes, and nucleotide binding to the Nef protein is thought to occur. These events are postulated to lead to 1) autophosphorylation of Nef, 2) alterations in the transduction pathway, exemplified by phosphorylation of cellular factor, e.g. a Nef-responsive factor, NRF (Fig. 7) and 3) subsequent alterations of regulatory events in the nucleus. A potential target of nuclear activity includes transcriptional suppression of the HIV-1 LTR (Guy et al., 1990a).

In view of the present controvery concerning many of the steps postulated for Nef expression and function, the challenges to establish its mode of action are immense. What are the diverse molecular forms of

Nef? Do they include other primary translation products? Are there alternatively spliced viral mRNAs including only portions of the Nef open reading frame, possibly fusing it with other open reading frames? Does proteolytic processing of Nef occur? What posttranslational modifications occur? What is the relationship of structural variants to differences in biochemical functions (e.g. phosphorylation), cell localization, and biological function (e.g. virus suppession)?

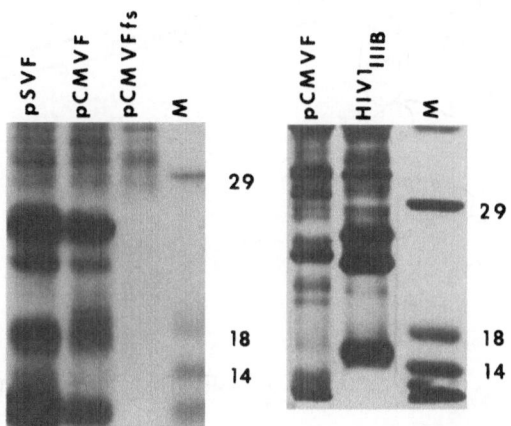

Fig. 6. Heterogeneity of HIV-1 Nef proteins. COS-1 cells were transfected with nef expression plasmids utilizing an SV40 promoter (pSVF) or a CMV promoter (pCMVF), or a frameshifted form of nef expressed from a CMV promoter (pCMVFfs). Lymphoid cells were infected with HIV-1-IIIB. Cells were labeled with [35S]-methionine and cysteine and immunoprecipitated with a specific anti-NEF antiserum. The molecular weights of marker proteins (M) are indicated in kDa to the right of each panel.

Does Nef bind nucleotides? If so, which ones? Does it utilize nucleotide triphosphates for autophosphorylation? If so, on which residues and with what effect on biochemical and biological function? If Nef is not a nucleotide binding protein, why is there so much sequence homology in the center of the protein in comparisons of different lentiviral proteins?

What is the relationship of cellular localization to Nef activity? What molecular forms are localized in each region of the cell? What effect does sequence variation play on cell localization?

Does Nef have pleiotropic effects? Does it down-regulate virus replication in some circumstances and up-regulate it in other contexts? What effect does cellular environment have on these effects? What is the relationship of each molecular form of Nef to these biological activities Is transcriptional suppression the only down-regulatory effect of Nef, or does down-regulation of CD4, as suggested by Guy et al. (1988), also play a role? What is the mechanism of up-regulation of virus expression by Nef, if in fact this does occur?

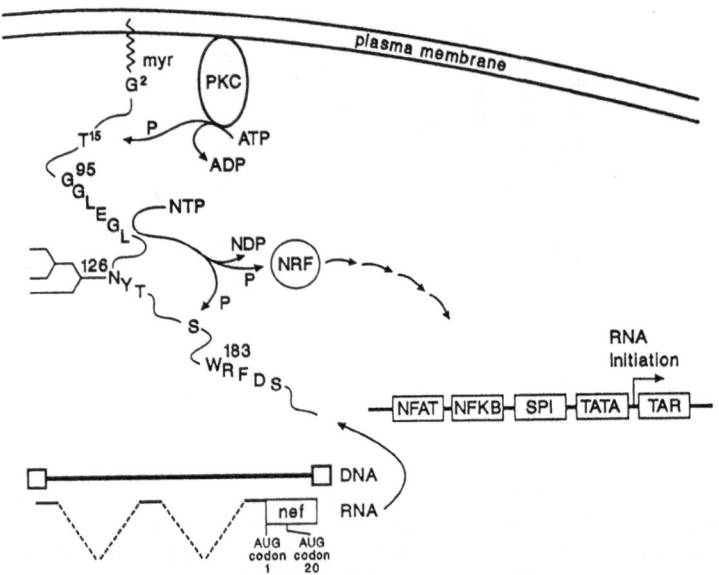

Fig. 7. Structure and function of NEF. Shown in the lower left hand portion is a multiply spliced nef mRNA expressed from HIV-1 proviral DNA. Protein products are initiated at both the first and second AUG codons. The larger protein product is shown as a protein myristoylated at a glycine residue at position 2 of the NEF precursor (G2), possibly mediating binding to the plasma membrane. Threonine at position 15 (T15) is phosphorylated by protein kinase C. A potential nucleotide binding domain is shown near residues 95-100. Cleavage of nucleotide triphosphate (NTP) may lead to autophosphorylation at a serine residue, as well as phosphorylation of a NEF-responsive factor (NRF) that mediates NEF activity in the nucleus. The asparagine at position 126 is glycosylated, and sequences near residue 183 may mediate GTPase activity (Guy et al, 1990b). Potential sites for action of NEF on LTR activity are shown at the lower right.

Perhaps more importantly, how does Nef serve as a pathogenicity factor in vivo? Does this information provide a clue to novel therapeutic approaches? Would viruses defective in Nef activity serve as attenuated mutants that might be useful for vaccine experiments?

ACKNOWLEDGEMENTS

This work was supported by contracts DAMD-87-17C-7107 and DAMD-90C-0125.

REFERENCES

Ahmad, N., and Venkatesan, S., 1988, Nef-protein of HIV-1 is a transcriptional repressor of HIV-1 LTR. Science 241:1481.

Ahmad, N., Maitra, R. K., and Venkatesan, S., 1989, Rev-induced modulation of Nef protein underlies temporal regulation of human immunodeficiency virus replication. Proc. Natl. Acad. Sci. U.S.A. 86:6111.

Allan, J. S., Coligan, J. E., Lee, T. H., McClane, M., Kanki, P., Groopman, J., and Essex, M., 1985, A new HIV antigen detected by antibodies from AIDS patients. Science 230:810.

Ameisen, J. C., Guy, B., Chamaret, S., Loche, M., Mouton, Y., Neyrinck, J. L., Khalife, J., Leprevost, C., Beaucaire, G., Boutillon, C., Gras-Masse, H., Maniez, M., Kieny, M. P., Laustriat, D., Bertheir, A., Mach, B., Montagnier, L., Lecocq, J.P., and Capron, A., 1989, Antibodies to nef protein and to nef peptides in HIV-1-infected seronegative individuals. AIDS Res. & Hum. Retorvir. 5:279.

Ameisen, J.-C., Guy, B., Lecocq, J.-P., Chamaret, S., Montagnier, L., Loche, M., Mach, B., Tartar, A., Mouton, Y., and Capron, A., 1989, Persistent antibody response to the HIV-1 negative regulatory factor in HIV-1 infected seronegative persons. N. Engl. J. Med. 320:251.

Arya, S. K., 1987, 3'orf and sor genes of human immunodeficinecy virus: in vitro transcription-translation and immunoreactive domains. Proc. Natl. Acad. Sci. U.S.A. 84:5429.

Bachelerie, F., Alcami, J., Hazan, U., Israel, N., Goud, B., Arenzana-Seisdedos, F., and Virelizier, J.-L., 1990, Constitutive expression of human immunodeficiency virus (HIV) nef protein in human astrocytes does not influence basal or induced HIV long terminal repeat activity, J. Virol 64:3059.

Bahraoui, E., Yagello, M., Billaud, J.-N., Sabatier, J.-M., Guy, B., Muchmore, E., Girard, M., and Gluckman, J.-C., 1990, Immunogenicity of the human immunodeficiency virus (HIV) recombinant nef gene product. Mapping of T-cell and B-cell epitopes in immunized chimpanzees, AIDS Res. & Hum. Retrovir. 6:1087.

Bryant, M. and Ratner, L., 1990, The biology and molecular biology of HIV, in "Pediatric AIDS: The challenge of HIV infection in infants, children, and adolescents," P. A. Pizzo and C. Wilfert, eds., Williams & Wilkins, Baltimore.

Chakrabarti, L., Guyader, M., Alizon, M., Daniel, M. D., Desrosiers, R. C., Tiollais, P., and Sonigo, P., 1987, Sequence of SIV from macaque and its relationship to other human or simian retroviruses, Nature (London) 328:543.

Cheingsong-Popov, R., Panagiotidi, C., Ali, M., Bowcock, S., Watikins, P., Aronstam, A., Wassef, M., and Weber, J., 1990, Antibodies to HIV-1 nef (p27): prevalence, significance, and relationship to seroconversion. AIDS Res. & Hum. Retrovir. 6:1099.

Cheng-Mayer, C., Lannello, P., Shaw, K., Luciw, P. A., and Levy, J. A., 1989, Differential effects of nef on HIV replication: implications for viral pathogenesis in the host, Science 246:1629.

Culmann, B., Gomard, E., Kieny, M.-P., Guy, B., Dreyfus, F., Saimot, A.-G., Sereni, D., and Levy, J. P., 1989, An antigenic peptide of the HIV-1 NEF protein recognized by cytotoxic T lymphocytes of seropositive indviduals in association with different HLA-B molecules. Eur. J. Immunol. 19:2383.

de Ronde, A., Reiss, P., Dekker, J., deWolf, F., Van den Hoek, A., Wolfs, T., Debouck, C., and Goudsmit, J., 1988, Seroconversion to HIV-1 negative factor. Lancet 2:574.

Fisher, A. G., Collalti, E., Ratner, L., Gallo, R. C., and Wong-Staal, F., 1985, A molecular clone of HTLV-III with biological activity. Nature (London) 316:262.

Fisher, A. G., Ratner, L., Mitsuya, H., Marselle, L. M., Harper, M. E., Broder, S., Gallo, R. C., and Wong-Staal, F. 1986, Infectious mutants of HTLV-III with changes in the 3'region and markedly reduced cytopathic effects. Science 233:655.

Franchini, G., Robert-Guroff, M., Wong-Staal, F., Ghrayeb, J., Kato, I., Chang, T., and Chang, N., 1986, Expression of the 3'orf of HTLV-III in bacteria: demonstration of its immuoreactivity with human sera. Proc. Natl. Acad. Sci. U.S.A. 83:5282.

Franchini, G., Robert-Guroff, M., Ghrayeb, J., Chang, N.T., and Wong-Staal, F., 1986, Cytoplasmic localization of the HTLV-III 3'orf protein in cultured T cells. Virol. 155:593.

Gallo, R., Wong-Staal, F., Montagnier, L., Haseltine, W., and Yoshida, M., 1988, HIV/HTLV nomenclature Nature (London) 333:504.

Gluckman, J. C., Fretz-Foucault, C., Rouzioux, C., Perret, P., Lopez, O., Bucquet, D., and Bahraoui, E., 1989, Lack of anti-p27nef antibody detection in HIV seronegative high-risk people. AIDS 3:855.

Gombert, F., O., Blecha, W., Tahtinen, M., Ranki, A., Pfeifer, S., Troger, W., Braun, R., Muller-Lantzsch, N., Jung, G., Rubsamen-Waigmann, H., and Krohn, K., 1990, Antigenic epitopes of NEF proteins from different HIV-1 strains as recognized by sera from patients with manifest and latent HIV infection. Virol. 176:458.

Guatelli, J. C., Gingeras, T. R., and Richman, D. D., 1990, Alternative splice acceptor utilization during human immunodeficiency virus type 1 infection of cultured cells. J. Virol. 64:4093.

Guy, B., Acres, R. B., Kiney, M. P., and Lecocq, J.-P., 1990a, DNA binding factors that bind to the negative regulatory element of the human immunodeficiency virus-1: regulation by nef, J. AIDS 3:797.

Guy, B., Kieny, M. P., Riviere, Y., Le Peuch, C., Dott, K., Girard, M., Montagnier, L., and Lecocq, J.P., 1987, HIV F/3'orf encodes a phosphorylated GTP-binding protein resembling an oncogene product. Nature (London) 330:266.

Guy, B., Riviere, Y., Dott, K., Regnault, A., and Kieny, M. P., 1990b, Mutational analysis of the HIV-1 nef protein, Virol. 176:413.

Hammes, S. R., Dixon, E. P., Malim, M. H., Cullen, B. R., and Greene, W. C., 1989, Nef protein of human immunodeficiency virus type 1: evidence against its role as a transcriptional inhibitor. Proc. Natl. Acad. Sci. U.S.A. 86:9549.

Kaminchik, J., Bashan, N., Pinchasi,D., Amit, B., Sarver, N., Johnston, M. I., Fischer, M., Yavin, Z., Gorecki, M., and Panet, A., 1990, Expression and biochemical characterization of human immunodeficiency virus type 1 nef gene product. J. Virol. 64:3447.

Kestler, H. W., III., Mori, K., Silva, D. P., Kodama, T., King, N. W., Daniel, M. D., and Desrosiers, R. C., 1990, Nef genes of SIV, J. Med. Primatol. 19:421.

Kim, S., Ikeuchi, K., Bryn, R., Groopman, J., and Baltimore, D., 1989, Lack a negative influence on viral growth by the nef gene of human immunodeficiency virus type 1. Proc. Natl. Acad. Sci. U.S.A. 86:9544.

Krohn, K., Ovod, V., Lagerstedt, A., Gobert, F., Jung, G., Hakkarainen, K., and Ranki, A., 1991, Expression kinetics and cellular localization of HIV-1 TAT and NEF proteins in relation to expression of structural viral mRNA and viral particles, in. "Vaccines 1991," F. Brown, R. Chanock, H. Ginsberg, and R. Lerner, eds., Cold Spring Harbor Press, Cold Spring Harbor, New York.

Luciw, P.A., Cheng-Mayer, C., and Levy, J.A., 1987. Mutational analysis of the human immunodeficiency virus: the orf-B region down-regulates virus replication. Proc. Natl. Acad. Sci. U.S.A. 84:1434.

Myers, G., Rabson, A.B., Berzofsky, J.A., Smith, T.A., and Wong-Staal, F., 1990, eds., "Human Retroviruses and AIDS", Los Alamos National Laboratory, Los Alamos, New Mexico.

Niederman, T.M.J., Hu, W., and Ratner, L., 1990, SIV Nef depresses virus replication and mRNA accumulation in COS cells, Submitted.

Niederman, T.M.J., Thielan, B.J., and Ratner, L, 1989, Human immunodeficiency virus type 1 negative factor is a transcriptional silencer. Proc. Natl. Acad. Sci. U.S.A. 86:1128.

Poulin, L. and Levy, J.A., 1989, In vitro expression of a functional HIV-1-nef gene product Fifth International Conference on AIDS Abstract No. T.C.P. 132.

Ranki, A., Krohn, M., Allain, J. P., Franchini, G., Valle, S. L., Antonen, J., Leuther, M., and Krohn, K., 1987, Long latency precedes overt seroconversion in sexually transmitted human immunodeficiency virus infection.- Lancet 2:589.

Ratner, L., Fisher, A., Jagodzinski, L., L., Mitsuya, H., Liou, R.-S., Gallo, R. C., and Wong-Staal, F., 1987, Complete nucleotide sequences of functional clones of the AIDS virus, AIDS Res. & Hum. Retrovir. 3:57.

Ratner, L., Haseltine, W., Patarca, R., Livak, K. J., Starcich, B., Josephs, S. F., Doran, E. R., Rafalski, J. A., Whitehorn, E. A., Baumeister, K., Ivanoff, L., Petteway, S. R., Pearson, M. L., Lautenberger, J. A., Papas, T. S., Ghrayeb, J., Chang, N. T., Gallo, R. C., and Wong-Staal, F., 1985a, Complete nucleotide sequence of the AIDS virus, HTLV-III, Nature (London) 313:277.

Ratner, L., Starcich, B., Josephs, S. F., Hahn, B. H., Reddy, E. P., Livak, K. J., Petteway, S. R., Jr., Pearson, M. L., Haseltine, W. A., Arya, S. K., and Wong-Staal, F., 1985b, Polymorphism of the 3'open reading frame of the virus associated with the acquired immune deficiency syndrome, human T-lymphotropic virus type III. Nucl. Acids Res. 13:8219.

Reiss, P., de Ronde, A., Lange, J. M. A., de Wolf, F., Dekker, J., Debouck, C., and Goudsmit, J., 1989, Antibody response to the viral negative factor (nef) in HIV-1 expression, AIDS 3:227.

Sabatier, J. M., Clerget-Raslain, B., Fontan, G., Fenouillet, E., Rochat, H., Granier, C., Gluckman, J. C., Van Rietschoten, J., Montagnier, L., and Bahraoui, E., 1989, Use of synthetic peptides for the detection of antibodies against the nef regulating protein in sera of HIV-infected patients. AIDS 3:215.

Samuel, K. P., Seth, A., Konopka, A., Lautenberger, J. A., and Papas, T. S., 1987, The 3'orf protein of human immunodeficiency virus shows structural homology with the phosphorylation domain of human interleukin-2 receptor and the ATP-bidning site of the protein kinase family, FEBS Letters 318:81.

Terwilliger, E., Sodroski, J. G., Rosen, C. A., and Haseltine, W.A., 1986. Effects of mutations within the 3'orf open reading frame region of human T-cell lymphotropic virus type III (HTLV-III/LAV) on replication and cytopathogenicity. J. Virol. 60:754.

Wieland, U., Kuhn, J. E., Jassoy, C., Rubsamen-Waigmann, H., Wolber, V., Braun, R. W., 1990, Antibodies to recombinant HIV-1 vif, tat, and nef proteins in human sera. Mol. Microbiol. Immunol. 179:1.

216

FUNCTIONAL ANALYSIS OF THE HIV-1

GENE PRODUCTS *vif* AND *vpu*

Klaus Strebel and Malcolm A. Martin

Laboratory of Molecular Microbiology
National Institute of Allergy and Infectious Diseases
Bethesda, MD

INTRODUCTION

Like other members of the lentivirus subfamily, HIV-1 contains several open reading frames (ORFs) in addition to the gag, pol, and env ORFs present in other retroviruses. Considerable effort has been put into the analysis of these extra genes to understand their significance for virus replication. We were focusing in the last few years on the functional analysis of the HIV-1 vif and vpu genes. The vif gene (previously also designated A, Q, P', ORF-1, or sor), partially overlaps the pol gene (see Fig. 1); its protein product has a relative molecular mass of 23.000 (M_r 23K) and is present in productively infected cells. The vif gene is highly conserved within HIV-1 and analoguous genes have been identified in HIV-2 and other lentiviruses suggesting that the gene function is an important one.

The vpu gene partially overlaps with the env gene (see Fig. 1) and was only recently shown to encode a 16 kDa protein that is expressed in cells infected with HIV-1 but not HIV-2 or SIV. The predicted amino acid sequence of vpu suggests an extremely hydrophobic N-terminus, while the C-terminal portion of the 81 amino acid protein is hydrophilic, containing a high number of charged amino acids. Both the size of vpu as well as its amino acid composition show similarities to some small membrane-associated proteins present in ortho- and paramyxovirus infected cells. The vpu gene is not as highly conserved as the vif gene and appears to be dispensable for virus replication in tissue culture. However, a significant number of HIV-infected individuals show antibody response to vpu [1] suggesting that it has an important function in vivo.

The HIV-1 vif gene

A first insight into the vif function was obtained about three years ago by studying the replication kinetics of a vif-mutant in a tissue culture system [2]. A vif-mutant was constructed from the infectious molecular clone, pNL43, by deleting a 621 bp NdeI/EcoRI fragment, mapping in the vif gene

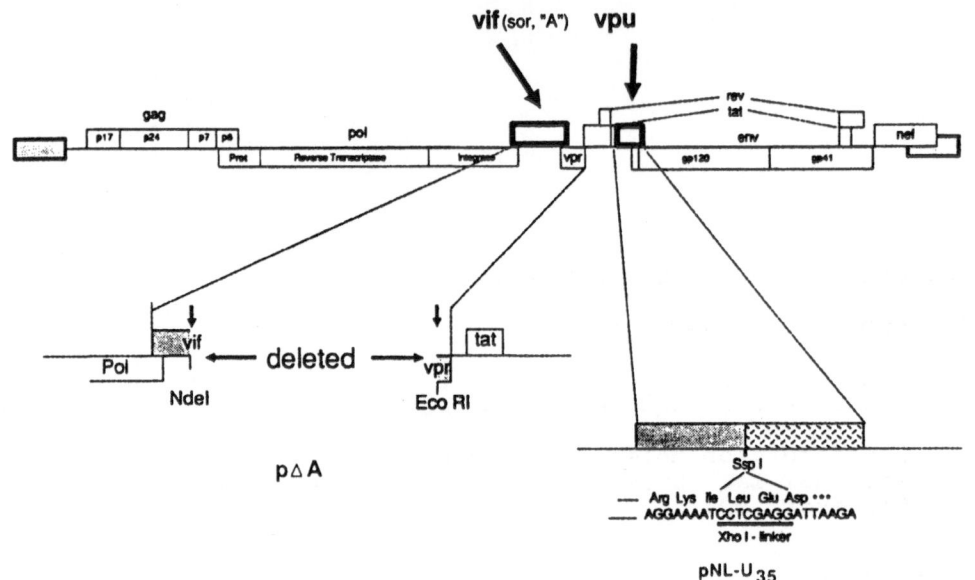

Figure 1. Genome organization of HIV-1 and structure of the vif mutant (pΔA) and the vpu mutant pNL-U₃₅. The location of the vif and vpu genes in the HIV-1 genome are highlighted.

region. DNA from both vif-mutant and wild type construct were independently transfected into a CD4⁺ T lymphoid cell line, A3.01, and virus production was measured by monitoring reverse transcriptase activity in the culture fluids (Fig. 2). Production of mutant virus was found to be significantly delayed and did not reach the level of wild type virus. The cultures producing mutant virus showed greatly reduced cytopathic effect and virus production could be measured for several months. This delay in virus spread lead us to suspect that vif might have either an effect on particle production or on the infectivity of HIV particles produced. To analyze a potential effect of vif on particle production, we transfected DNA of wild type and vif mutant constructs into a CD4-negative colon carcinoma cell line, SW480, and measured transient expression of HIV genes by western blotting. Both the relative amount of proteins and the protein patterns obtained were indistinguishable ruling out a function of vif on viral gene expression. To identify an effect of vif on viral infectivity we used equal amounts of cell-free virus (normalized for RT activity) to infect A3.01 cells (Fig. 3). In this experiment only virus from the wild type culture was able to establish a productive infection indicating that vif indeed affects the infectivity of virions. In contrast, when transfected SW480 cells were directly cocultivated with A3.01 cells, successful virus transmission was observed with both wild type and vif mutant cultures. When using concentrated virus stocks, vif mutant virus could be passaged onto A3.01 cells even as cell-free virus and subsequent titration experiments suggested that the loss of infectivity induced by the deletion of the vif gene was not absolute but in the order of 100 to 1000 fold. These experiments suggested that the HIV-1 vif gene while not affecting virus replication in transfected cells has a major impact on particle mediated virus transmission.

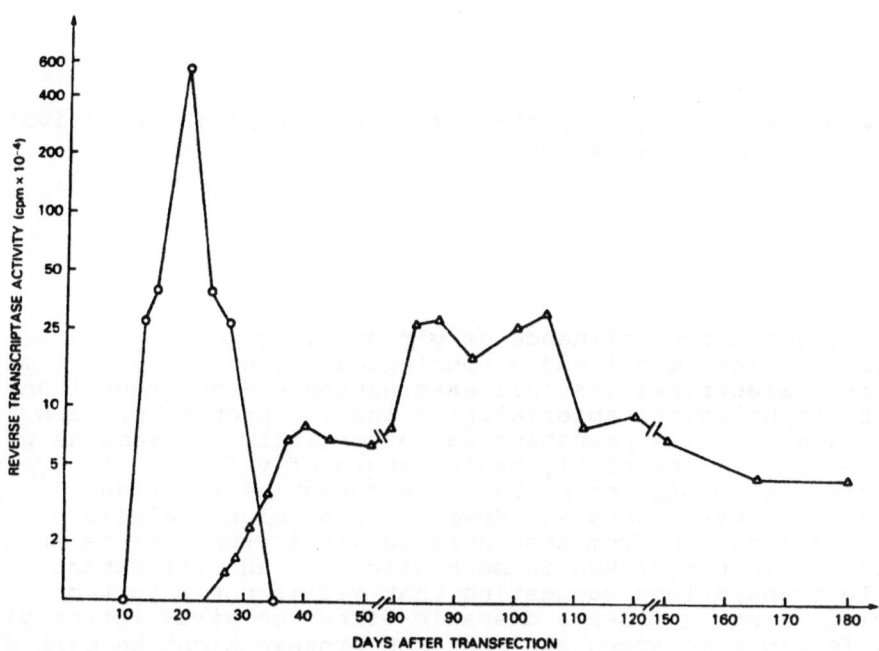

Figure 2. Virus production (measured as reverse transcriptase activity) after transfection of wild type (circles) or vif mutant DNA (triangles) into A3.01 cells.

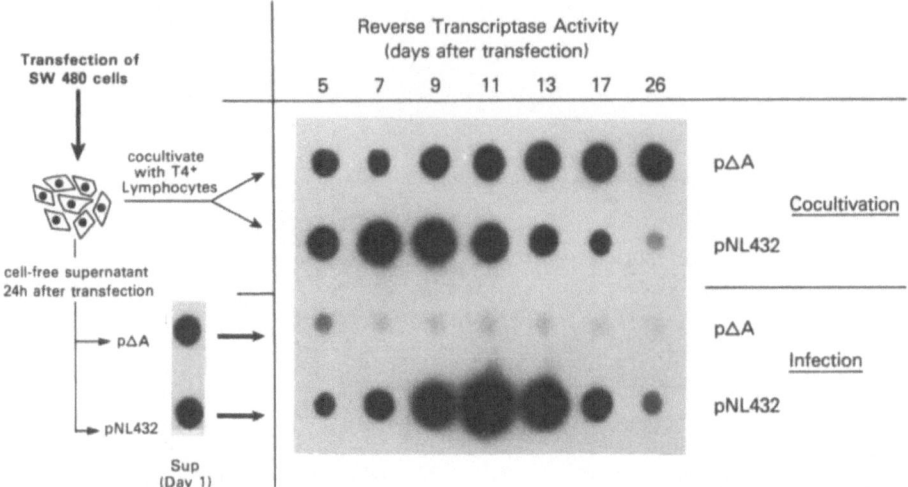

Figure 3. The activity of the vif mutant (p A) in cocultivation and particle-mediated infection.

To understand the influence of vif on virus infectivity we studied the biochemical and morphological structure of vif mutant particles. Electronmicroscopic examination did not reveal any obvious morphological abnormalities and the protein pattern of concentrated viral supernatant was essentially the same as wild type. Because of the highly basic nature of vif which is typical for nucleic acid binding proteins, a potential function of vif in RNA packaging was discussed. However, slot blot analysis of virion RNA obtained from concentrated virus preparations showed comparable amounts of RNA in both wild type and vif mutant particle preparations suggesting that vif is not involved in RNA packaging. Since vif has a dramatic effect on virus infectivity it was feasable to speculate that the protein might be part of the virion catalyzing an essential step in the viral infection cycle such as penetration, uncoating or as a cofactor in the integration reaction. We therefore attempted to identify the localization of vif by immunoprecipitating ^{35}S-methionine-labelled protein from cell lysates and cell-free supernatants with a vif-specific antiserum. While the 23 kDa vif protein was clearly identified in the cell lysate we were unable to detect vif in culture supernatants even after prolonged exposure of the gel. Thus, our inability to detect vif in virus particles under conditions where other low copy virion proteins such as integrase or reverse transcriptase were readily detected suggests that vif is not a component of HIV particles. To explain the effect of vif on virus infectivity we therefore have to assume that vif very likely exerts an enzymatic activity within an infected cell, acting most likely at the post-translational level, for example in stabilizing the proper conformation of structural proteins. We are currently conducting studies to investigate the nature of such a potential enzymatic activity and are further trying to narrow down the step(s) in the viral replication cycle which are affected by the absence of vif.

The HIV-1 vpu gene

The vpu gene was originally identified when sequence
information of the HIV genome became first available as an open
reading frame (ORF) with a coding capacity of 81 amino acids.
However, since the vpu ORF was not strictly conserved between
different HIV-1 isolates and furthermore no counterpart was found
in HIV-2 or SIV it was initially considered non-functional.
Finally, after the presence of an immune response to vpu was
detected in about one third of a randomly selected set of HIV-
posite human sera it became obvious that vpu was indeed actively
synthesized in virus producing cells [3]. To study the function
of vpu, we constructed a vpu mutant, pNL-U$_{35}$ from the infectious
molecular clone pNL43 by introducing an 8 bp XhoI DNA linker
(CCTCGAGG) at an SspI site (pos. 6189) located within the vpu
gene (Fig. 1). Insertion of the XhoI linker at this position
generated a translational frame shift resulting in premature
termination 35 codons from the NH$_2$-terminus of the vpu protein.
DNA of both pNL-U$_{35}$ and pNL43 was transfected into SW480 cells
and virus containing cell-free culture fluid was used to infect
A3.01 cells. Virus production was monitored at 2 day intervals
and analyzed for RT activity. Cultures infected with vpu-
defective virus produced less detectable progeny virus than wild
type virus (Fig. 4). In contrast, when intracellular protein
levels were analyzed by western blotting, cells producing vpu-
mutant virus consistently showed higher levels of viral proteins
than the corresponding wild type control [4]. The total amount of
viral protein produced in the two systems was estimated to be
approximately the same, suggesting that the absence of vpu did
not affect viral protein synthesis per se but rather influenced
the release of virus from infected cells. This observation was
further confirmed by pulse-chase experiments. While the kinetics
of processing of viral precursor proteins were unaffected in
these experiments, release of mature virion proteins from the
cells was significantly delayed [5]. In addition, electron
microscopic analysis of vpu-mutant cultures showed multiple
budding structures on the surface of infected cells and, at the
same time, extensive budding at internal membranes was observed
[4]. While internal budding has been described in the literature
for wild type virus in monocyte/macrophage systems, its
appearance in a T-lymphoid system was highly unusual. It is
presently unclear whether internal budding is a direct
consequence of the vpu defect or is indirectly caused by the
accumulation of budding structures at the cell surface,
redirecting budding to secondary membranes. To get more insight
into the function of vpu,we attempted to localize this protein in
virus producing cells. Pulse-chase experiments such as the one
shown in Figure 5 indicated that vpu is not released into the
culture supernatant in detectable quantities and is thus unlikely
to be virion-associated. Indirect immunofluorescence further
demonstrated that vpu is predominantly localized in a perinuclear
region, most likely the endoplasmic reticulum or the golgi
apparatus [4]. In vitro studies showed that vpu is an integral
membrane protein [5] with its hydrophobic N-terminus acting as
the membrane anchor. The membrane association of vpu makes it
attractive to speculate that it might have a role in assisting
the transport of viral proteins, such as gag or env, to the cell
surface. However, this would make it difficult to explain why vpu
is not packaged into virions itself. Alternatively, vpu might
somehow change the structure of the cell membrane and indirectly
facilitate detachment of maturing particles. Such a function

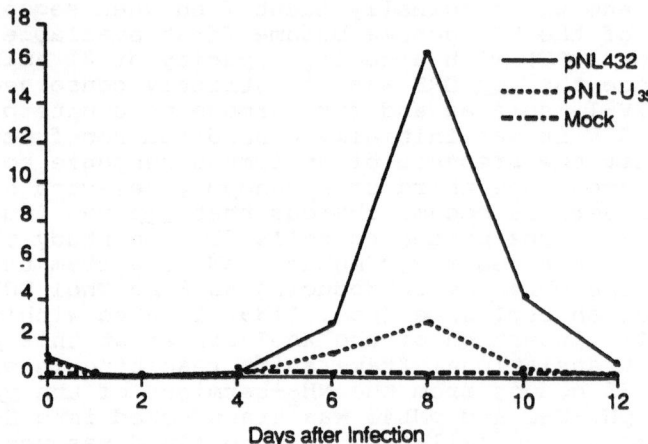

Figure 4. Virus production in A3.01 cells after infection with wild type or vpu mutant virus.

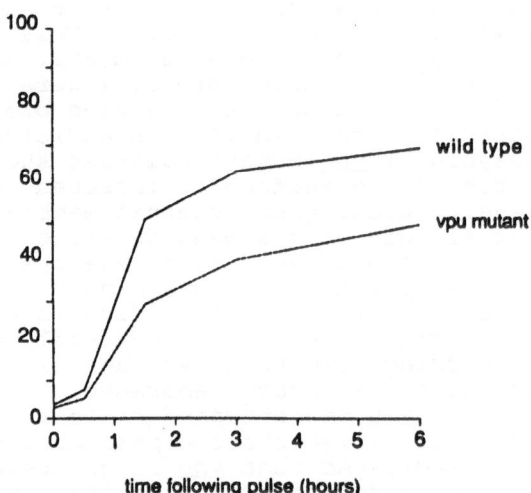

Figure 5. Release of wild type or vpu-mutant virus (measured as p24gag) from infected cells.

would be compatible with the observation that <u>vpu</u> is dispensible for virus replication, at least in vitro. Studies to determine a potential protein/protein interaction of <u>vpu</u> with other viral or cellular factors are currently in progress.

SELECTED PAPERS

[1] T. Schneider et al., AIDS Res Hum Retroviruses
     6:943-950 (1990)
[2] K. Strebel et al., Nature 328: 728-730 (1987)
[3] K. Strebel, T. Klimkait, and M.A. Martin
     Science 241: 1221-1223 (1988)
[4] T. Klimkait et al., J. Virol. 64: 621-629 (1990)
[5] K. Strebel et al., J. Virol. 63: 3784-3791 (1989)

ENVELOPE PSEUDOTYPES OF HIV AND HTLV

Nathaniel R. Landau*, Kathleen A. Page*,
and Dan R. Littman*#

*Department of Microbiology and Immunology and
#Howard Hughes Medical Institute, University of
California, San Francisco
San Francisco, California 94401

INTRODUCTION

The course of viral infections can be influenced by a variety of host factors as well as by co-infection of the host with one or more other viruses. Studies in cells co-infected with the human immunodeficiency virus (HIV) and a second virus have demonstrated transactivation of transcription from the HIV LTR by products of the second virus. Several classes of viruses, including papovaviruses (6), herpesvirus (8, 13, 15) and hepatitis B virus (23) have been shown to have this transactivating property. A second virus can also activate transcription of HIV proviruses by indirect mechanisms, such as by inducing the transcription factor NFkB, which acts on the HIV LTR (14). The human T cell leukemia viruses (HTLV-I and HTLV-II) may enhance HIV replication by a different means; HTLV particles have a direct mitogenic effect on T cells (possibly by binding to a cell surface growth factor receptor) and thus stimulating production of HIV from infected T cells (5, 28).

Infection with a second virus may influence not only the replication rate, but also the cellular tropism of HIV. By contributing its envelope glycoprotein to HIV particles, the second virus might cause the production of phenotypically mixed virions whose cellular tropism is determined by both envelope glycoproteins. Phenotypic mixing, in which the envelope glycoprotein of one virus is incorporated into the envelope of another virus, has been shown for several classes of viruses and results in production of hybrid particles whose cellular tropism is contributed by both viral *env* glycoproteins (for review see reference 1). These hybrid viruses, called pseudotypes, occur in cultured cells doubly infected with amphotropic (25) or xenotropic (11) MLV and HIV. In this review, we will discuss our results on the formation of pseudotypes by HTLV and HIV in cultured cells. These findings may be of clinical importance, since they suggest that such pseudotypes may form *in vivo* and thus alter the course of disease. In addition, the viral pseudotypes may be

Advances in Molecular Biology and Targeted Treatment for AIDS
Edited by A. Kumar, Plenum Press, New York, 1991

useful tools for studying cellular factors that participate in HIV infection and the molecular basis of envelope-core interactions.

EXPERIMENTAL APPROACH AND RESULTS

HIV tropism is limited to primate cells and, with a few exceptions, to those expressing the viral receptor, CD4. We wished to determine whether pseudotyping of HIV could broaden its host range, enabling the virus to infect CD4⁻ cells as well as cells of species other than human. To study phenotypic mixing in the absence of viral interactions that could complicate the interpretation of experimental results, pseudotypes were produced in the absence of viral replication. This was accomplished by transfecting COS cells with HIV or HTLV genomes, whose *env* genes had been replaced by selectable marker genes, thus generating defective virions that could be complemented with various retroviral envelope glycoproteins (10, 16). Cells infected with such viral particles are detected by their ability to form colonies in selective media.

HIV-gpt has a substitution of the SV-gpt selectable marker gene in the *env* gene of an otherwise replication competent HIV provirus (16). Cotransfection of this plasmid with an envelope glycoprotein expression vector encoding an HTLV-I, ecotropic, or amphotropic murine leukemia virus (E- or A-MLV) envelope glycoprotein resulted in production of HIV particles whose host range was determined by the product of the transfected *env* glycoprotein gene (Figure 1, panel A). HIV bearing the A-MLV *env* glycoprotein infected both human and mouse cells while HIV bearing the E-MLV *env* glycoprotein infected only 3T3 cells, in accordance with the known distribution of both receptors. HIV bearing the HTLV-I *env* glycoprotein infected human cells and, with significantly lower efficiency, murine cells, regardless of the presence of CD4 on these cells. The low level of infectivity of murine cells parallels the results of Sommerfelt et al., who determined the distribution of HTLV receptor using VSV(HTLV) pseudotypes (24).

In experiments analogous to those with HIV, we tested the ability of HTLV to be pseudotyped by MLV *env* glycoproteins. The vector HTLV-II-neo, in which the SV-neo selectable marker gene replaces the *env* gene of an otherwise replication competent HTLV-II provirus, was transfected together with the MLV *env* glycoprotein vectors. The tropism of viruses produced reflected the specificities of the two envelope glycoproteins, respectively (Figure 1B). Neither the HIV nor HTLV pseudotypes required species-specific cellular factors for steps in the viral life cycle up to proviral integration, as shown by their ability to infect murine cells.

One unusual aspect of HTLV replication is the extremely low infectivity of cell-free virion preparations (3, 27). In general, propagation of HTLV has been successful only by cocultivation of producer cells with target cells, suggesting a strong preference for spread of the virus by cell-cell contact. It is therefore of interest to note that both HTLV-II and HIV pseudotypes bearing the HTLV-1 *env* glycoprotein were substantially more infectious when tested in the cocultivation rather than the cell-free assay. This result supports the notion that the cocultivation preference of HTLV is a property of its *env* glycoprotein. This notion is

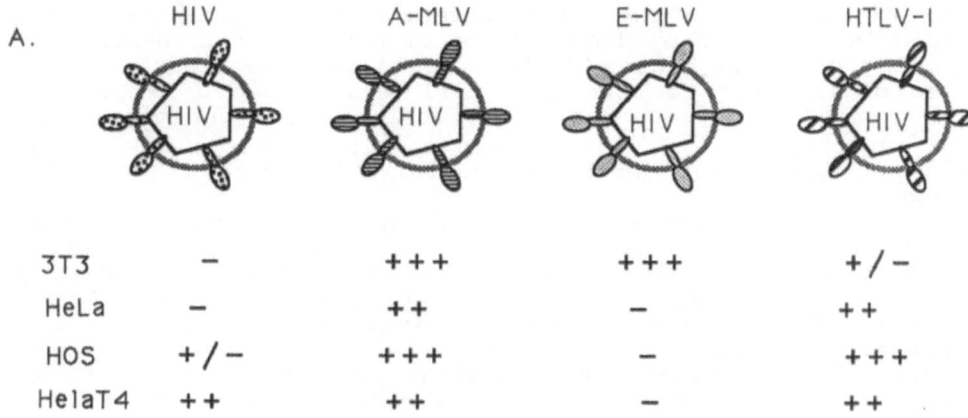

A.

	HIV	A-MLV	E-MLV	HTLV-I
3T3	–	+++	+++	+/–
HeLa	–	++	–	++
HOS	+/–	+++	–	+++
HelaT4	++	++	–	++

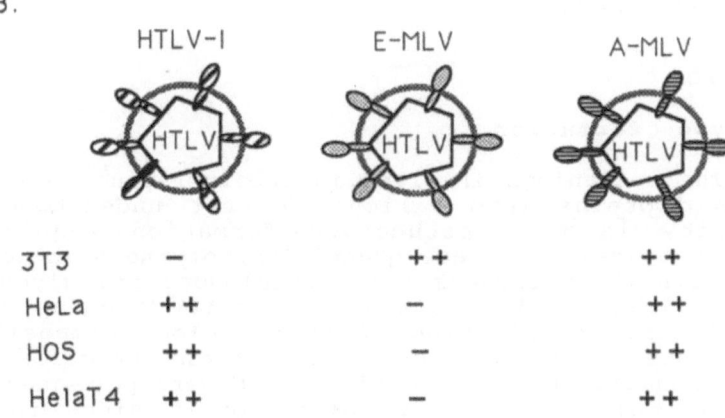

B.

	HTLV-I	E-MLV	A-MLV
3T3	–	++	++
HeLa	++	–	++
HOS	++	–	++
HelaT4	++	–	++

Figure 1.

Cellular Tropism of HIV and HTLV-II Pseudotypes. Pseudotypes were
produced by transfection of COS cells with the retroviral vectors
HIV-gpt (panel A) or HTLV-II-neo (panel B) and envelope glycoprotein
expression vectors for HIV, HTLV-I, A-MLV, E-MLV. Murine fibroblast
(3T3) or human cervical carcinoma (Hela and Hela-T4) or human
osteosarcoma (HOS) cells were infected by cell-free COS cell
supernatants (panel A) or by cocultivation with transfected,
mitomycin-C treated COS cells. Infected cells were selected
for resistance to mycophenolic acid (panel A) or G418 (panel B).

further supported by the ability of HTLV-II(MLV) pseudotypes
to overcome the block to cell-free infectivity. The
cocultivation preference is apparently not a property of the
HTLV core particle, but rather of the env glycoprotein. This
property may be a reflection of instability of the env
glycoprotein, gp61, either due to dissociation of the surface
protein, gp46, from the transmembrane protein, gp21, or to its
rapid inactivation by denaturation or proteolysis. Hence only
budding or newly budded virions would be infectious. It is
also of interest to note that cell-cell spread of HTLV does
not proceed through an env glycoprotein-independent mechanism,
but requires the env glycoprotein for infectivity; env⁻ virus
was not infectious, even in cocultivation assays (data not
shown).

Whether this apparent preference for spread by cell-
cell contact reflects a selective advantage to the survival
of HTLV in the population or whether it simply reflects a
lack of selective pressure to achieve higher cell-free
infectivity is not clear. Restriction to cell-cell
transmission may function to limit the rate of HTLV
replication, increasing the survival time of the host; or it
may limit the effectiveness of the host's immune response
against the virus. Unlike HIV, in which the env glycoprotein
appears to rapidly diverge to escape neutralization, the HTLV
env glycoprotein is highly conserved between isolates (21) and
may therefore be more readily subject to inactivation by the
host immune system.

DISCUSSION

Pseudotype formation

The mechanisms involved in incorporating heterologous
env glycoproteins into virions are not understood. One
possibility is that pseudotype formation requires the
presence, at least in trace quantities, of the homologous env
glycoprotein which directs the heterologous env glycoprotein
into the viral envelope and initiates the budding reaction
(26). Initial examination of the envelope composition of
VSV(MLV) pseudotypes showed that a small fraction of the
virions produced from a mixed infection were pure pseudotypes
containing only the MLV env glycoprotein since they were
resistant to neutralization with antibody directed against the
VSV-G protein. However, further analysis showed that
treatment with anti-VSV-G antibody and complement could remove
most of the previously non-neutralizable fraction suggesting
the presence of at least a small quantity of the homologous
envelope glycoprotein (26). Furthermore, analysis of
VSV(Sendai) envelope composition formed under conditions in
which it was possible to vary the amount of VSV-G protein
present in the cell supported a role for the homologous env
glycoprotein in virion budding and assembly (12). As the
amount of cellular VSV-G protein was decreased, the amount of
budded virion diminished and the number of Sendai env
glycoprotein per virion increased, but the number of VSV-G
protein molecules per virion remained fixed. This suggested
that the Sindbis env glycoprotein did not substitute for VSV-G
in the assembly and budding process and that its incorporation
into virions could be due to trapping between VSV-G protein
molecules in the virus envelope.

The results of our experiments argue against such a mechanism for pseudotyping of HIV and HTLV since the homologous *env* glycoprotein was not introduced into the virus producing cells. It is unlikely that *env* glycoproteins produced from endogenous genes play a role in the pseudotype formation as these do not appear to exist in the primate genome. Sequences homologous to HIV and HTLV do exist in the genome, but they exhibit homology to the *gag/pol* genes and not *env* (19). Our experiments also rule out the possibility that incorporation of heterologous *gag* proteins into the virion core particle is involved in directing the incorporation of heterologous *env* glycoprotein, since these *gag* proteins were not expressed in the packaging cells.

Although the homologous *env* glycoprotein was not required, it remained possible that it would influence, either positively or negatively, the efficiency of pseudotype formation. Mixed pseudotypes were therefore prepared to test this possibility. Transfection with a mixture HIV and MLV or HTLV *env* glycoprotein expression vectors did not affect the amount of pseudotype virus produced (10). The homologous *env* glycoprotein did not stimulate the formation of pseudotypes, as might have been expected if it nucleates the assembly process; nor did it appear to compete away the heterologous *env* glycoprotein, as might have been expected if it were to bear an assembly recognition sequence more compatible with the homologous core proteins.

The experiments discussed up to this point could lead to the conclusion that phenotypic mixing occurs in a completely promiscuous fashion, such that any *env* glycoprotein would form pseudotypes with any core particle. However, this does not appear to be the case. Experiments in which we expressed MLV *gag* and *pol* gene products in COS cells with HIV or HTLV *env* glycoproteins yielded either undetectable or low titers, respectively (N.R. Landau and D.R. Littman, unpublished observations). There is no obvious explanation for this nonreciprocity. We are currently investigating whether the *env* glycoproteins were incorporated into the virion but failed to function or whether they were not incorporated. It was not possible to overcome this block to pseudotype formation by using chimeric HIV or HTLV *env* glycoproteins in which the cytoplasmic and transmembrane domains were substituted with the analogous regions of MLV p15E, arguing against the existence of a simple core recognition sequence present in the transmembrane or cytoplasmic domains of the *env* glycoprotein (N.R. Landau and D.R. Littman, unpublished observations). However, it is difficult to rule out the possibility that these chimeric *env* glycoproteins were improperly folded and therefore non-functional.

Models for pseudotype formation

1. _Conservation of a specific signal for *env* glycoprotein incorporation_. In the simplest form of this model, a short stretch of amino acids in the cytoplasmic tail or transmembrane domain of the *env* glycoprotein is recognized by the *gag* matrix protein that mediates incorporation of *env* glycoprotein into the budding virion. However, such a conserved sequence is not obvious from inspection of several cytoplasmic and transmembrane domain sequences. Sequences in the cytoplasmic tail of the Rous sarcoma virus envelope

glycoprotein do not contain such a signal, since deletion of this region does not affect incorporation of the gp85^{env} into virions (18). Furthermore, chimeric *env* glycoproteins containing HTLV, HIV, or RSV *env* glycoprotein ectodomains and MLV transmembrane and cytoplasmic domains failed to be functionally incorporated into MLV virions. This argument is weakened by the possibility that the chimeric *env* glycoproteins may be incorrectly folded.

2. **Budding from a restricted site in the cell membrane**. Budding could occur at specialized plasma membrane sites from which most other membrane proteins are excluded. Such a mechanism would require that both the *env* glycoprotein and the *gag* proteins contain signals for transport to this site. This would allow for pseudotyping of any two viruses that bud from such a site. This mechanism cannot be ruled out, but seems unlikely due to the apparently random cell surface distribution of budding viruses.

3. **Heterologous gag incorporation into virions results in heterologous *env* glycoprotein incorporation**. The extent to which *gag* proteins mix in double infections has not been investigated for retroviruses due to the difficulty of separating virions from one another. Incorporation of even small amounts heterologous gag containing signals for incorporation of *env* glycoprotein might mediate pseudotype formation. This mechanism is argued against by the results discussed above in which pseudotypes were formed in the absence of heterologous *gag* proteins.

4. **Incorporation of homologous *env* glycoprotein nucleates budding**. Although this mechanism may operate for the rhabdovirus pseudotypes discussed above, it is argued against for the HIV and HTLV pseudotypes by our experiments in which pseudotyping was unaffected by the presence or absence of homologous *env* glycoprotein.

5. **Conservation of a specific property of the *env* glycoprotein**. This model is similar to model 1, but in this case the conserved signal for *env* incorporation is not a specific sequence of amino acids but a structural motif. For example, the two dimensional array of *env* glycoproteins at the cell surface might orient the transmembrane and cytoplasmic domains of the *env* glycoprotein molecules so that they form appropriate contacts with the gag matrix proteins of the virion core. These contacts need not be highly specific due to the polymeric nature of the interaction between the envelope and the core. This model is consistent with all the data presented.

6. **Random incorporation of cell surface proteins**. This model is unlikely to be correct due to the apparent exclusion of many membrane proteins from retroviral particles (2). In addition, *env* glycoprotein is not present at high concentration on the surface of virus producing cells.

Clinical and Laboratory Implications

It has yet to be shown that pseudotypes are formed in animals infected with two unrelated viruses . However, pseudotyping between closely related viruses is well established, as in the case of the pathogenic, defective viruses that are dependent on a replication competent helper virus to spread in animals. The critical determinant in formation of pseudotypes *in vivo* is likely to be the number of cells that are doubly infected. This will be influenced by the fraction of cells infected by each virus and by the tropism of the two viruses. HIV and HTLV as well as several other viruses often present in HIV infected individuals have overlapping tropism and therefore some cells could be doubly infected. The fraction of such cells will be small, however, because each virus infects only a small number of available cells. Suggestive evidence for pseudotyping *in vivo* has been obtained in the feline system (17). Doubly infected cats showed increased spread and broadened tissue distribution of FIV as a result of coinfection with FeLV. Pseudotyping of FIV by FeLV could explain these findings, but other interpretations are possible.

In doubly infected patients pseudotyping of HIV could play a role in allowing HIV to infect CD4⁻ cells. This could explain why virus is found in several CD4⁻ brain cell types in patients with neurological symptoms (20). Presence of HIV pseudotypes could also account for some of the laboratory results showing CD4-independent infection of cell lines or primary cells (7, 9). Most HIV infected individuals are also infected with CMV (4) and many are infected with hepatitis B virus or human herpes virus-6 (22). It will be of interest to determine which other viruses are capable of pseudotyping HIV.

In addition to the possible significance of pseudotyping in doubly infected patients, pseudotypes prepared in the laboratory serve as valuable tools for the study of cellular factors involved in determining viral host range as well as assembly mechanisms of retroviruses.

REFERENCES

1.   **Boettiger, D.** Animal Virus Pseudotypes. Prog. Med. Virol. **25:** 37-68, 1979.

2.   **Calafat, J. H., H. Janssen, P. Demant, H. Hilgers and J. Zavada.** Specific Selection of host Cell glycoproteins During Assembly of Murine Leukaemia Virus and Vesicular Stomatitis Virus: Presence of Thy-1 Glycoprotein and Absence of H-2, Pgp-1 and T-200 Glycoproteins on the Envelopes of these Virus Particles. J. Gen. Virol. **64:** 1241-1253, 1983.

3.   **Chen, I. S. Y., S. G. Quan and D. W. Golde.** Human T-Cell Leukemia Virus Type II Transforms Normal Human Lymophocytes. Proc. Natl. Acad. Sci. USA. **80:** 7006-7009, 1983.

4.   **Fiala, M., L. A. Cone, C. Chang and E. Mocarski.** Cytomegalovirus Viremia Increases with Progressive Immune Deficiency in Patients Infected with HTLV-III. AIDS Research. **2:** 175-181, 1986.

5.   **Gazzolo, L. and M. D. Dodon.** Direct Activation of Resting T Lymphocytes by Human T-Lymphotropic Virus Type I. Nature. **326:** 714-717, 1987.

6.   Gendelman, H. E., W. Phelps, L. Feigenbaum, J. M. Ostrove, A. Adachi, G. Howley, G. Khoury, H. S. Ginsberg and M. A. Martin. Transactiviation of the Human Immunodeficiency Virus Long Terminal Repeat Sequence by DNA Viruses. Proc. Natl. Acad. Sci., USA. **83**: 9759-9763, 1983.

7.   Harouse, J. M., C. Kunsch, H. T. Hartle, M. A. Laughlin, J. A. Hoxie and B. Wigdahl. CD4-Independent Infection of Human Neural Cells by Human Immunodeficiency Virus Type 1. J. Virol. **19**: 19-25, 1989.

8.   Horvat, R. T., S. Wood and N. Balachandran. Transactivation of Human Immunodeficiency Virus Promoter by Human Herpesvirus-6. J. Virol. **63**: 970-973, 1989.

9.   Kunsch, C., H. T. Hartle and B. Wigdahl. Infection of Human Fetal Dorsal Root Ganglion Glial Cells with Human Immunodeficiency Virus Type 1 Involves an Entry Mechanism Independent of the CD4 T4A Epitope. J. Virol. **63**: 5054-5061, 1989.

10.   Landau, N. R., K. A. Page and D. R. Littman. Pseudotyping with HTLV-I Broadens the HIV Host Range. J. Virol. In Press.

11.   Lusso, P., F. di Marzo Veronese, B. Ensoli, G. Franchini, C. Jemma, S. E. DeRocco, V. S. Kalyararaman and R. C. Gallo. Expanded HIV-1 Cellular Tropism by Phenotypic Mixing with Murine Endogenous Retroviruses. Science. **247**: 1990.

12.   Metsikko, K. and H. J. Garoff. Role of Heterologous and Homologous Glycoprotiens in Phenotypic Mixing Between Sendai Virus and Vesicular Stomatitis Virus. J. Virlol. **63**: 5111-5118, 1989.

13.   Mosca, J. D., D. P. Bednarik, N. B. Raj, C. Rosen, J. G. Sodroski, W. A. Haseltine and P. M. Pitha. Herpes Simplex Virus Type-1 Can Reactivate Transcription of Latent Human Immunodeficiency Virus. Nature. **325**: 67-70, 1987.

14.   Nabel, G. and D. Baltimore. An Inducible Transcription Factor Activates Expression of Human Immunodeficiency Virus in T Cells. Nature. **326**: 711-713, 1987.

15.   Ostrove, J. M., J. Leonard, K. E. Weck, A. B. Rabson and H. E. Gendelman. Activation of the Human Immunodeficiency Virus by Herpes Simplex virus Type 1. Virology. **12**: 3726-3732, 1987.

16.   Page, K. A., N. R. Landau and D. R. Littman. Construction and Use of an HIV Vector for Analysis of Viral Infectivity. J. Virol. In Press.

17.   Pedersen, N. C., M. Torten, B. Rideout, E. Sparger, T. Tonachini, P. A. Luciw, C. Ackley, N. Levy and J. Yamamoto. Feline Leukemia Virus Infection as a Potentiating Cofactor for the Primary and Secondary Stages of

Experimentally Induced Feline Immunodeficiency Virus Infection. J. Virol. **64**: 598-606, 1990.

18.   **Perez, L. G., G. L. Davis and E. Hunter**. Mutants of the Rous Sarcoma Virus Envelope Glycoprotein that Lack the Transmembrane Anchor and Cytoplasmic Domains: Analysis of Intracellular Transport and Assembly into Virions. J. Virol. **61**: 2981-2988, 1987.

19.   **Perl, A., J. D. Rosenblatt, I. S. Chen, J. P. DiVincenzo, R. Bever, B. J. Poiesz and G. N. Abraham**. Detection and Cloning of New HTLV-Related Endogenous Sequences in Man. Nucl. Acids Res. **17**: 6841-6854, 1989.

20.   **Price, R. W., B. Brew, J. Sidtis, M. Rosenblum, A. C. Schneck and P. Cleary**. The Brain in AIDS: Central Nervous System HIV-1 Infection and AIDS Dementia Complex. Science. **239**: 5586-5591, 1988.

21.   **Ratner, L**. "Molecular Variation of Human T-Lymphotropic Viruses and Clinical Associations." Human Retrovirology HTLV. Blattner ed. 1990 Raven Press. New York.

22.   **Salahuddin, S. Z., D. V. Ablashi, P. D. Markham, S. F. Josephs, B. Kramarsky and R. C. Galo**. Isolation of a New Virus, HBLV, in Patients with Lymphoproliferative Disorders. Science. **234**: 596-601, 1986.

23.   **Seto, E., Y. Benedict, T. S. Yen, B. M. Peterlin and J. Ou**. Transactivation of the Human Immunodeficiency Virus Long Terminal Repeat by the Hepatitis B Virus X Protein. Proc. Natl. Acad Sci. USA. **85**: 8286-8290, 1988.

24.   **Sommerfelt, M. A., B. P. Williams, P. R. Clapham, E. Solomon, P. N. Goodfellow and R. Weiss**. Human T Cell Leukemia Viruses Use a Receptor Determined by Human Chromosome 17. Science. **242**: 1557-1559, 1988.

25.   **Spector, D. H., E. Wade, D. A. Wright, V. Koval, C. Clark, D. Joquish and S. A. Spector**. Human Immunodeficiency Virus Pseudotypes with Expanded Cellular and Species Tropism. J. Virol. **64**: 2298-2308, 1990.

26.   **Witte, O. N. and D. Baltimore**. Mechanism of Formation of Pseudotypes Between Vesicular Stomatitis Virus and Murine Leukemia Virus. Cell. **11**: 505-511, 1977.

27.   **Yammamoto, N., M. Okada, Y. Koyanagi, M. Kannagi and Y. Hinuma**. Transformation of Human Leukocytes by Cocultivation with an Adult T Cell Leukemia Virus Producer Cell Line. Science. **217**: 737-739, 1982.

28.   **Zack, J. A., A. J. Cann, H. P. Lugo and I. S. Chen**. HIV-1 Production from Infected Peripheral Blood T Cells after HTLV-I Induced Mitogenic Stimulation. Science. **240**: 1026-1029, 1988.

CD4-*PSEUDOMONAS* EXOTOXIN: A STRATEGY FOR AIDS THERAPY BASED ON

SELECTIVE KILLING OF HIV-INFECTED CELLS

Per Ashorn, Bernard Moss and Edward A. Berger

Laboratory of Viral Diseases
National Institute of Allergy and Infectious Diseases
National Institutes of Health
Bethesda, MD   20892

INTRODUCTION

Therapeutic Concepts Based on Derivatives of CD4

The receptor for human immunodeficiency virus (HIV) is CD4, a 55 kD glycoprotein expressed on the surface of certain human lymphoid and monocytic cell types[1]. The finding[2-7] that recombinant soluble forms of CD4 retain the capacity for high affinity binding to gp120 (the external subunit of the HIV envelope glycoprotein) has suggested potential therapeutic uses of CD4 derivatives. For example, soluble truncated forms of CD4 (sCD4) are able to neutralize HIV infectivity *in vitro*[2-6], suggesting they may have specific anti-viral activity in HIV-infected individuals. Several "second-generation" applications of this neutralizing concept are under development, including attachment of the CD4 to immunoglobulin constant region sequences[8-12], or presentation of the CD4 in association with erythrocytes[13]. These modifications provide the advantages of increased plasma half-life compared to sCD4, and possibly enhanced efficiency of neutralization due to multivalency.

An alternative concept is to use the CD4 moiety as a targeting agent to direct the effector activities of novel hybrid molecules to selectively kill HIV-infected cells. This can be achieved with molecules containing CD4 linked to a toxin[14,15], which should selectively bind to and kill HIV-infected cells which express gp120 at their surface (Fig. 1). The aforementioned CD4-immunoglobulin hybrid proteins also have the potential to selectively kill HIV-infected cells by virtue of the Fc regions which can mobilize normal immune system functions, such as complement-mediated lysis and antibody-dependent cellular cytotoxicity. CD4 can also be inserted into liposomes[16], suggesting the potential to selectively deliver cytotoxic compounds to HIV-infected cells. The high affinity interaction between CD4 and gp120 has thus spawned a diversity of strategic concepts for treatment of HIV infection. This report focuses on one of these approaches.

Advances in Molecular Biology and Targeted Treatment for AIDS
Edited by A. Kumar, Plenum Press, New York, 1991

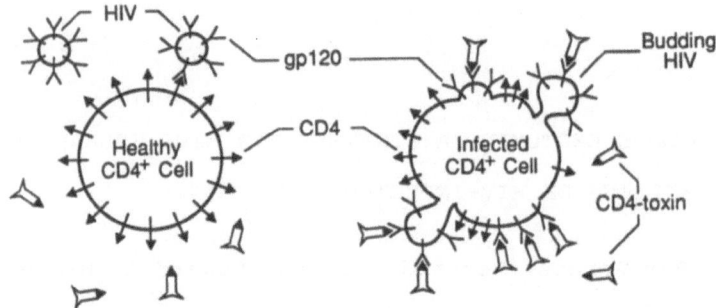

Fig. 1. Selective killing of HIV-infected cells by a hybrid
protein containing CD4 linked to a toxin. The CD4 moiety
directs the hybrid toxin to selectively bind to and kill the
HIV-infected cell, which expresses gp120 at its surface. The
uninfected cell lacks gp120 and is thus spared from killing.

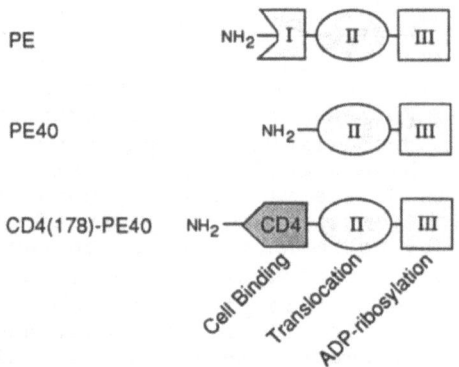

Fig. 2. Structure/function domains of PE and its derivatives.
Domain I of native PE binds to receptors present on a wide
variety of mammalian cell types, giving the toxin broad
specificity; domain II is involved in translocation of the
toxin to the cytoplasm; domain III ADP-ribosylates elongation
factor 2, thereby inactivating protein synthesis and killing
the cell. PE40 is a recombinant construct lacking domain I,
and serves as a control for non-specific toxicity. In
CD4(178)-PE40, the CD4 region consists of the N-terminal 178
amino acids, representing the first two domains of CD4; the
gp120-binding site is contained within the first domain.

CD4-*Pseudomonas* Exotoxin Hybrid Protein

As illustrated schematically in Fig. 2, the domain stucture of *Pseudomonas aeruginosa* exotoxin A (PE) makes this an ideally suited molecule for the construction of recombinant derivatives containing alternative ligands in place of the normal cell binding domain[17]. We therefore genetically engineered hybrid proteins consisting of portions of CD4 linked to the translocation and ADP-ribosylation domains of PE. The molecule we have focused on, designated CD4(178)-PE40, contains the first two domains of CD4, including gp120-binding site located within the amino-terminal domain[1]. CD4(178)-PE40 selectively binds to cells expressing gp120[14], and is presumably translocated to the cytoplasm where it ADP ribosylates elongation factor 2, resulting in inactivation of protein synthesis and consequent cell death[17].

We have used two types of systems to measure the anti-HIV activity of CD4(178)-PE40[14,18-21]. In the first, we measure direct killing of a cell population in which most of the cells express surface HIV envelope glycoprotein. This can be achieved using cells expressing recombinant envelope glycoproteins, e.g. by using vaccinia virus vectors[14,18]; alternatively, cell lines which are chronically infected with HIV can be tested as targets[14,18,19]. In the second system, we measure the effects of CD4(178)-PE40 on the spread of HIV throughout a population of susceptible cells[18,20,21]. This system allows assessment of the ability of the hybrid toxin to protect a susceptible population from virus-mediated killing, as well as the toxin's ability to suppress virus production. In such assays, appropriate controls must be included to demonstrate that the protective effects observed are truly due to selective killing of the infected cells, rather than to a simple neutralization effect by the CD4 moiety of the hybrid toxin. In both the direct killing and HIV spread systems, we have used MTT oxidation as a convenient, highly reproducible measure of the relative numbers of viable cells.

RESULTS AND DISCUSSION

Selective Killing of HIV-Infected Cells by CD4(178)-PE40

We have assayed the activity of CD4(178)-PE40 on chronically infected human T-cell lines that constitutively produce virus[14,18,19]. The hybrid toxin kills two such cell lines with IC_{50} values in the range of 100 pM, but has no effect on the corresponding uninfected parental cell lines (Table 1). By contrast, native PE kills both the infected and uninfected cells (not shown), indicating that the resistance of the parental lines to CD4(178)-PE40 is not simply due to their inherent insensitivity to PE derivatives. Control experiments verify that the specificity of CD4(178)-PE40 for HIV-infected cells is provided by the CD4 moiety: i) PE40, which lacks a cell binding domain, has no effect on infected or uninfected cells, and ii) the activity of the hybrid toxin is neutralized by anti-CD4 monoclonal antibodies, by recombinant gp120, and by sCD4, none of which affect native PE[19]. The selective killing of HIV-infected cells is not due to mere interaction of the

Table 1. Sensitivity of HIV-1 infected and uninfected T-cell lines to killing by CD4(178)-PE40 or control proteins.
   Two T-cell lines chronically infected with HIV-1 were tested; the corresponding uninfected parental cell lines served as controls. After 4-5 days of culture in the presence of the indicated proteins, the relative numbers of viable cells were determined by the MTT assay. IC_{50} is the concentration giving 50% cell killing. CD4-PE40$_{asp553}$ is a recombinant protein with greatly reduced cytotoxicity due to a mutation abolishing ADP-ribosylation activity. In some experiments lys-PE40, a genetically engineered version of PE40 with an additional lysine residue at the N-terminus, was used instead of PE40. *no toxicity at this concentration; n.d. = not determined.

| | | IC_{50} (nM) | | | |
Cell Line	HIV-1 Infected	CD4-PE40	PE40	sCD4	CD4-PE40$_{asp553}$
8E5	YES	0.08	>10*	>10*	>10*
A3.01	NO	>10*	>10*	>10*	>10*
H9/HTLVIIIB	YES	0.1	>8*	n.d.	n.d.
H9	NO	>10*	>8*	n.d.	n.d.

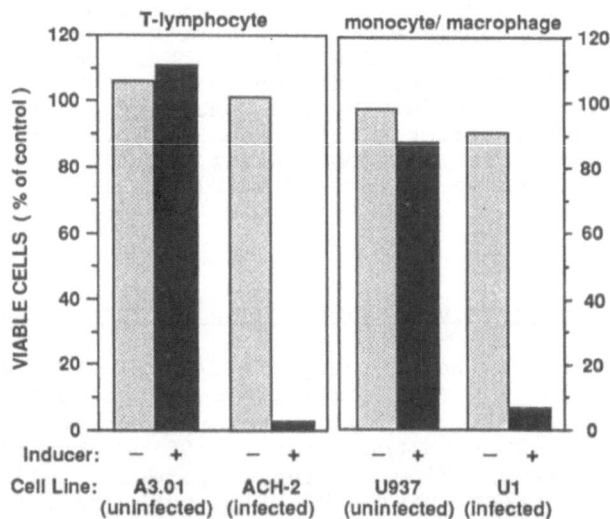

Fig. 3. Effects of CD4(178)-PE40 on chronically infected cell lines inducible for HIV-1 expression. The T-lymphocyte lines were cultured 5 days with 4 nM CD4(178)-PE40, in the absence or presence of inducer (a crude cytokine mix with TNF-alpha as the major active component). The monocyte/macrophage lines were treated for 24 hr without or with inducer (10 nM PMA), then cultured 4 days with 100 nM CD4(178)-PE40. Relative viable cell numbers are expressed as the % of the MTT values obtained for the corresponding cultures in the absence of CD4(178)-PE40.

238

hybrid toxin with the HIV envelope glycoprotein, as judged by the failure of CD4 derivatives lacking an enzymatically active domain (i.e. sCD4 or CD4(178)-PE40asp553) to promote killing.

The lengthy time interval between initial HIV infection and progression to severe immunodeficiency and disease raises the possibility that activation of latently infected cells may play a critical role in the transition from the asymptomatic to the diseased state[22]. It was therefore of interest to test the sensitivity of latently HIV-infected cells to CD4(178)-PE40, under conditions of latency vs. active virus production. Fig. 3 summarizes results with latently infected cell lines of both the T-lymphocyte[19] and monocyte/macrophage[21] lineages. In both cases, the infected cells are readily killed by the hybrid toxin, but only when virus expression is induced; the parental uninfected cell lines are not killed under either condition. Extrapolating to the clinical situation, we predict that CD4(178)-PE40 would be most effective at eliminating the subpopulation of infected cells which are actively producing virus. This may have important implications for the timing of treatment protocols involving this agent.

Inhibition of Spreading HIV Infection by CD4(178)-PE40

The therapeutic potential of CD4(178)-PE40 depends on its ability to inhibit HIV spread by selectively killing the infected cells. We have tested for this activity using several systems, including continuous T-cell lines[18,20] and primary cultures of human T-cells[20,21] and monocyte/macrophages[21]. As summarized in Table 2, the hybrid toxin greatly suppresses virus production in each of these systems. Furthermore, in both the continuous and primary T-cell cultures, the virus-mediated killing of the target cell population is significantly delayed (not shown). Both of these protective effects are due mainly to selective killing of the target cells rather than simply to neutralization by the CD4-moiety of the hybrid toxin, since CD4 derivatives lacking an active toxin moiety are much less effective[18,20,21]. However, despite the suppression of virus production and delay of target cell killing with CD4(178)-PE40, the infection eventually spreads and eliminates most of the target cell population[18,20,21]. This is not surprising, since the hybrid toxin can exert its selective killing effect only after some envelope glycoprotein has been expressed on the surface of the target cell; complete inhibition of virus spread can be achieved only if the infected cell is killed before any free virus release or cell-cell transmission occurs. We conclude that under these experimental conditions, CD4(178)-PE40 does not kill the infected cell rapidly enough to completely prevent spread of infection. However, the infected cells are killed by the toxin before producing the amount of virus that they would normally produce before succumbing to the cytopathic effect of the virus; this accounts for the dramatic suppression of virus production.

Synergistic Action of CD4(178)-PE40 and Reverse Transcriptase (RT) Inhibitors

There is a growing appreciation that effective treatment of HIV infection may require combination therapy with agents

Table 2. Inhibition of virus producton by CD4(178)-PE40 during
spreading HIV infection in various cell types. The effects of
the indicated proteins are expressed as the % of total virus
produced in control cultures with no additions. For the T-cell
line experiment[18], chronically infected H9/HTLVIIIB cells were
mixed with uninfected A3.01 cells at a ratio of 1:1000.
Cultures were maintained in the absence or presence of the
indicated CD4 derivative at 10 nM. Every two days the RT
activity was measured and the cultures were split 1:5. The
results shown are based on the total RT measured during 18
days, at which point nearly all of the target cell population
had been eliminated. For the primary T-lymphocyte
experiment[21], PHA-stimulated PBMCs were infected with HIV-1
(LAV isolate) and cultured with IL-2 in the absence or presence
of the indicated CD4 derivative at 50 nM. The results shown
are based on the p24 produced from days 3-6, which represented
at least 80% of the total which could be produced in this
system. For the primary macrophage experiment[21], primary
monocyte/macrophages obtained by elutiation were stimulated
with recombinant CSF-1. The adherent cells were infected with
HIV-1$_{AD-87(M)}$, then cultured in the absence or presence of the
indicated CD4 derivative at 10 nM. The results shown are based
on the total RT produced during 15 days.

| Human Target Cell | Virus Produced (% of control) | |
	CD4(178)-PE40	sCD4
T-cell line (A3.01)	7	55
Primary T-lymphocytes	8	95
Primary macrophages	17	120

that attack different aspects of the infection process[23].
Particularly appealing would be the use of an agent that blocks
the virus replicative cycle (thereby preventing spread of the
infection to healthy target cells), plus another that
selectively kills those cells which are already infected. The
RT inhibitors AZT and ddI represent drugs of the former
category[24], whereas CD4(178)-PE40 exemplifies the latter
category. In experiments with normal primary T-cells infected
with HIV *in vitro*, we have shown that strong mutual
potentiation of the anti-viral effects are observed with
combinations of CD4(178)-PE40 and either AZT or ddI[20]. These
two classes of drugs thus act synergistically to inhibit HIV
spread. A particularly dramatic consequence of this
synergistic action is the finding that the combination of
CD4(178)-PE40 and AZT can completely eliminate infectious HIV
from cultures of human T-cell lines[20]. In the experiment shown
in Fig. 4, A3.01 cells acutely infected with HIV were cultured
for 31 days in the presence of the indicated drug combinations,
then for an additional 3 weeks following termination of drug
treatment. Viable cell number and RT activity in the
supernatant were used as indicators of HIV spread. In the
absence of drugs, most of the cells are killed by the virus

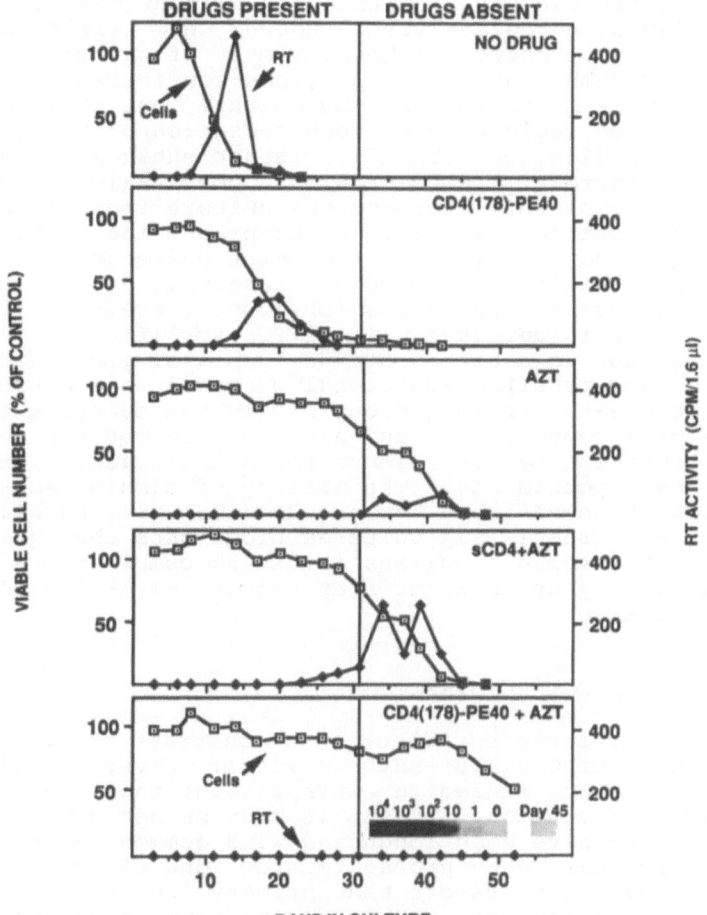

Fig. 4. Elimination of infectious HIV by combination treatment
with CD4(178)-PE40 plus AZT. A3.01 cells were acutely infected
with HIV-1 (LAV isolate), then cultured in the presence of the
indicated drugs [CD4(178)-PE40 and sCD4 at 10 nM; AZT at 1mM].
The cultures were diluted 1:12 every 2-3 days with medium
containing drugs at the original concentrations. Beginning at
day 31, the drugs were omitted from the dilution medium. Virus
production was monitored by RT activity in the supernatant.
Relative viable cell numbers were determined by the MTT
procedure, and are expressed as the percent of the value
obtained with a control culture containing no HIV. The gradual
decline in relative cell number after cessation of drug
treatment in the lowest panel is due to variation introduced by
the repeated dilution process, not to HIV spread. The
autoradiograph in the lowest panel shows the assay for HIV-1
DNA using a PCR-amplification method. On the left is a series
of signals from amplification reaction mixtures, each
containing the indicated numbers of HIV-1 proviruses, obtained
by serial dilution of a lysate from chronically infected U1
cells. The experimental sample on the right shows the absence
of a signal using a volume of lysate corresponding to 75% of
the cells remaining at day 45.

within the first two weeks, with a concomitant peak of RT. CD4(178)-PE40 alone delays virus-induced cell killing and suppresses virus production; however most of the cells eventually succumb. AZT alone provides considerable protection as long as it is maintained in the culture, but cell death and virus production rapidly ensue upon cessation of treatment. Inclusion of sCD4 along with AZT does not enhance the antiviral effect. In contrast, treatment with a combination of CD4(178)-PE40 and AZT completely protects the culture from HIV spread. Thus, virus-mediated cell killing and production of free virus are prevented, both during the treatment phase and after termination of treatment. Moreover, analysis of the cells remaining two weeks after cessation of drug treatment indicates that no HIV DNA is detected using a PCR amplification technique sensitive enough to detect provirus from a single infected cell (Fig. 4, bottom panel). Additional controls were performed to confirm the absence of infectious HIV in the cultures treated with this drug combination, and also to rule out the possiblity that the treatment merely selects for a CD4-negative, HIV-resistant cell population. The ability of simultaneous treatment with an RT inhibitor and CD4(178)-PE40 to eliminate infectious HIV from T-cell cultures highlights the therapeutic potential of treatment regimens involving combinations of a virostatic drug plus an agent that selectively kills HIV-infected cells.

CONCLUSIONS AND FUTURE DIRECTIONS

The findings presented herein demonstrate the potent anti-HIV properties of CD4(178)-PE40 *in vitro*. Several additional factors must be considered in extrapolating these results to the clinical situtation. First, it must be noted that all our studies to date have been conducted with common laboratory HIV isolates which have been propagated for long periods *in vitro*. In view of the recent report that primary isolates from infected individuals are much less sensitive than laboratory isolates to neutralization by sCD4[25], it is of great importance to test the effects of CD4(178)-PE40 against fresh patient isolates. Furthermore, the report that plasma from infected individuals contains anti-gp120 antibodies capable of inhibiting binding of sCD4[26] makes it imperative to test for possible neutralizing activity against CD4(178)-PE40 in patient serum. Problems arising from immunogenicity and non-specific toxicity, as previously observed in animals with another PE derivative[27], may restrict the time period over wich the drug can be administered. Nevertheless, the promising results obtained thus far with CD4(178)-PE40 suggest that the hybrid toxin, particularly in combination with other anti-HIV agents, could potentially eliminate a large fraction of infected cells in the body, thereby reducing the viral burden and providing a significant therapeutic effect to HIV-infected individuals.

ACKNOWLEDGEMENTS

We thank P. B. Robbins for excellent technical assistance. This work was supported in part by the National Institutes of Health Intramural AIDS Targeted Antiviral Program.

REFERENCES

1. D. R. Klatzman, J. S. McDougal, and P. J. Maddon, Immunodef. Rev. 2:43 (1990).
2. D. H. Smith, R. A. Byrn, S. A. Marsters, T. Gregory, J. E. Groopman, and D. J. Capon, Science 238:1704 (1987).
3. R. A. Fisher, J. M. Bertonis, W. Meier, V. A. Johnson, D. S. Costopoulos, T.Liu, R. Tizard, B. D. Walker, M. S. Hirsch, R. T. Schooley, and R. A. Flavell, Nature 331:76 (1988).
4. R. E. Hussey, N. E. Richardson, M. Lowalski, N. R. Brown, H.-C. Chang, R. F. Siliciano, T. Dorfman, B. Walker, J. Sodroski, and E. L. Reinherz, Nature 331:78 (1988).
5. K. C. Deen, J. S. McDougal, R. Inacker, G. Folena-Wasserman, J. Arthos, J. Rosenberg, P. J. Maddon, R. Axel and R. W. Sweet, Nature 331:82 (1988).
6. A. Traunecker, W. Luke, and K. Karjalainen, Nature 331:84 (1988).
7. E. A. Berger, T. R. Fuerst and B. Moss, Proc. Natl. Acad. Sci., USA 85:2357 (1988).
8. D. J. Capon, S. M. Chamow, J. Mordenti, S. A. Marsters, T. Gregory, H. Mitsuya, R. A. Byrn, C. Lucas, F. M. Wurm, J. E. Groopman, and D. H. Smith, Nature 337:525 (1989).
9. A. Traunecker, J. Schneider, H. Kiefer, and K. Karjalainen, Nature 339:68 (1989).
10. T. Mizukami, C. D. Smith, E. A. Berger, and B. Moss, Int. Conf. AIDS 5:556 (1989).
11. R. A. Byrn, J. Mordenti, C. Lucas, D. Smith, S. A. Marsters, J. S. Johnson, P. Cossum, S. M. Chamow, F. M. Wurm, T. Gregory, J. E. Groopman, and D. J. Capon, Nature 344:667 (1990).
12. G. Zettlmeissl, J. P. Gregersen, J. M. Duport, S. Mehdi, G. Reiner, and B. Seed, DNA Cell. Biol. 9:347 (1990).
13. Y. Mouneimne, P. F. Tosi, R. Barhoumi, and C. Nicolau, Biochim. Biophys. Acta 1027:53 (1990).
14. V. K. Chaudhary, T. Mizukami, T. R. Fuerst, D. J. FitzGerald, B. Moss, I. Pastan, and E. A. Berger, Nature 333:369 (1988).
15. M. A. Till, V. Ghetie, T. Gregory, E. J. Patzer, J. P. Porter, J. W. Uhr, D. J. Capon, and E. S. Vitetta, Science 242:1166 (1988).
16. A. Cudd, C. A. Noonan, P. F. Tosi, J. L. Melnick, and C. Nicolau, J. Acquir. Immune Defic. Syndr. 3:109 (1990).
17. I. Pastan and D. FitzGerald, J. Biol. Chem. 264:15157 (1989).
18. E. A. Berger, K. A. Clouse, V. K. Chaudhary, S. Chakrabarti, D. J. FitzGerald, I. Pastan, and B. Moss, Proc. Natl. Acad. Sci. USA 86:9539 (1989).
19. E. A. Berger, V. K. Chaudhary, K. A. Clouse, D. Jaraquemada, J. A. Nicholas, K. L. Rubino, D. J. FitzGerald, I. Pastan, and B. Moss, AIDS Res. Hum. Retroviruses 6:795 (1990).

20. P. Ashorn, B. Moss, J. N. Weinstein, V. K. Chaudhary, D. J. FitzGerald, I. Pastan, and E. A. Berger, Proc. Natl. Acad. Sci. USA 87:8889 (1990).
21. P. Ashorn, G. Englund, M. A. Martin, B. Moss, and E. A. Berger, submitted.
22. Z. F. Rosenberg and A. S. Fauci, AIDS Res. Hum. Retroviruses 5:1 (1989).
23. V. A. Johnson and M. S. Hirsch, in: "Antiviral Chemotherapy: New Directions for Clinical Applications and Research", J. Mills and L. Corey, eds., Elsevier, New York, Vol. 2:275 (1989).
24. R. Yarchoan, H. Mitsuya, C. E. Myers, and S. Broder, N. Engl. J. Med. 321:726 (1989).
25. E. S. Daar, X. L. Li, T. Moudgil, and D. D. Ho, Proc. Natl. Acad. Sci., USA 87:6574 (1990).
26. L. N. Callahan and M. A. Norcross, Lancet 734 (1989).
27. H. Loberboum-Galski, L. V. Barrett, R. L. Kirkman, M. Ogata, M. C. Willingham, D. J. FitzGerald, and I. Pastan, Proc. Natl. Acad. Sci. USA 86:1008 (1989).

BINDING OF A REDUCED-PEPTIDE INHIBITOR AND A STATINE-CONTAINING

INHIBITOR TO THE PROTEASE FROM THE HUMAN IMMUNODEFICIENCY VIRUS

Paula M. D. Fitzgerald, Brian M. McKeever, Jody F. VanMiddlesworth
and James P. Springer

Merck Sharp & Dohme Research Laboratories
P. O. Box 2000, Ry80M203
Rahway, New Jersey 07065 USA

INTRODUCTION

In the replication cycle of human immunodeficiency virus, viral mRNA is translated as a polyprotein precursor. Cleavage by a viral-encoded protease is required for release of mature viral structural proteins and enzymes from the precursor; the key role of the protease in the maturation of this virus, the causative agent in acquired immune deficiency syndrome (AIDS), makes the it an attractive target for therapeutic intervention in the treatment of AIDS. To aid in the design of safe and effective inhibitors, we have studied the structure of the protease from human immunodeficiency virus type 1 (HIV-1), both in its native state[1] and in complex with a statine-containing inhibitor, acetyl-pepstatin[2]. We report here the determination of the structure of the protease in complex with a reduced-peptide inhibitor, L-365,862 (Ser-Gln-Asn-PheΨ(CH$_2$-N)Pro-Ile-Val-Gln). The amino acid sequence of this peptide corresponds to the cleavage site between the membrane-associated and capsid proteins in the polyprotein (with the substitution of Phe for Tyr); the replacement of the scissile bond in this octapeptide with a reduced-peptide linkage generates an inhibitor that has been used as an affinity ligand in the purification of the protease[3]. Although the structure of this complex is still being refined, a preliminary comparison to the structure of the complex between HIV-1 protease and acetyl-pepstatin reveals considerable difference in the length of hydrogen bonds between the protein and the inhibitor. The conformation of the backbone of most of the inhibitor is extended, but the backbone has a turn conformation at the position of the amino-terminal serine residue; turn-forming at this position may facilitate binding between the protease and its natural substrates.

METHODS

Recombinant HIV-1 protease (NY5 strain) was prepared as described previously[3]. The inhibitor (dissolved in DMSO) was added to a 6 mg/ml protein stock solution to achieve a final inhibitor to active site ratio of 2 to 1. Crystals were grown by the hanging drop method; each crystallization experiment consisted of mixing 2 µl of inhibited protein stock solution with 2 µl of precipitation solution and equilibrating this drop over a well containing 1 ml of the precipitating solution. Crystals grew from precipitation solutions that contained 800-1200 mM sodium chloride, 100 mM acetic acid/sodium acetate buffer, pH 5.5-6.0, and 10 mM trans-4,5-dihydroxy-1,2-dithiane.

Diffraction data were measured on a Nicolet multiwire area detector mounted on a Rigaku rotating anode X-ray generator operating at 50 kV and 60 mA. Data were measured to a nominal resolution of 2.5 Å from a single crystal that measured 0.5 x 0.3 x 0.04 mm. The unit cell of this crystal (a=58.36, b=87.02, c=46.56 Å, space group P2$_1$2$_1$2) was approximately isomorphous to that of crystals of the previously determined complex of HIV-1 pro-

Advances in Molecular Biology and Targeted Treatment for AIDS
Edited by A. Kumar, Plenum Press, New York, 1991

Table 1. Summary of current restrained least-squares refinement model

Root mean square deviation	σ	Value in current model	Number of constraints total	> 2 σ	> 3 σ
From ideal distances (Å)					
Bond distances	0.020	0.033	1615	30	4
Angle distances	0.030	0.051	2194	169	50
Planar 1-4 distances	0.040	0.057	517	21	8
From ideal planarity (Å)	0.020	0.015	265	0	0
From ideal chirality (Å^3)	0.150	0.240	263	14	5
From permitted contact distances (Å)					
Single torsion contacts	0.500	0.262	542	0	0
Multiple torsion contacts	0.500	0.336	654	4	0
Possible H-bonds	0.500	0.275	98	0	0
From ideal torsion angles (°)					
Planar groups (0 or 180)	3.0	3.1	210	4	1
Staggered groups (±60 or 180)	15.0	25.6	302	47	28
Orthonormal groups (±90)	20.0	17.9	13	0	0

Resolution range (Å)	Number of Reflections	R
8.00 - 6.01	355	0.261
6.01 - 5.02	417	0.190
5.02 - 4.40	491	0.140
4.40 - 3.96	533	0.148
3.96 - 3.63	595	0.153
3.63 - 3.37	611	0.157
3.37 - 3.16	594	0.188
3.16 - 2.99	540	0.193
2.99 - 2.84	450	0.218
2.84 - 2.71	333	0.206
2.71 - 2.60	326	0.218
2.60 - 2.50	305	0.200
8.00 - 2.50	5550	0.177

tease and acetyl-pepstatin. Coordinates from an advanced stage of the refinement of that structure, modified by removal of the inhibitor, the solvent model, and all side chains modeled in alternate conformations, were used as the starting point in creating a model for the structure of the complex between HIV-1 protease and L-365,862. An electron density map calculated with coefficients $2F_o-F_c$ revealed clearly defined electron density for the reduced octapeptide inhibitor, and a model for the inhibitor was built to that density. Refinement of the structure proceeded by alternating 10 cycles of least-squares refinement using program PROLSQ[4] and sessions of manual rebuilding of the model using program FRODO[5]. A solvent model was gradually introduced by inspection of electron density maps calculated with coefficients F_o-F_c. The current restrained least-squares refinement model, which is described in Table 1, contains 88 solvent molecules and 8 amino acid side chains modeled in two partially-occupied conformations. Although modeling of alternative conformations might seem inappropriate at this resolution, electron density maps clearly indicated alternate conformations and each side chain so modeled had been previously observed to adopt alternate conformations in the 2.0 Å refined structure of the complex between HIV-1 protease and acetyl-pepstatin.

RESULTS

Hydrogen-bonding interactions

Although the pattern of hydrogen bonds between the inhibitor L-365,862 and HIV-1

Table 2. Interatomic distances corresponding to potential hydrogen bonding interactions in complexes between HIV-1 protease and inhibitors

Atom in protein or solvent	Atom in Acetyl-pepstatin	Distance in Å	Atom in L-365-862	Distance in Å
Residues in body of protein				
Asp-230 $O^{\delta 1}$			Ser-1 N	3.0
Asp-229 $O^{\delta 1}$	Val-2 N	3.3	Gln-2 N	3.5
Asp-229 N	Val-2 O	3.3	Gln-2 O	3.1
Gly-227 O	Sta-4 N	3.0	Phe-4 N	2.4
Asp-225 $O^{\delta 1}$	Sta-4 OH	2.5		
Asp-225 $O^{\delta 2}$	Sta-4 OH	2.6		
Asp-25 $O^{\delta 2}$	Sta-4 OH	3.6		
Asp-25 $O^{\delta 1}$	Sta-4 OH	3.9		
Gly-27 O	Ala-5 N	3.1	Ile-6 N	3.0
Asp-29 N	Ala-5 O	3.1	Ile-6 O	3.2
Asp-29 $O^{\delta 1}$			Gln-8 N	3.4
Asp-30 $O^{\delta 1}$	Sta-6 $O^{\tau 1}$	2.2	Gln-8 $O^{\tau 1}$	3.5
Residues in flap				
Gly-248 N	Ac-1 O	3.1	Ser-1 O	3.0
Gly-248 O	Val-3 N	2.9	Asn-3 N	2.9
Solvent atom	Val-3 O	2.9	Asn-3 O	3.7
Solvent atom	Sta-4 O	2.3	Pro-5 O	3.6
Gly-48 O	Sta-6 N	2.7	Val-7 N	4.2
Gly-48 N	Sta-6 OH	3.0	Val-7 O	3.9

protease is similar to that reported previously for complexes of the protease with the reduced peptide inhibitor MVT-101[6] and with acetyl-pepstatin[2], the lengths of some of the hydrogen-bonding interactions are substantially different. As is illustrated in Table 2, the hydrogen-bonding distances are of equivalent magnitude between atoms in the amino-terminal portion of the inhibitor and hydrogen-bonding partners in the protein, but the distances are significantly longer between atoms in the carboxyl-terminal portion of the inhibitor and atoms in the protein. The potential hydrogen bonds that involve atoms in residue Gln-8 are at the long end of the range that is usually considered to represent hydrogen bonding, and the distances between atoms in residue Val-7 and potential hydrogen-bonding partners in the flap of the protein are too long to be considered hydrogen bonds at all.

The geometry of the hydrogen bonds involving the solvent atom that bridges between the inhibitor and residues in the flap is also considerably different in this complex. This solvent atom, which is assumed to be the oxygen of a water molecule, accepts hydrogen bonds from the amide nitrogens of residues Ile-50 and Ile-250 in the flaps, and donates hydrogen bonds to two carboxyl oxygens in the inhibitor. In the previously determined structures, the hydrogen bond distances are 2.5, 2.7, 3.0, and 3.3 Å (for MVT-101) and 2.4, 2.9, 2.9, and 3.0 Å (for acetyl-pepstatin). In the L-365,862 complex the distances are significantly longer (2.6, 3.4, 3.6, and 3.7 Å). In the acetyl-pepstatin complex, the hydrogen bonding is nearly tetrahedral at the solvent atom, with angles in the range 95-111°; in the L-365,862 complex the geometry is more distorted, with angles in the range 81-116°.

Conformation of the inhibitor

The conformation of the inhibitor is illustrated in Fig. 1. As seen in other complexes of HIV-1 protease with inhibitors, the conformation of the backbone of L-365,862 is largely extended. The replacement of the amide linkage between residues Phe-4 and Pro-5 with the CH_2-N linkage allows the peptide torsion angle to deviate from planarity; the value of 151° observed for this angle is similar to the value of 142° observed for the corresponding angle in the reduced peptide inhibitor MVT-101[6]. The backbone of the inhibitor deviates from an extended conformation at residue Ser-1, where the conformational torsion angles are those of a β-turn.

In the current model for the structure of the complex of L-365,862 and HIV-1 protease, the electron density is consistent with only a single binding mode for the inhibitor. A similar unidirectional binding has been reported for MVT-101[6], whereas the binding of acetyl-pepstatin is bidirectional[2]. The protease is chemically a dimer and the structure of the native enzyme is twofold symmetric. However, in the complexes with inhibitors the protease dimer shows significant structural deviation from twofold symmetry. Despite this asymmetry, acetyl-pepstatin binds to the enzyme in two, roughly twofold symmetric, orientations. The electron density maps for the L-365,862 complex have been examined carefully, and while a twofold related binding mode is not clearly indicated, there is weak unexplained density adjacent to the side chain of Pro-5. This density may correspond to a minor population of inhibitor molecules bound in the twofold related direction (the position occupied by Pro-5 in the current model would be occupied by the larger Phe-4 in the alternate binding mode). Further refinement of the structure should resolve this issue.

DISCUSSION

The structure of the complex of L-365,862 and HIV-1 protease displays a weakened pattern of hydrogen bonding when compared to the structures of the protease in the MVT-101 and acetyl-pepstatin complexes. It is tempting to correlate this loosened hydrogen-bonding pattern with the potency of the compound as an inhibitor, but such correlations must be made with extreme caution. The measured K_i for L-365,862 is 1 μM[3], which is weaker than the K_i of 0.78 μM reported for MVT-101[6], but this difference may not be large enough to be significant. The difference is greater between the K_i for L-365,862 and the value of 20 nM reported for acetyl-pepstatin[7], but the K_i for acetyl-pepstatin is highly dependent on pH and ionic strength. Indeed, each of these inhibition constants has been measured under different conditions, thereby making correlation of structure and potency of inhibitor difficult.

With the exception of the reduced-peptide linkage at the scissile bond, L-365,862 is a natural substrate for the protease. Although the natural substrates of retroviral proteases show great diversity, certain patterns have been noted[8]. One of these is the preference for the amino acids Ser, Thr, Pro, Ala, and Gly, amino acids commonly observed in β-turns in protein structures, at the P4 and P4' positions of the substrate sequence. HIV-1 protease is a small enzyme, and model building of substrates to the protease indicates only six substrate binding sites (three on each side of the scissile bond) on the active-site face of the enzyme.

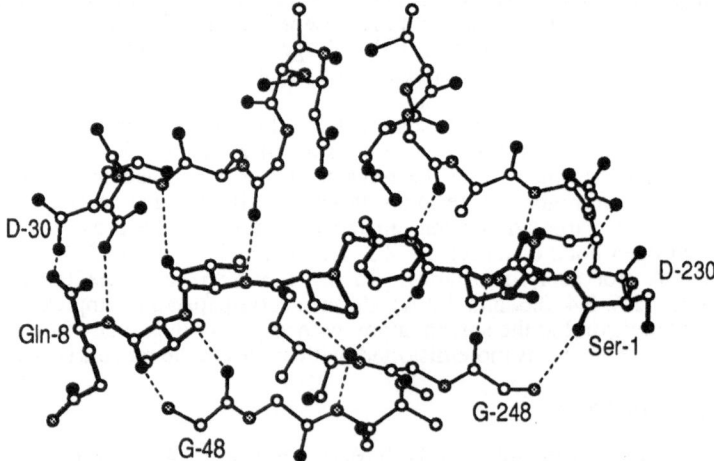

Fig. 1. The conformation of the inhibitior L-365,862 as bound to HIV-1 protease. Carbon atoms are indicated by open circle, nitrogen atoms by grey circles and oxygen atoms by black circles. Potential hydrogen bonds are indicated by dashed lines. Bonds between atoms in the inhibitor are drawn as thick lines; atoms in the protein are connected with thin lines.

The natural substrates of the enzyme are large polyproteins, and if more than eight residues were to interact with the protease there would have to be a turn at the P4 and P4' residues. In the structure of the complex between HIV-1 protease and L-365,862, we find that residue Ser-1 (in the P4 position) does indeed adopt a turn conformation. The density for residue Gln-8 (in the P4' position) is much less well resolved, and although the current model for the structure has been built in a turn conformation, the alternate extended conformation cannot be ruled out.

ACKNOWLEDGEMENTS

We would like to thank Paul Darke, Jill Heimbach, Chih-Tai Leu, and Wayne Herber of the Department of Molecular Biology, Merck Sharp and Dohme Research Laboratories, West Point, Pennsylvania for preparation of HIV-1 protease and Victor Garsky, James Guare and Ken Stauffer of the Department of Medicinal Chemistry, Merck Sharp & Dohme Research Laboratories, West Point, Pennsylvania for preparation of the inhibitor L-365,862.

REFERENCES

1. M. A. Navia, P. M. D. Fitzgerald, B. M. McKeever, C.-T. Leu, J. C. Heimbach, W. K. Herber, I. S. Sigal, P. L. Darke, and J. P. Springer, Three-dimensional structure of aspartyl-protease from human immunodeficiency virus HIV-1, *Nature* 337:615 (1989).
2. P. M. D. Fitzgerald, B. M. McKeever, J. F. VanMiddlesworth, J. P. Springer, J. C. Heimbach, C.-T. Leu, W. K. Herber, R. A. F. Dixon, and P. L. Darke, Crystallographic Analysis of a Complex between Human Immunodeficiency Virus Type 1 Protease and Acetyl-Pepstatin at 2.0-Å Resolution, *J. Biol. Chem.* 265:14209 (1990).
3. J. C. Heimbach, V. M. Garsky, S. R. Michelson, R. A. F. Dixon, I. S. Sigal, and P. L. Darke, Affinity Purification of the HIV-1 Protease, *Biochem. Biophys. Res. Commun.* 164:955 (1989).
4. W. A. Hendrickson, Stereochemically Restrained Refinement of Macromolecular Structures, *Meth. Enzymol.* 115:252 (1985).
5. T. A. Jones, A Graphics Model Building and Refinement System for Macromolecules, *J. Appl. Crystallogr.* 11:268 (1978).
6. M. Miller, J. Schneider, B. K. Sathyanarayana, M. V. Toth, G. R. Marshall, L. Clawson, L. Selk, S. B. H. Kent, and A. Wlodawer, Structure of Complex of Synthetic HIV-1 Protease with a Substrate-Based Inhibitor at 2.3 Å Resolution, *Science* 246:1149 (1989).
7. A. D. Richards, R. Roberts, B. M. Dunn, M. C. Graves, and J. Kay, Effective blocking of HIV-1 proteinase activity by characteristic inhibitors of aspartic proteinases, *FEBS Lett.* 247:113 (1989).
8. L. E. Henderson, R. E. Benveniste, R. Sowder, T. D. Copeland, A. M. Schultz, and S. Oroszlan, Molecular Characterization of *gag* Proteins from Simian Immunodeficiency Virus (SIV$_{Mne}$), *J. Virol.* 62:2587 (1988).

A T CELL CLONE WHICH REFLECTS THE FUNCTIONAL DEFECTS

OBSERVED IN THE T CELLS OF AIDS PATIENTS

Perez, V.L. and T.M. Folks

Retrovirus Diseases Branch, Division of Viral and
Rickettsial Diseases, Center for Infectious
Diseases, Centers for Disease Control, Public Health
Service, U.S. Department of Health and Human
Services, 1600 Clifton Road N.E., Atlanta, Georgia
U.S.A.

INTRODUCTION

Infection of T cells by the Human Immunodeficiency Virus
(HIV) renders the body susceptible to attack by opportunistic
infections as these cells of the immune system are not only
killed by the virus, but are also rendered defective in their
function (1,2,3,4). It is well documented that T cell defects
associated with HIV infection in AIDS patients include reduced
production of the lymphokine, IL-2, and an inability to
mobilize Ca^{++} following mitogenic or antigenic stimulation. T
cell dysfunction has also been attributed to downmodulation of
CD3 and CD4, possibly as a result in their uncoupling (5,6,7).
Since many of these events that lead to T cell activation are
under some of the same controls as events involved in the
activation of HIV (8-9), a T cell model system to study
physiologic influences on HIV expression and the effects of the
virus on the cell's activation signals would be of great value.

T-CELL MODEL DEVELOPMENT

Cellular models of HIV infection are useful for in vitro
studies, especially when they contain the entire proviral
genome and respond to physiological stimuli. Cytokine and
antigen-induced modification of HIV expression have been
observed in both acutely and chronically HIV infected cellular
models (10-11). Many experiments involving HIV infected normal
PBL are difficult to evaluate because of individual-to-
individual variation, randomly mixed infections, and the
unavailability of infected clones. Because of these
limitations, tumor cell models are often desirable for in vitro
HIV T cell studies.

However, there are limited number of TcR/CD3$^+$ bearing CD4$^+$
tumor cell lines susceptible to HIV infection. This, of
course, makes TcR activation studies of HIV quite difficult.
The CD3$^+$ tumor line, Jurkat (12), is most extensively used as a

human cell line model for T cell activation studies, especially when TcR-CD3 crosslinking studies are involved (13,14). Because of this we chose the Jurkat cell line as our potential model. After infecting Jurkat cells with HIV-1, we cloned by limiting dilution the surviving cells from the acute cytopathic infection. These survivor clones were then screened for reverse transcriptase (RT) production. From these populations, we further subcloned one HIV infected Jurkat clone, J1.1, that contained a latent expressing HIV genome.

1) EXPRESSION OF CD3 ON J1.1

Reports have varied concerning the effects HIV infection has on the expression of the CD3 molecule (11.5) Using the J1.1 cell line as a representative chronically HIV infected clone, we tested for surface expression of CD3 and compared this to CD3 expression of 5 uninfected Jurkat clones by Flow Cytometry (FCM).

Clone-to-clone variability was found to be remarkably small (less than 5 channels of fluorescence intensity). Even though the J1.1 cell was positive for surface expression of CD3 as compared to an unstained control, all of the uninfected Jurkat clones showed a median fluorescence intensity of approximately 10-fold greater than that of the HIV infected J1.1 line.

2) EFFECTS OF CROSSLINKING CD3 ON Ca^{++} MOBILIZATION

The finding of reduced surface expression by the HIV-1 infected J1.1 clone suggested that this clone might be defective in additional functions linked to TcR signaling. Previous reports indicate that T-cells from HIV-1 infected individuals are defective in their capacity to mobilize Ca^{++} (15,7). Therefore, uninfected Jurkat cells and J1.1 cells were treated with anti-CD3 monoclonal antibody and immediately tested for Ca^{++} mobilization by the indo-loading technique. Uninfected Jurkat, when stimulated with anti-CD3, could mobilize intracellular Ca^{++}, while J1.1 showed approximately a 60% decrease following similar treatment. This data and those of Linnette et al. support the concept that Ca^{++} mobilization is defective in HIV infected T cells.

There are several possible mechanisms which may explain the Ca^{++} defect in the J1.1 clone. As reported, the J1.1 clone has a reduced surface density of CD3. This could imply too few CD3 molecules available for effective functional crosslinking or a potential uncoupling of CD3 with other transducing molecules responsible for transmembrane signaling. Since CD4 is completely modulated from the surface of the J1.1 clone (most likely due to gp-120/CD4 complexes), it is conceivable that CD3-CD4 interactions are critical during anti-CD3 stimulation for maximal Ca^{++} mobilization and cellular signaling.

It is also very difficult to rule out other viral gene products different from gp120 which might contribute to Ca^{++} pathway defects. Ruegg and Strand (17) have recently reported suppression of anti-CD3 induced Ca^{++} mobilization by an HIV gp41 peptide. It was interesting in their study that the levels of one of the second messenger stimulants of Ca^{++}, inositol triphosphate (IP_3), were unaffected by the peptide. This indicates that the inhibition observed must be distal to IP_3

formation or affecting a co-stimulatory signal. Indeed, their findings further showed that protein kinase C (PKC), a stimulant of Ca^{++} export by way of a Ca^{++}-ATPase, was also inhibited by the peptide from phosphorylating the CD3 gamma-chain. It will now be important to determine if the Ca^{++} defect observed in the J1.1 clone is also linked to a similar defect in PKC activity.

3) EFFECTS OF CROSSLINKING CD3 ON IL-2 PRODUCTION

Secretion of IL-2 has also been described to be defective in the T-cells from AIDS patients (15,2). Therefore, the J1.1 clone was further investigated for its ability to produce IL-2 in response to TcR stimulation. After solid phase crosslinking of cells with anti-CD3 for 48 hours, supernatant

culture fluids were added to HT-2 indicator cells to test for IL-2 activity. J1.1 cells were found to secrete approximately 80% less IL-2 activity than uninfected Jurkat cells after 48 hours of anti-CD3 crosslinked activation.

Since IL-2 secretion is dependent on CD3 activation and Ca^{++} mobilization, it was not unsuspected that J1.1 might also be defective in this function.

Table 1. T-cell defects seen in the HIV-1 infected J1.1 clone

	CD3 Expression	Ca^{++} mobilization	IL-2 production
Jurkat	Normal	800nM of Ca^{++} mobilization	100% maximal
J1.1 HIV+ clone	Decreased 10-fold	Decreased 60% (300nM)	Decreased 80%

CONCLUSION

The HIV infected Jurkat clone, J1.1, is a new model which demonstrates many of the HIV induced T cell defects observed in the T cells of AIDS patients. These defects can be observed when functional and phenotypic comparisons are carried out between J1.1 and uninfected Jurkat cells.

In addition, pathways have been elucidated where by induction of physiological T cell signals can be studied in the context of HIV infection. Cellular and biochemical dissection of these pathways should contribute to a better understanding of the influence of HIV on T cell function and may give rise to new ideas that will help to develop more effective therapeutic approaches to AIDS.

REFERENCES

1. Fauci, A.S. 1988. The human immunodeficiency virus: infectivity and mechanisms of pathogenesis. Science 239:617-622.

2. Lane, H.C., J.M. Depper, W.C. Greene, G. Whalen, T.A. Waldman and A.S. Fauci. 1985. Qualitative loss of immune function in patients with the acquired immunodeficiency syndrome. Evidence for a selective defect in soluble antigen recognition. N.Eng. J. Med. 313:79-84.

3. Folks, T.M., J. Kelly, S. Benn, A. Kinter, J. Justement, J. Gold, J. Redfield, K. Sell and A.S. Fauci. 1986. Susceptibility of normal human lymphocytes to infection with HTLV-III/LAV. J. Immunol. 136:4049-4053.

4. Giorgi, J.V., J.L. Fahey, D.C. Smith, L.E. Hultin, H.L. Chang, R.T. Mitsuyasu, R. Detels. 1987. Early effects of HIV on CD4 lymphocytes in vivo. J. Immunol. 138:3725-3730.

5. Stevenson, M., X. Zhang, and D.J. Volsky. 1987. Downregulation of cell surface molecules during noncytopathic infection of T cells with human immunodeficiency virus. J. Virol. 61:3741-3748.

6. Hoxie, J.A., B.S. Haggarty, J.L. Rackowski, N. Pillsbury, and J.A. Levy. 1985. Persistent noncytopathic infection of normal human T lymphocytes with AIDS-associated retrovirus. Science 229:1400-1402.

7. Linette, G.P., R.J. Hartzman, J.A. Ledbetter, and C. June. 1988. HIV-1-infected T cells show a selective Signaling defect after perturbation of CD3-Antigen receptor. Science 241:573-576.

8. Harada, S., Y. Koyanagi, H. Nakashima, N. Kobayashi, and N. Yamamoto. 1986. Tumor promoter, TPA, enhances replication of HTLV-III/LAV. Virology 154:249-250.

9. Siekevitz, M., S.F. Josephs, M. Dukovich, N. Peffer, F. Wong-Staal, W.C. Greene. 1987. Activation of the HIV-1 LTR by T cell mitogens and the transactivator protein of HTLV-I. Science 238:1575-1578.

10. Clouse, K.A., D. Powell, I. Washington, G. Poli, K. Strebel, W. Farrar, P. Barstad. J. Kovacs, A.S. Fauci, and T.M. Folks. 1989. Monokine regulation of human immunodeficiency virus-1 expression in a chronically infected human T cell clone. J. Immunol. 142:431-438.

11. Lawrence, J., S.M. Friedman, E.K. Chartash, M.K. Crow and D.N. Posnett. 1989. Human immunodeficiency virus infection of helper T cell clones. J. Clin. Invest. 83:1843-1848.

12. Tong-Starksen, S., P.A. Luciew, and B.M. Peterlin. 1989. Signaling through T Lymphocyte proteins, TCR/CD3 and CD28, activates the HIV-1 long terminal repeat. J. Immunol. 142:702-707.

13. Jin, Y-J., D.R. Kaplan, M. White, G.C. Spagnoli, T.M. Roberts, and E.L. Reinherz. 1990. Stimulation via CD3-Ti but not CD2 induces rapid tyrosine phosphorylation of a 68kDa protein in the human Jurkat T cell line. J. Immunol. 144:647-652.

14. Mauger, B., A. Weiss, J. Imboden, T. Laing, and J.D. Stobo. 1987. The role of protein kinase C in transmembrane signaling by the T-cell antigen receptor complex:effects of stimulation with soluble or immobilized CD3 antibodies. J. Immunol. 139:2755.

15. Gupta, S., and B. Vayuvegula. 1987. Human immunodeficiency virus-associated changes in signal transduction. J. Clin. Imunol. 7:486-489.

16. Murray, H.W., B.Y. Rubin, H. Masur, and R.B. Roberts. 1984. Impaired production of lymphokines and immune (gamma) interferon in the acquired immunodeficiency syndrome. N. Eng. J. Med. 310:883-889.

17. Ruegg, C.L., and Strand, M. 1990. Inhibition of protein kinase C and anti-CD3 induced Ca^{++} influx in Jurkat T-cells by a synthetic peptide with sequence identity to HIV-1 gp41. J. Immunol. 144:3928-290.

ROLES OF NUCLEOCAPSID CYSTEINE ARRAYS IN RETROVIRAL ASSEMBLY AND
REPLICATION: POSSIBLE MECHANISMS IN RNA ENCAPSIDATION

Robert J. Gorelick[1], Stephen M. Nigida, Jr.[1], Larry O. Arthur[1],
Louis E. Henderson[1], and Alan Rein[2*]

[1]AIDS Vaccine Program, PRI/DynCorp and [2]Laboratory of Molecular
Virology and Carcinogenesis, ABL-Basic Research Program, NCI-
Frederick Cancer Research and Development Center, Frederick
MD 21702

SUMMARY

The nucleocapsid (NC) proteins of all retroviruses contain either
one or two copies of a sequence motif, $C-X_2-C-X_4-H-X_4-C$, which has been
termed the "cysteine array" or Cys-His box. Studies from several
laboratories have shown that mutants in this motif direct the production
of virus particles; these particles are structurally normal in many
respects, but are either partially or completely deficient in genomic
RNA. Further, even when some RNA is encapsidated, the particles are
almost completely noninfectious.

We review some of the known properties of these particles and of
particles lacking genomic RNA for other reasons. Based on these
properties, we propose that only dimeric RNA is encapsidated, and that
the cysteine array(s) are involved in a positive selection or "search"
for this dimeric RNA during virus assembly. Further speculations on the
nature of the protein-RNA interactions in RNA packaging are also
presented.

INTRODUCTION

Experiments in a variety of retroviral systems have shown clearly
that expression of the Gag polyprotein precursor is sufficient for
assembly of virus particles (e.g., Shields et al., 1978; Gheysen et al.,
1989). Thus, all of the functions required for virus assembly must be
performed by this protein.

Following virus assembly and release, the Gag polyprotein is
cleaved by the viral protease into a series of smaller proteins. One of
these cleavage products is termed the nucleocapsid (NC) protein. NC
proteins are small, basic proteins which are apparently associated with
genomic RNA in the mature virion and which exhibit nonspecific binding
to single-stranded nucleic acids in vitro. It seems possible that one
function performed by this protein (and by the corresponding domain of
the Gag polyprotein before cleavage) is nonspecific association with the
viral RNA, perhaps like that of histones with DNA, and that the binding
observed in vitro is a manifestation of this functional activity.
However, it is now clear that NC performs a complex series of functions
which are only partially understood at present.

---

*Corresponding author

Advances in Molecular Biology and Targeted Treatment for AIDS
Edited by A. Kumar; Plenum Press, New York, 1991

In all retroviruses, the NC protein has either one or two copies of a sequence motif, $C-X_2-C-X_4-H-X_4-C$, which has been termed the "cysteine array" or "Cys-His box" (Henderson et al., 1981; Oroszlan and Copeland, 1985; Covey, 1986). The ubiquity of this motif in retroviral NC proteins suggested that it might perform an essential role in retroviral replication. In addition, the conserved sequence bears a striking resemblance to the "zinc finger" motif implicated in recognition of specific DNA sequences by a wide variety of eukaryotic transcription factors (Berg, 1986; Evans & Hollenberg, 1988). This similarity raises the possibility that the cysteine arrays might be involved in specific nucleic acid recognition during retroviral assembly. We and others have begun to analyze the function(s) of this motif by generating and characterizing mutants in Moloney murine leukemia virus (M-MuLV), avian sarcoma virus, and human immunodeficiency virus-1 (HIV-1).

The results of these studies point to a number of interesting conclusions concerning retrovirus assembly and replication. In the present report, we review some of these findings and speculate on their possible implications. We propose that the cysteine arrays function during virus assembly in a "search" or positive selection for a specific sequence or structure present in the genomic RNA of the virus. Some of the ramifications of this idea are discussed below.

RESULTS AND DISCUSSION

Mutants of M-MuLV

The sequence of the single cysteine array of the M-MuLV NC protein is as follows.

Cys-Ala-Tyr-Cys-Lys-Glu-Lys-Gly-His-Trp-Ala-Lys-Asp-Cys.
26   27   28   29   30   31   32   33   34   35   36   37   38   39

We generated point mutations in an infectious molecular clone of M-MuLV by oligonucleotide-directed mutagenesis, individually changing each of the underlined residues to serine. We also generated one double mutant in which the cysteines at both position 26 and position 29 were changed to serine.

As we have reported (Gorelick et al., 1988), each of these mutants exhibited a remarkable phenotype: when expressed in mammalian cells, they gave rise to virus particles which appeared to be structurally normal, but contained little or no detectable genomic RNA (Fig. 1). Thus, the cysteine array appears to function in the specific recognition and/or packaging of genomic RNA during virus assembly. Further, this function is evidently exquisitely sensitive to changes in the cysteine array, since merely replacing a sulfur with an oxygen atom, as in C26S, C29S, and C39S, prevents efficient RNA packaging.

Cysteine-array Mutants are Recessive in Mixed Infection

It was of interest to determine whether the hypothetical "search" function could successfully be performed by wild-type protein in the presence of mutant protein, or whether cysteine-array mutants would be dominant negative mutants. We therefore superinfected cells containing the mutants with the 10A1 isolate of wild-type MuLV. Supernatants were harvested from these dually infected cells five days later, and the titer of replication-competent MuLV in these fluids was determined by the S^+L^- focus assay. As is shown in Table 1, the wild-type virus grew to virtually the same titer in the cells containing the mutants as in control cells.

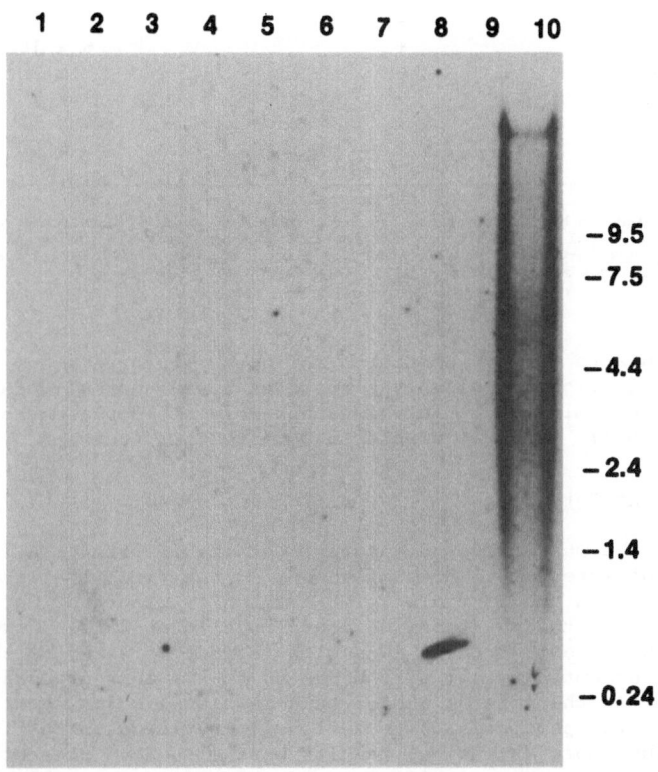

Fig. 1.   Northern analysis of mutant and wild-type viral RNA with an
M-MuLV probe.   Samples were adjusted for equal amounts of
p30^{CA} content.   An M-MuLV 2.1-kilobase nick-translated ^{32}P-
labeled *gag-pol* probe was used.   Lane 1, negative control,
pGCcos3neo; lane 2, mutant C26S; lane 3, mutant C26S/C29S;
lane 4, mutant C29S; lane 5, mutant Y28S; lane 6, mutant
W35S; lane 7, mutant C39S; lanes 8 and 9, 1:100 and 1:10
dilutions of wild-type M-MuLV, respectively; lane 10, wild-
type M-MuLV.   (Reproduced from Gorelick et al., 1988).

Table 1.   M-MuLV NC Mutants are Recessive in Mixed Infection.

Virus	Superinfection	Progeny Virus Titer
C26S	-	<4 x 10^0 FIU/ml
	+	1.7 x 10^3
C29S	-	<4 x 10^0
	+	4 x 10^2
C26S/C29S	-	<4 x 10^0
	+	2 x 10^2
C39S	-	<2 x 10^0
	+	6 x 10^2
Wild Type	-	2 x 10^4
	+	3 x 10^4
None	-	<4 x 10^0
	+	5 x 10^2

CHO cultures which had been stably transfected with M-MuLV NC mutants (Gorelick et al., 1988) were superinfected with 10A1 MuLV, which can efficiently infect hamster cells (Ott et al., 1990). Culture fluids were harvested five days after superinfection, and assayed for replication-competent MuLV by the S^+L^- focus assay (Bassin et al., 1971). FIU, focus-inducing units.

Thus the mutants we have tested do not appear to have a dominant effect in mixed infection; if wild-type cysteine arrays normally function in a search for genomic RNA during virus assembly, then the mutants behave here like "null" mutants, unable to perform this function.

RNA in Mutant Particles

Since the mutants do not package detectable viral RNA, it was obviously of interest to determine whether the viral RNA is replaced by cellular RNAs in the particles. We originally reported (Gorelick et al., 1988) that the particles do contain cellular RNAs. However, we have now characterized these RNAs with respect to size; as shown in Fig. 2, lane 2, essentially all of the RNA in the virions produced by the C39S mutant is the size of tRNA. tRNA is also a major constituent of wild-type virus particles (Fig. 2, lane 1; reviewed in Coffin, 1984). (Similar [but more extensive] results have also been presented by Meric and Goff [1989]). The significance of these findings will be considered below.

Nature of Viral RNA Packaged by Some Cysteine-array Mutants: Is Viral RNA Always Dimeric?

Certain cysteine array mutants package easily detectable, albeit reduced, levels of genomic RNA. This is evident in the case of mutant Y28S (see Fig. 1), and was also observed by Meric and Goff (1989) in mutant Y28G. As discussed below, other examples include mutants of HIV-1 and Rous sarcoma virus (RSV), and also Kirsten sarcoma virus (KiSV) particles rescued by the M-MuLV mutants. This observation suggests that some alterations of the cysteine array impair, but do not completely eliminate, the ability of the viral protein to recognize and package the viral genome during virus assembly. In turn, this finding may provide us with some preliminary clues as to the nature of the protein structures required for the packaging function.

The properties of the RNA encapsidated by these mutants may also shed some light on the nature of the packaging function. Meric and Goff (1989) showed that the viral RNA packaged in several M-MuLV cysteine

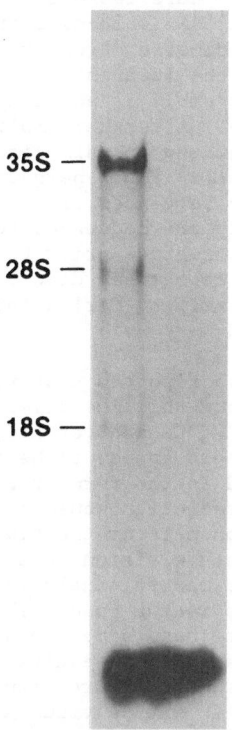

35S —

28S —

18S —

Fig. 2. Analysis of RNAs contained in mutant and wild-type viral particles. CHO cells, transfected with the C39S mutant and wild-type plasmid clones, were labeled with ^{32}P-ortho-phosphate. Virus was pelleted and viral RNA was isolated. The RNA was fractionated on a formaldehyde agarose gel and the gel was dried and autoradiographed. Lane 1, wild type; lane 2, C39S.

array mutants is in the form of a dimer, as it is in all wild-type retroviral particles. Thus each mutant particle which contains viral RNA contains two monomers linked in a dimeric structure. Since these mutants package far less than one genomic RNA molecule per particle, this result would be impossible if RNA monomers were packaged individually and then associated with each other in the assembled particle to form a dimer. Thus, this finding strongly suggests that viral RNA is packaged as a dimer.

A similar result was also reported a number of years ago by Levin et al. (1974). When cells producing wild-type MuLV are treated with actinomycin D, they continue to release virus particles for a number of hours. Ultimately, the virus particles produced under these conditions contain no detectable genomic RNA. However, in the early stages of actinomycin treatment, these virions have a low content of genomic RNA. In this case, as in the case of the cysteine-array mutants described above, the RNA present in these particles is in the dimeric form (Levin et al., 1974). Again, this observation strongly suggests that RNA monomers are packaged in pairs, rather than individually.

Very little is known about how monomers of retroviral genomic RNA are linked in the dimeric structure found in virus particles. If it is true, as we suggest here, that RNA molecules are packaged as dimers, then it is possible that the dimeric structure is required for packaging. That is, perhaps the "search" function postulated above for RNA packaging during virus assembly is actually a search for the distinctive dimeric structure. A further implication of this idea is that sequences in the RNA required for dimerization would hence be necessary for packaging as well. Thus, perhaps the "Ψ" sequences, which are defined experimentally as sequences required for packaging (e.g., Mann et al., 1983), are really the sequences required for association of the monomers into a dimeric structure prior to packaging. Data supporting a close relationship between dimerization and encapsidation have also been presented recently by Darlix and colleagues (Bieth et al., 1990; Prats et al., 1990).

Several laboratories have reported that very fresh virus appears to contain monomeric RNA, and that this RNA dimerizes upon incubation of the virus (Cheung et al., 1972; Canaani et al., 1973; Korb et al., 1976). These observations would appear to be incompatible with the proposal that RNA is packaged in the form of a dimer. One way of resolving this apparent contradiction would be to propose that the initial association of the monomers at the time of packaging is different from that in the mature virion, and is more easily disrupted during extraction. Indeed, a careful analysis by Stoltzfus and Snyder (1975) demonstrates that this is the case. They showed that fresh virus does contain dimers, but that these dimers dissociate more readily than those present in mature virus. Their results are consistent with the idea that dimers in both fresh and mature virus are held together by base-pairing, but that those in mature virus are more extensively base-paired than those in fresh virus.

Virus particles containing some monomeric RNA have also been described by Meric and colleagues (Meric and Spahr, 1986; Dupraz et al., 1990), in their studies of cysteine array mutants of avian sarcoma virus. It seems very plausible that, as in the case of fresh wild-type virus particles (Stoltzfus and Snyder, 1975), these "monomers" are really fragile dimers. The observations with cysteine array mutants (Meric and Spahr, 1986; Dupraz et al., 1990) thus raise the intriguing possibility that the cysteine arrays are involved in stabilizing the dimeric structure in the particle. Monomeric RNA has also been observed in a protease mutant of avian leukosis virus (Stewart et al., 1990). It would clearly be of great importance to understand the structure of the dimer in greater detail than at present.

As noted above (Fig. 2; Meric and Goff, 1989), cysteine array mutants of M-MuLV contain tRNA despite the absence of genomic RNA. It is not completely clear, from the data currently available, whether the viral RNA is replaced by additional tRNA in the mutant particles. However, in the particles released by actinomycin D-treated cells (Levin et al., 1974), the total amount of RNA is apparently the same as that in normal, infectious particles (Levin and Seidman, 1979). Thus in this case the genomic RNA is apparently replaced by additional tRNA. One hypothesis which seems consistent with these data is that during virus assembly, the Gag proteins normally "search" for RNA with some degree of double-stranded character; if no genomic RNA is available for dimerization and encapsidation, then cellular tRNA is incorporated in its place. (It should be noted that while reverse transcriptase is not required for encapsidation of tRNA, it is clearly involved in the selective incorporation into the virion of the primer tRNA species [Sawyer and Hanafusa, 1979; Peters and Hu, 1980; Levin and Seidman, 1981]).

Properties of Kirsten sarcoma virus (KiSV) Particles Rescued by Cysteine-array Mutants

In our initial studies of the cysteine-array mutants (Gorelick et al., 1988), it seemed important to establish that they function in *trans*. To investigate this question, we tested the ability of our mutants to package a second, MuLV-related RNA, i.e., the genome of KiSV. This genome contains terminal segments from MuLV, additional sequences from retrovirus-related endogenous elements termed rat VL30s, and coding sequences from the *K-ras* gene (Ellis et al., 1981). Thus, it does not encode any viral structural proteins, and is efficiently rescued by wild-type MuLVs. When cells containing the KiSV genome were transfected with plasmids containing the M-MuLV cysteine array mutants, they produced virus particles, as expected. These particles contained

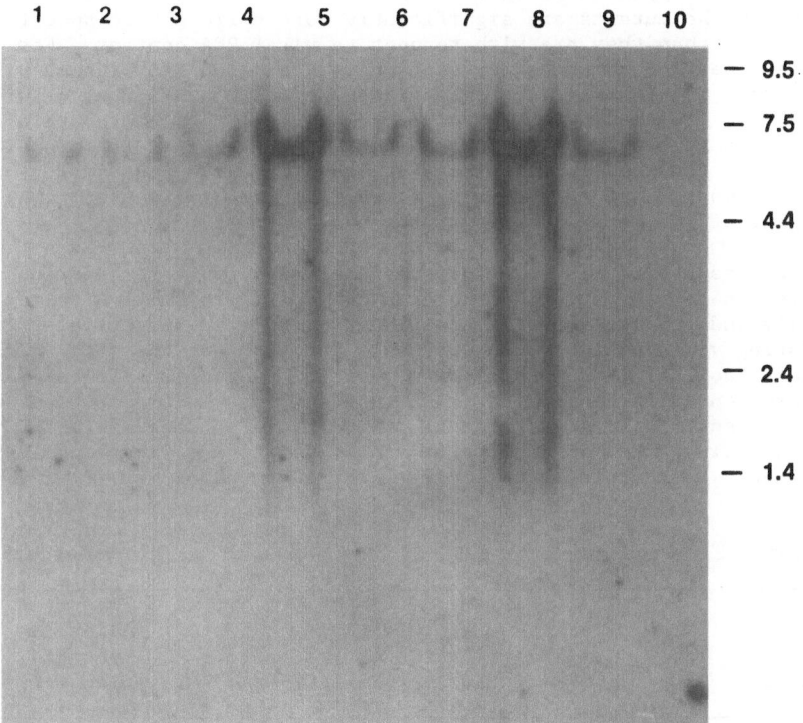

Fig. 3.   Northern analysis of RNA from virus contained in culture
          fluids from KiSV-transformed rat cells. Viral RNA was
          hybridized using a nick-translated ^{32}P-labeled Kirsten
          ras probe. Each lane contains the same amount of virus
          particles, as measured by immunoblotting for p30^{CA}.
          Lane 1, negative control, pGCcos3neo; lane 2, mutant
          C26S; lane 3, mutant C26S/C29S; lane 4, mutant C29S;
          lane 5, mutant Y28S; lane 6, mutant W35S; lane 7, mutant
          C39S; lane 8, wild-type M-MuLV; lanes 9 and 10, 1:10 and
          1:100 dilutions of wild-type M-MuLV, respectively.
          (Reproduced from Gorelick et al., 1988)

significantly less KiSV RNA than the wild-type controls (Fig. 3). This finding confirms that the mutations act in *trans*. However, it is also striking that each of the mutants packages easily detectable levels of KiSV RNA. Thus, each of the alterations of the cysteine array of M-MuLV tested here reduces, but does not eliminate, the ability to package KiSV RNA, even though the majority of them do prevent packaging of detectable levels of M-MuLV RNA.

It seems likely that the "packaging" sequence of KiSV RNA is derived, at least in part, from rat VL30 sequences, and is different from that of M-MuLV. The data suggest that the altered cysteine arrays of the mutants are more effective in packaging RNA containing the KiSV sequence than the M-MuLV sequence, although it is also possible that the hypothetical "search" function of the cysteine array is not required for packaging KiSV.

When the infectivity titers of the KiSV stocks characterized in Fig. 3 were determined, a remarkable result was obtained. As shown in Table 2, the stocks contained almost no infectious KiSV. Comparing the data in Fig. 3 with those in Table 2, we see that the KiSV particles rescued by the mutants are significantly more deficient in specific infectivity than they are with respect to viral RNA content. For example, the C26S mutant stock contains ~3% as much KiSV RNA as wild-type virus, but its specific infectivity is only ~0.01% that of wild-type virus. This quantitative discrepancy shows that many particles in these virus stocks contain KiSV RNA but do not register as infectious KiSV. Thus the mutations in the cysteine array have apparently impaired a second function of this domain, as well as reducing the efficiency of packaging of genomic RNA during virus assembly.

Our results on KiSV provide experimental support for the idea that the cysteine array performs more than one critical function during virus assembly and replication. In addition to a role in selection of genomic RNA during virus assembly, the possible functions of the array might include stabilizing the association between RNA monomers in the mature particle; annealing the primer tRNA to the genome in preparation for reverse transcription (Prats et al., 1988); and/or cleavage of NC protein early in the infectious process (Roberts and Oroszlan, 1989). Experiments to test these possibilities are now under way.

Table 2. Properties of KiSV Stocks "Rescued" by $p10^{gag}$ Mutants

| Virus | p30 | Infectivity | | KiSV specific infectivity |
		FFU/ml	CPFU/ml	
C26S	14	2×10^0	$<1 \times 10^0$	0.00009
C29S	17	4×10^0	$<1 \times 10^0$	0.0001
C26S/C29S	33	2×10^1	1×10^0	0.0004
Y28S	37	2×10^3	3×10^1	0.03
W35S	60	5×10^2	1×10^0	0.005
C39S	25	4×10^1	$<1 \times 10^0$	0.001
Wild type	44	7×10^4	4×10^4	(1)

K-NRK cells were transfected with plasmids containing the indicated viral genomes, and stable G418-resistant transfectants were isolated. Culture fluids were assayed as indicated. p30 values are from protein immunoblotting results expressed in arbitrary densitometric units. Infectivity is focus-forming units (FFU) of KiSV per ml (Rein, 1982) or complementation plaque-forming units (CPFU) per ml (Rein and Bassin, 1978). KiSV specific infectivity value is the FFU/p30 ratio of the mutant divided by the FFU/p30 ratio of the wild-type virus. (Reproduced from Gorelick et al., 1988).

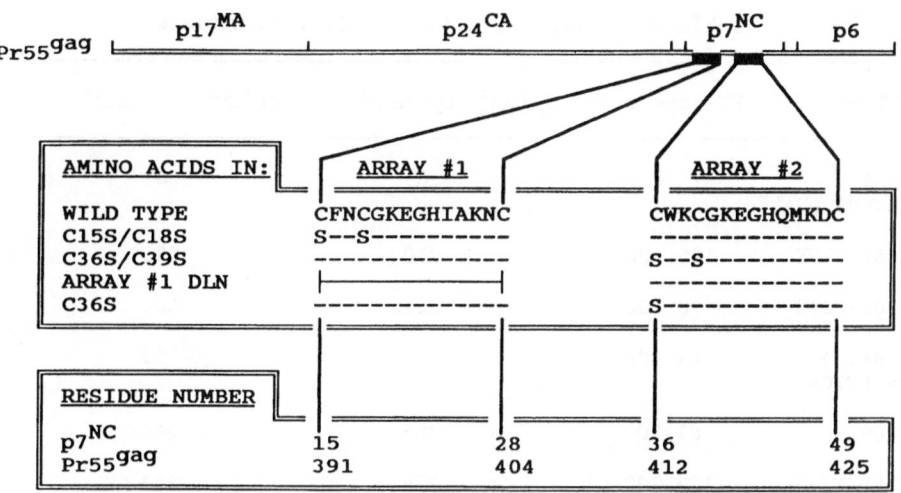

Fig. 4. Mutants of HIV-1 nucleocapsid (p7^{NC}). The top line shows
the cleavage pattern of Pr55^{gag} (the Gag precursor
polyprotein) and the positions of the two cysteine arrays
within p7^{NC}. Beneath this line are shown the amino acid
sequences of the two arrays in the pNL4-3 clone of HIV-1
(Adachi et al., 1986) and in the mutants described here.
(Reproduced with permission from Gorelick et al., 1990).

Cysteine Array Mutants of HIV-1

We have now extended our studies of the function of the cysteine
array by generating mutants in HIV-1 (Gorelick et al., 1990). HIV-1
differs from MuLV in that the NC protein contains two, rather than one,
cysteine arrays. We generated the mutants shown in Fig. 4 in the pNL4-3
clone of HIV-1 (Adachi et al., 1986) using oligonucleotide-directed
mutagenesis. We have studied the properties of these mutants by
transfecting plasmid clones of the intact viral genomes into HeLa cells
and harvesting the supernatants of these HeLa cultures 2-5 days later.

As in the case of MuLV, each of the mutants gave rise to virus
particles with essentially the same efficiency as the wild-type. As
indicated in Fig. 5 and Table 3, the particles appear to contain normal
levels of viral proteins, and the Gag precursor is processed properly.
Northern analysis (Fig. 6) showed that each mutant packages less genomic
RNA than the wild-type (each lane in the Figure contains the same amount
of virus particles, as determined by reverse transcriptase [RT]
activity); however, easily detectable levels of viral RNA are present in
each of the mutant stocks. Using the serial dilutions of the wild-type
sample as a "calibration" of the Northern blot (Fig. 6), we estimate the
genomic RNA contents of the mutant particles to be: C15S/C18S, ~2%;
array one deletion, ~3%; C36S, ~6%; and C36S/C39S, ~20%, taking wild-
type as 100%.

Table 3.   Characteristics of HIV-1 NC Mutants.

MUTANT	RT (cpm/mL)[a]	p24^{CA} (pg/mL)[b]	gp120^{SU} (pg/mL)[b]
CALF THYMUS DNA[c]	0	0	0
C15S/C18S	125,900	710	338
C36S/C39S	70,200	361	404
ARRAY #1 DELETION	44,200	260	153
C36S	67,200	383	244
WILD TYPE	122,700	779	557

[a]CPM ^{3}H-TTP incorporated/mL of culture fluid.
[b]Picograms of p24^{CA} or gp120^{SU}/mL of culture fluid determined by competition radioimmunoassay.
[c]Cells transfected with carrier DNA alone. A background of 420 cpm/mL has been subtracted from the RT values shown. (Reproduced with permission from Gorelick et al., 1990).

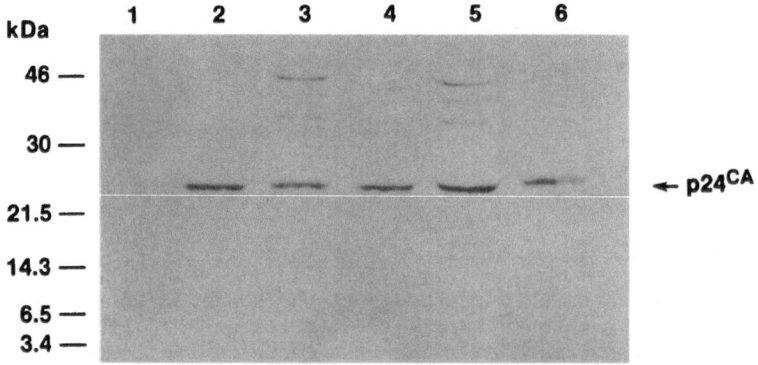

Fig. 5.   Protein immunoblotting analysis of mutant and wild-type virus particles.   Particles were analyzed for p24^{CA} by immunoblotting. Lane 1, HeLa cells transfected with calf thymus DNA; lane 2, C15S/C18S; lane 3, C36S/C39S; lane 4, array one deletion; lane 5, C36S; land 6, wild type. (Reproduced with permission from Gorelick et al., 1990).

To test the mutant particles for infectivity, we incubated H9 cells with the same virus preparations used above, and analyzed the culture fluids for RT activity (Fig. 7A). H9 cultures exposed to wild-type virus showed a rapid rise in RT activity. However, no RT activity was observed in the H9 cultures incubated with the mutant particles, even after passage for one month. Negative results with the mutants were also obtained in similar tests using human peripheral blood lymphocytes instead of H9 cells, and in experiments using direct cocultivation of the transfected HeLa cells with permissive cells.

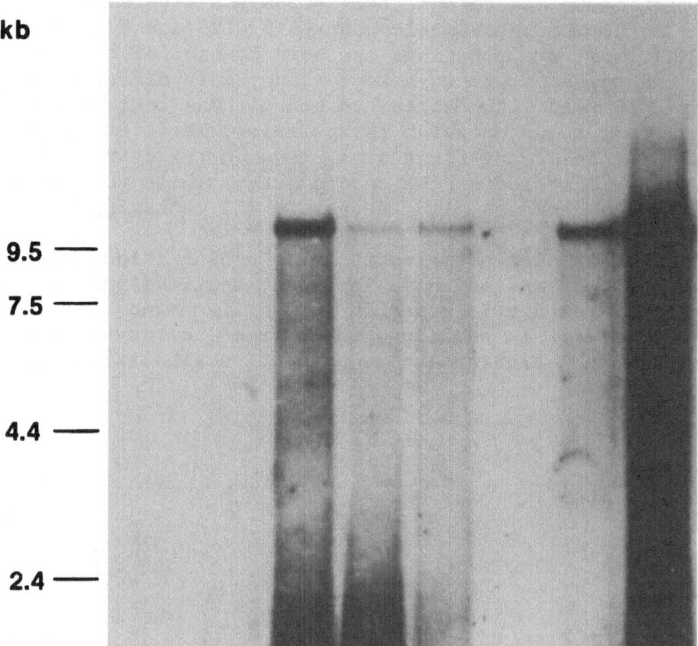

Fig. 6. Northern blot analysis of mutant and wild-type virus particles.
Particles were analyzed for HIV-1 genomic RNA using nick-
translated ^{32}P-labeled pNL4-3 plasmid as a probe. Lane 1,
HeLa cells transfected with calf thymus DNA; lane 2, C15S/C18S;
lane 4, C36S/C39S; lane 4, array one deletion; lane 5, C36S;
lane 6, 1:100 dilution of wild type; lane 7, 1:10 dilution of
wild type; lane 8, wild type. (Reproduced with permission from
Gorelick et al., 1990).

A quantitative assessment of the relative infectivities of mutant
and wild-type virus particles was performed by an endpoint dilution
assay (Fig. 7B). Infectious virus was detected in a 1:100,000 dilution
of wild type stock, while undiluted C15S/C18S mutant stock (adjusted to
the same RT activity as wild type) failed to show any infectious virus
(Fig. 7B). We conclude that the mutants are at least 100,000 times less
infectious than an equivalent amount of wild type virus.

These results on HIV-1 lead to the following conclusions: (1) as in
murine and avian retroviruses, intact cysteine arrays are required for
efficient packaging of genomic RNA during virus assembly. (Experiments
in support of this conclusion have also been reported by Aldovini and
Young [1990]). (2) Just as in the case of KiSV rescued by cysteine-
array mutants of M-MuLV, some of the particles of HIV-1 cysteine-array
mutants do contain genomic RNA; however, these particles are
nevertheless noninfectious. Thus disruption of the cysteine array
apparently introduces another defect, in addition to its effect on
packaging of viral RNA. The nature of the second defect is unknown,
but, as noted above, might involve dimer stabilization, primer
annealing, or a late cleavage event. (3) The two cysteine arrays of
HIV-1 are not functionally redundant: mutations in either array are
sufficient to reduce the packaging of viral RNA and to eliminate any
detectable infectivity in the particles.

It is interesting to note that alterations in the first array
(C15S/C18S and the array one deletion) reduce RNA encapsidation somewhat
more than mutations in the second array (C36S and C36S/C39S). This
difference is particularly dramatic when C15S/C18S and C36S/C39S are
compared, since these two mutations are both changes of the first two
cysteines in the respective arrays to serines; they differ by
approximately 7-10-fold with respect to genomic RNA content (see Fig.
6). (In a previous study on avian retroviruses, Meric et al. [1988]
found that a deletion of the first array reduced infectivity more
drastically than a deletion of the second array, while the two mutants
exhibited similar phenotypes with respect to RNA packaging.)

Because of their near-normal structure, noninfectious HIV-1 mutants
of the type described here may have important applications as non-
hazardous reagents for use in research and as immunogens in vaccine and
immunotherapy studies. In addition, the extreme conservation of the
cysteine array among retroviruses may present possibilities for

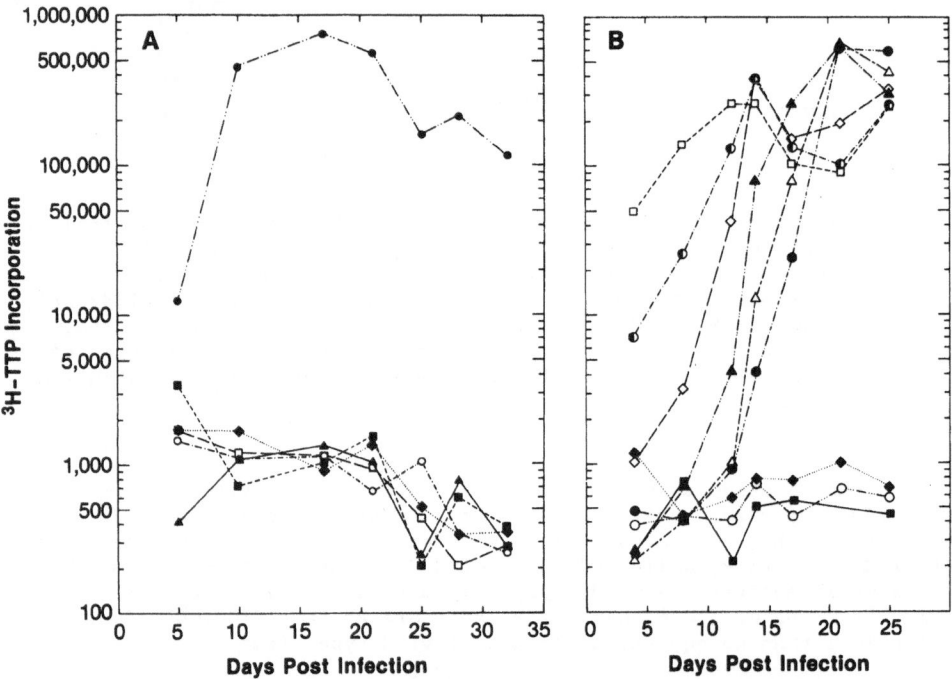

Fig. 7. Infectivity assays of mutant and wild-type virus particles.
(A) Infectivity assays of mutants. HeLa supernatants were
adjusted for equal amounts of RT activity. H9 cultures were
infected, passaged, and monitored for particle-associated RT
activity. Wild type (●), C15S/C18S (■), C36S/C39S (□),
array #1 deletion (O), C36S (◆) and calf thymus DNA negative
control (▲). (B) Dilution endpoint assay of wild-type HIV-1.
HeLa supernatants containing wild-type (□) and the C15S/C18S
(◆) mutant viruses were adjusted for equal RT activity.
Supernatant containing wild-type virus was also serially
diluted 10 (⦿), 100 (◇), 1,000 (▲), 10,000 (△), 100,000
(●) and 1,000,000 (O) fold; calf thymus DNA negative control
(■). H9 cells were infected and tested as in (A). (Reproduced
with permission from Gorelick et al., 1990).

developing universal retroviral diagnostic reagents, while the exquisite sensitivity of the virus to changes in the array suggests that the array is a potential target for the development of antiviral drugs.

CONCLUDING REMARKS

We have postulated here that the cysteine array(s) participate in a positive "search" function for viral RNA during virion assembly. It is striking that a number of mutants in which conserved cysteine residues are replaced with serine residues package some viral RNA. (Similarly, an RSV mutant in which the conserved histidine residue in the first array is replaced with a proline packages viral RNA with only slightly reduced efficiency [Dupraz et al., 1990]). These observations suggest that the ability to perform this hypothetical "search" function is impaired, but not eliminated, by these changes in the conserved residues of the cysteine array.

It has been suggested by us and others (Berg, 1986; Gorelick et al., 1988; South et al., 1990) that the cysteine array may co-ordinate zinc ions, in direct analogy to the zinc fingers found in a number of eukaryotic transcription factors. While retroviral NC proteins clearly have the capability to co-ordinate zinc (Schiff et al., 1988; Jentoft et al., 1988; Roberts et al., 1989; South et al., 1990; Green and Berg, 1990), little if any zinc is detected in purified virus particles (Jentoft et al., 1988); thus it is not yet clear whether or when zinc binding might be involved in cysteine array function. If co-ordination of zinc by this protein motif is indeed involved in encapsidation of viral RNA, then the partial ability to package viral RNA exhibited by the mutant proteins might imply that they retain some ability to co-ordinate zinc and to fold into some semblance of their wild-type conformation. This idea may not be completely implausible: Green and Berg (1990) have recently analyzed metal binding by peptides representing wild-type and mutant MuLV cysteine arrays. They reported that some mutant peptides bound cobalt with dissociation constants only slightly higher than that of the wild-type. Perhaps the functions of the cysteine array will not be understood until the genetic data are complemented by three-dimensional structural studies of mutant and wild-type NC proteins.

ACKNOWLEDGMENTS

We wish to thank Janet Hanser and Emily Keene for technical assistance; Judith Levin, Patricia Powell, and Michael Summers for many helpful discussions; Marilyn Powers for oligonucleotides; and Carol Shawver for preparation of the manuscript. This research was supported in part by the National Cancer Institute, DHHS, under contract no. N01-CO-74101 with ABL, and under contract no. N01-CO-74102 with PRI/DynCorp, and Public Health Service Individual National Research Service Award No. 5 F32 AIO 7684-02 from the National Instututes of Health.

LITERATURE CITED

Adachi, A., Gendelman, H. E., Koenig, S., Folks, T., Willey, R., Rabson, A., Martin, M. A., 1986, Production of acquired immunodeficiency syndrome-associated retrovirus in human and nonhuman cells transfected with an infectious molecular clone, _J. Virol._, 59:284-291.

Aldovini, A. and Young, R. A., 1990, Mutations of RNA and protein sequences involved in human immunodeficiency virus type 1 packaging result in production of noninfectious virus, J. Virol., 64:1920-1926.

Bassin, R. H., Tuttle, N., and Fischinger, P. J., 1971, Rapid cell culture assay for murine leukaemia virus, Nature (London), 229:564-566.

Berg, J. M., 1986, Potential metal-binding domains in nucleic acid binding proteins, Science, 232:485-487.

Bieth, E., Gabus, C., and Darlix, J.-L., 1990, A study of the dimer formation of Rous sarcoma virus RNA and of its effect on viral protein synthesis in vitro, Nucleic Acids Res., 18:119-127.

Canaani, E., Helm, K. V. D., and Duesberg, P., 1973, Evidence for 30-40S RNA as precursor of the 60-70S RNA of Rous sarcoma virus particles, Proc. Natl. Acad. Sci. USA, 72:401-405.

Cheung, K.-S., Smith, R. E., Stone, M. P., and Joklik, W. K., 1972, Comparison of immature (rapid harvest) and mature Rous sarcoma virus particles, Virology, 50:851-864.

Coffin, J., 1984, Structure of the retroviral genome, in: "RNA Tumor Viruses", 2nd ed., R. Weiss, N. Teich, H. Varmus, and J. Coffin, eds., Cold Spring Harbor Laboratory, Cold Spring Harbor, NY, pp. 261-368.

Covey, S. M., 1986, Amino acid sequence homology in gag region of reverse transcribing elements and the coat protein gene of cauliflower mosaic virus, Nucleic Acids Res., 14:623-633.

Dupraz, P., Oertle, S., Meric, C., Damay, P., and Spahr, P.-F., 1990, Point mutations in the proximal Cys-His box of Rous sarcoma virus nucleocapsid protein, J. Virol., 64:4978-4987.

Ellis, R. W., Defeo, D., Shih, T. Y., Gonda, M. A., Young, H. A., Tsuchida, N., Lowy, D. R., and Scolnick, E. M., 1981, The p21[src] genes of Harvey and Kirsten sarcoma viruses originate from divergent members of a family of normal vertebrate genes, Nature (London), 292:506-511.

Evans, R. M. and Hollenberg, S. M., 1988, Zinc fingers: Gilt by association, Cell, 52:1-3.

Gheysen, D., Jacobs, E., de Foresta, F., Thiriart, C., Francotte, M., Thines, D., and De Wilde, M., 1989, Assembly and release of HIV-1 precursor Pr55[gag] virus-like particles from recombinant baculovirus-infected insect cells, Cell, 59:103-112.

Gorelick, R. J., Henderson, L. E., Hanser, J. P., and Rein, A., 1988, Point mutants of Moloney murine leukemia virus that fail to package viral RNA: Evidence for specific RNA recognition by a "zinc finger-like" protein sequence, Proc. Natl. Acad. Sci. USA, 85:8420-8424.

Gorelick, R. J., Nigida, S. M., Jr., Bess, J. W., Jr., Arthur, L. O., Henderson, L. E., and Rein, A., 1990, Noninfectious human immunodeficiency virus type 1 mutants deficient in genomic RNA, J. Virol., 64:3207-3211.

Green, L. M. and Berg, J. M., 1990, Retroviral nucleocapsid protein-metal ion interactions: Folding and sequence variants, Proc. Natl. Acad. Sci. USA, 87:6403-6407.

Henderson, L. E., Copeland, T. D., Sowder, R. C., Smythers, G. W., and Oroszlan, S., 1981, Primary structure of the low-molecular-weight nucleic acid binding proteins of murine leukemia viruses, J. Biol. Chem., 256:8400-8406.

Jentoft, J. E., Smith, L. M., Fu, X., Johnson, M., and Leis, J., 1988, Conserved cysteine and histidine residues of the avian myeloblastosis virus nucleocapsid protein are essential for viral replication but are not "zinc-binding fingers", Proc. Natl. Acad. Sci. USA, 85:7094-7098.

Korb, J., Travnicek, M., and Riman, J., 1976, The oncornavirus maturation process: Quantitative correlation between morphological changes and conversion of genomic virion RNA, Intervirology, 7:211-224.

Levin, J. G., Grimley, P. M., Ramseur, J. M., and Berezesky, I. K., 1974, Deficiency of 60 to 70S RNA in murine leukemia virus particles assembled in cells treated with actinomycin D, J. Virol., 14:152-161.

Levin, J. G. and Seidman, J. G., 1979, Selective packaging of host tRNA's by murine leukemia virus particles does not require genomic RNA, J. Virol., 29:328-335.

Levin, J.G. and Seidman, J. G., 1981, Effect of polymerase mutations on packaging of primer tRNA^{Pro} during murine leukemia virus assembly, J. Virol., 38:403-408.

Mann, R., Mulligan, R. C., and Baltimore, D., 1983, Construction of a retrovirus packaging mutant and its use to produce helper-free defective retroviruses, Cell, 33:153-159.

Méric, C. and Goff, S. P., 1989, Characterization of Moloney murine leukemia virus mutants with single-amino-acid substitutions in the Cys-His box of the nucleocapsid protein, J. Virol., 63:1558-1568.

Méric, C., Gouilloud, E., and Spahr, P.-F., 1988, Mutations in rous sarcoma virus nucleocapsid protein p12 (NC): Deletions of Cys-His boxes, J. Virol., 62:3328-3333.

Méric, C. and Spahr, P.-F., 1986, Rous sarcoma virus nucleic acid-binding protein p12 is necessary for viral 70S RNA dimer formation and packaging J. Virol., 60:450-459.

Oroszlan, S. and Copeland, T. D., 1985, Primary structure and processing of gag and env gene products of human T-cell leukemia viruses HTLV-Icr and HTLV-IATK, in: "Current Topics in Microbiology and Immunology," P. K. Vogt, ed., Springer-Verlag, New York, pp. 221-233.

Ott, D., Friedrich, R., and Rein, A., 1990, Sequence analysis of amphotropic and 10A1 murine leukemia viruses: Close relationship to mink cell focus-inducing viruses, J. Virol., 64:757-766.

Peters, G.G. and Hu. J., 1980, Reverse transcriptase as the major determinant for selective packaging of tRNA's into avian sarcoma virus particles, <u>J. Virol.</u>, 36:692-700.

Prats, A.-C., Christine, R., Wang, P., Erard, M., Housset, V., Gabus, C., Paoletti, C., and Darlix, J.-L., 1990, *cis* elements and *trans*-acting factors involved in dimer formation of murine leukemia virus RNA, <u>J. Virol.</u>, 64:774-783.

Prats, A. C., Sarih, L., Gabus, C., Litvak, S., Keith, G., and Darlix, J. L., 1988, Small finger protein of avian and murine retroviruses has nucleic acid annealing activity and positions the replication primer tRNA onto genomic RNA, <u>EMBO J.</u>, 7:1777-1783.

Rein, A., 1982, Interference grouping of murine leukemia viruses: a distinct receptor for the MCF-recombinant viruses on mouse cells, <u>Virology</u>, 120:251-257.

Rein, A. and Bassin, R. H., 1978, Replication-defective ecotropic murine leukemia viruses: detection and quantitation of infectivity using helper-dependent XC plaque formation, <u>J. Virol.</u>, 28:656-660.

Roberts, M. M. and Oroszlan, S., 1989, The preparation and biochemical characterization of intact capsids of equine infectious anemia virus, <u>Biochem. Biophys. Res. Commun.</u>, 160:486-494.

Roberts, W. J., Pan, T., Elliott, J. I., Coleman, J. E., and Williams, K. R., 1989, p10 single-stranded nucleic acid binding protein from murine leukemia virus binds metal ions via the peptide sequence $Cys^{26}-X_2-Cys^{29}-X_4-His^{34}-X_4-Cys^{39}$, <u>Biochemistry</u>, 28:10043-10047.

Sawyer, R. C. and Hanafusa, H., 1979, Comparison of the small RNAs of polymerase-deficient and polymerase-positive Rous sarcoma virus and another species of avian retrovirus, <u>J. Virol.</u>, 29:863-871.

Schiff, L. A., Nibert, M. L., and Fields, B. N., 1988, Characterization of a zinc blotting technique: Evidence that a retroviral gag protein binds zinc, <u>Proc. Natl. Acad. Sci. USA</u>, 85:4195-4199.

Shields, A., Witte, O. N., Rothenberg, E., and Baltimore, D., 1978, High frequency of aberrant expression of Moloney murine leukemia virus in clonal infections, <u>Cell</u>, 14:601-609.

South, T. L., Blake, P. R., Sowder, R. C., III, Arthur, L. O., Henderson, L. E., and Summers, M. F., 1990, The nucleocapsid protein isolated from HIV-1 particles binds zinc and forms retroviral-type zinc fingers, <u>Biochemistry</u>, 29:7786-7789.

Stewart, L., Schatz, G. and Vogt, V. M., 1990, Properties of avian retrovirus particles defective in viral protease, <u>J. Virol.</u>, 64:5076-5092.

Stoltzfus, C.M. and Snyder, P. N., 1975, Structure of B77 sarcoma virus RNA: Stabilization of RNA after packaging, <u>J. Virol.</u>, 16:1161-1170.

REGULATED PROTEOLYTIC PROCESSING WITHIN MATURE RETROVIRAL CAPSIDS

Michael M. Roberts, Eugene Volker[†], Terry D. Copeland,
Kunio Nagashima[*], M. Beth Cassell, Carlton J. Briggs and
Stephen Oroszlan

Laboratory of Molecular Virology and Carcinogenesis, ABL-
Basic Research Program, [*]Electron Microscopy/Cell Biology
Laboratory, Program Resources, Inc., NCI-Frederick Cancer
Research and Development Center, Frederick, MD 21702

SUMMARY

Capsid particles were prepared from equine infectious anemia virus
(EIAV) as a model retrovirus for the human immunodeficiency virus (HIV).
There is a stepwise cleavage of the nucleocapsid (NC) protein (p11) and
integrase (IN) (p32) during incubation of EIAV capsids at 37°C in 10 mM
Tris 1 mM EDTA (TE Buffer) at pH 7.6. The viral protease cleaves the NC
protein after the first Cys residue of both conserved (C X_2 C X_4 H X_4 C)
regions. The p11 → p6 cleavage occurs at the first Cys array. The p6
is then cleaved at the second Cys array, resulting in three main peptide
fragments appearing as a 4 KDa band on an SDS gel. The cleavage of a 6
KDa C-terminal fragment from IN starts when all the p11 is cleaved to
p6, and therefore occurs during the final fragmentation of the NC
protein. Capsids from other retroviruses also show NC protein cleavage
when incubated under similar conditions. It has been postulated that
proteolytic processing of the NC protein occurs *in vivo* during the early
stages of the viral life-cycle and may be required for replication.

INTRODUCTION

The viral aspartic proteinase has a well-established role during
the maturation stage of the viral life cycle. For the production of
fully infectious viral progeny following budding from the host cell, the
Gag and Gag-Pol polyproteins are cleaved into their individual protein
components (1). These cleavages are accompanied by a morphological
transition from the circular immature virion core into a collapsed core
surrounded by the viral envelope (2). In the lentiviruses, the mature
core is a cone shaped structure which consists of the ribonucleoprotein complex and active enzymes contained within the viral capsid.

Conditions were developed for the preparation of capsids of equine
infectious anemia virus (EIAV) with the primary aim of isolating a
macromolecular assembly that can be well-defined by physical and
chemical methods (3). Since extensive sequence homology has been
demonstrated between the *gag*-encoded structural proteins and the *pol*-
encoded enzymes of EIAV and the human immunodeficiency virus (HIV) (4),
and both viruses have similar morphology (5), this investigation of EIAV
capsids was intended to be a model study for HIV.

[†]Permanent address: Dept. of Chemistry, Shepherd College, Shepherdstown, WV, 25443.

Advances in Molecular Biology and Targeted Treatment for AIDS
Edited by A. Kumar, Plenum Press, New York, 1991

PREPARATION AND CHARACTERIZATION OF CAPSIDS

The capsid preparation scheme is outlined in Fig. 1. This involved stripping of the envelope from the whole virus with detergent followed by two cycles of sedimentation in Ficoll density gradients (3). In addition to the viral RNA, the protein components of the EIAV capsid are shown in Table 1. The p11 nucleocapsid (NC) and p26 capsid (CA) proteins are the main structural components, present in a 1:1 molar ratio. The minor components are p22, which remains to be characterized, and the *pol*-encoded proteins, one of which is a newly isolated 15 KDa polypeptide (Roberts, Copeland, and Oroszlan, to be published) located between the reverse transcriptase (RT) and integrase (IN) in the Gag-Pol polyprotein. The CA protein forms the exterior of the capsid particle, as demonstrated by immunoaggregation in the presence of p26 antibody (Özel, Roberts, Oroszlan, unpublished observation). Uninfected cells injected with freshly prepared capsid particles were able to produce virus after two weeks (Oroszlan, Boyd, Cassell, and Roberts, to be published). Therefore, all the components required for infection are encapsidated by the p26 CA protein.

NC PROTEIN CLEAVAGE

Purified capsid preparations stored in 10mM Tris 1mM EDTA (TE) buffer will show cleavage of the NC protein in the pH range of 6.0-7.6 (Fig. 2). As seen on SDS gels, the p11 is cleaved to p6 both at room temperature and at 37°C. This is followed by cleavage of the integrase

(IN) at 32 KDa to give a 6 KDa C-terminal fragment (Roberts, Copeland & Oroszlan, to be published). Whole virus incubated under the same conditions does not show cleavage, indicating that the virion has to be de-enveloped to allow the cleavage to proceed. The addition of

TABLE 1. Protein components of EIAV capsids

Protein	Coding gene	Protein nomenclature	Identification method
p11	*gag*	Nucleocapsid (NC)	SDS-PAGE
p26	*gag*	Capsid (CA)	SDS-PAGE, Western blot
p22	*gag*		SDS-PAGE, Western blot, N-terminal sequencing
p12	*pol*	Protease (PR)	RP-HPLC, SDS-PAGE, Western blot, enzymatic activity
p15	*pol*		RP-HPLC, SDS-PAGE, amino acid analysis, sequencing data
p32	*pol*	Integrase (IN)	SDS-PAGE, Western blot, sequencing data
p51/p66	*pol*	Reverse transcriptase (RT)	SDS-PAGE, endogenous activity

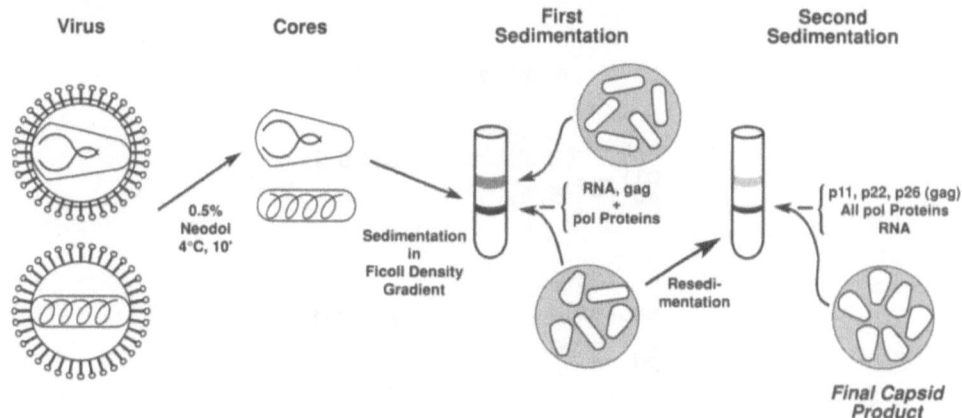

| Virus | Cores | First Sedimentation | Second Sedimentation |

0.5% Neodol 4°C, 10'

Sedimentation in Ficoll Density Gradient

RNA, gag + pol Proteins

Resedimentation

p11, p22, p26 (gag) All pol Proteins RNA

Final Capsid Product

Fig. 1.   Preparation scheme for EIAV capsids.

pepstatin to the incubated capsids will inhibit this cleavage, supporting the viral aspartic proteinase as the agent responsible for this process (3). There is an additional cleavage of the p6 which goes to completion in the presence of 10mM DTT, resulting in peptides which appear as a p4 band. This proceeds in parallel with the IN cleavage. Selective proteolysis therefore occurs in two stages, with p6 as an intermediate (6,7). The peptides resulting from the proteolysis were separated and purified by RP-HPLC. Chemical analysis of these peptides revealed the cleavage sites in p11 which occur after the first cysteine of each zinc binding domain.

The NC protein cleavage appears to be an *in situ* process in that it occurs only inside the capsid. It may require the viral RNA, since increasing the salt concentration to 0.3M will inhibit the cleavage (Fig. 3), and at this ionic strength the p11-RNA interactions are largely diminished (8). Also, the presence of EDTA in the buffer appears to be a requirement for cleavage to proceed. This indicates that some divalent cation may be regulating the proteolysis. In fact, the cleavage sites are located at the Cys residue that is likely to be one of the ligands coordinating to zinc to form the zinc finger structure (9). This would suggest that the NC protein, in the zinc-bound form, contains cysteine that is unable to fit into the active site cleft of the viral proteinase and form the extended conformation required for cleavage (10). Preliminary analyses using atomic absorption spectroscopy indicate that there are two zinc atoms per NC protein, in agreement with the number of conserved $CX_2CX_4HX_4C$ domains in the p11 sequence. The stoichiometry of zinc in retroviruses containing one of these domains remains to be determined in order to establish the endogenous association of zinc with the NC proteins as a general feature. In capsid preparations of Friend murine leukemia virus (FMuLV), which contains one zinc binding domain, a peptide has been recovered that begins with the sequence Ala-Tyr, indicating that cleavage at the first Cys has occurred, as in EIAV. This supports the prediction that cleavage is likely to occur at analogous sites in other retroviruses (7).

With the biochemistry of these cleavages established, their physiological significance remains to be demonstrated. We have described the *in vitro* removal of the viral envelope with detergent, releasing the capsid particles (3), and their introduction into a reducing environment at physiological pH and temperature. To a certain degree, these conditions resemble the events occuring during the entry

275

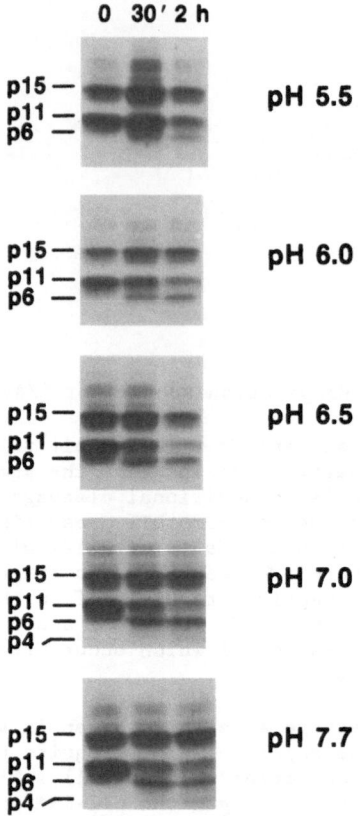

Fig. 2.   SDS gel showing the effect of pH on p11 cleavage in capsids.
Capsids were prepared in 10mM Tris-maleate buffer at pH 9.0,
then incubated in the same buffer containing 1mM EDTA and
20mM DTT at the above pH values at 37°C, then aliquoted at
time intervals of 0, 30 mins and 2hr.   There is about 6 μg
of total protein loaded per gel lane.

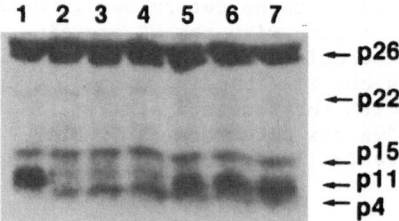

Fig. 3.   SDS gel of a capsid preparation (lane 1, 0hr) incubated at
NaCl concentrations of 0.0, 0.1, 0.2, 0.3, 0.4, and 0.5M
(lanes 2-7) for 24 hr at 37°C in pH 7.6 TE, 10mM DTT.

of virus into the uninfected cell. It remains to be determined whether the cleavage of the NC protein by the viral proteinase is important for viral replication. Site directed mutagenesis, at the NC cleavage sites, which are now in progress, should give a clearer insight into the biological relevance of this proteolytic process.

PROTEOLYSIS AND REVERSE TRANSCRIPTION

The addition of 6mM $MgCl_2$ to the capsid preparation will completely inhibit the NC protein cleavage. However, when added together with 200 μM each of dATP, dCTP, dGTP and dTTP as one would during an RT reaction, the cleavage proceeds (Fig. 4). These nucleotides are therefore interacting in some way with the components in the capsid to allow the cleavage by the viral proteinase to occur. A high resolution electron micrograph of one capsid particle (Fig. 5) shows some features of the capsomeric structure. The capsomers appear to consist of cylindrical assemblies 9nm in size with a central cavity of 3nm. This organization will allow the entry of nucleotide substrates into the capsid during reverse transcription. At the same time, the growing DNA strand will remain encapsidated by the CA protein. This supports the suggestion that reverse transcription takes place in an organized structure in the cytoplasm after the virus enters the cell (11,12).

The p11 → p6 cleavage will occur in capsid particles during reverse transcription. The DNA produced has been extracted and measured in electron micrographs by Pt/C shadowing to a length of 7.9 KB (Fig. 6), close to full-length EIAV DNA. The DNA-p26 complexes made band to 1.33 gcm^{-3} in CsCl gradients (Fig. 4). These could be analogous to the p30-DNA complexes isolated from the cytoplasm of cells freshly infected with MuLV (for review see ref. 11). The *in situ* proteolysis proceeds

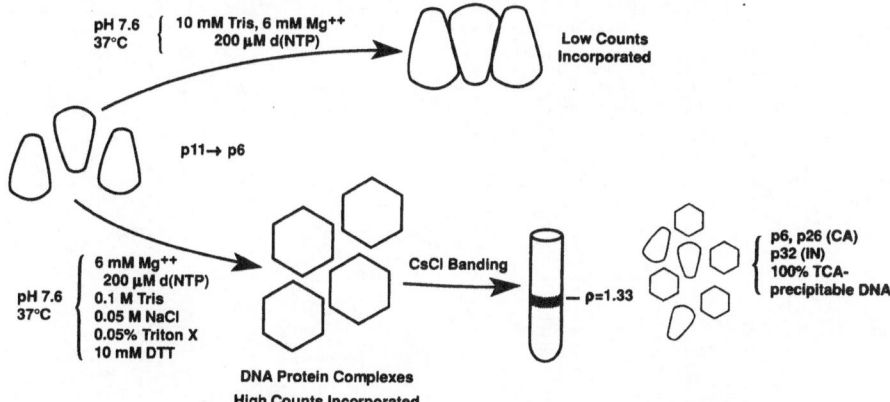

Fig. 4.   The cocktail ingredients required for endogenous reverse trans- cription in capsids. The NC protein cleavage, inhibited in the presence of 6 mM MgCl2, will proceed with the addition of 200 μM each of dATP, dCTP, dGTP and dTTP. However, extra ingredients such as salt, detergent and reducing agent are required for full incorporation of ^{3}H-dTTP labeled counts (lower half of figure). The morphology of the protein-DNA complexes made is not established, although the cone-shape appears to predominate in electron micrographs (Ozel, Roberts & Oroszlan, unpublished results) These complexes can be banded to a density of 1.33 gcm^{-3} in CsCl gradients.

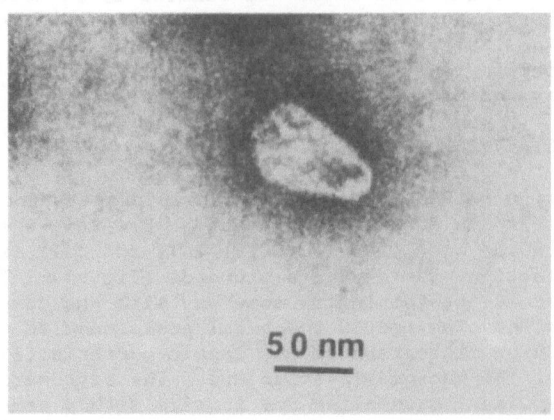

Fig. 5.   Electron micrograph of EIAV capsid particle (300,000X)
            negative stained with phosphotungstate.

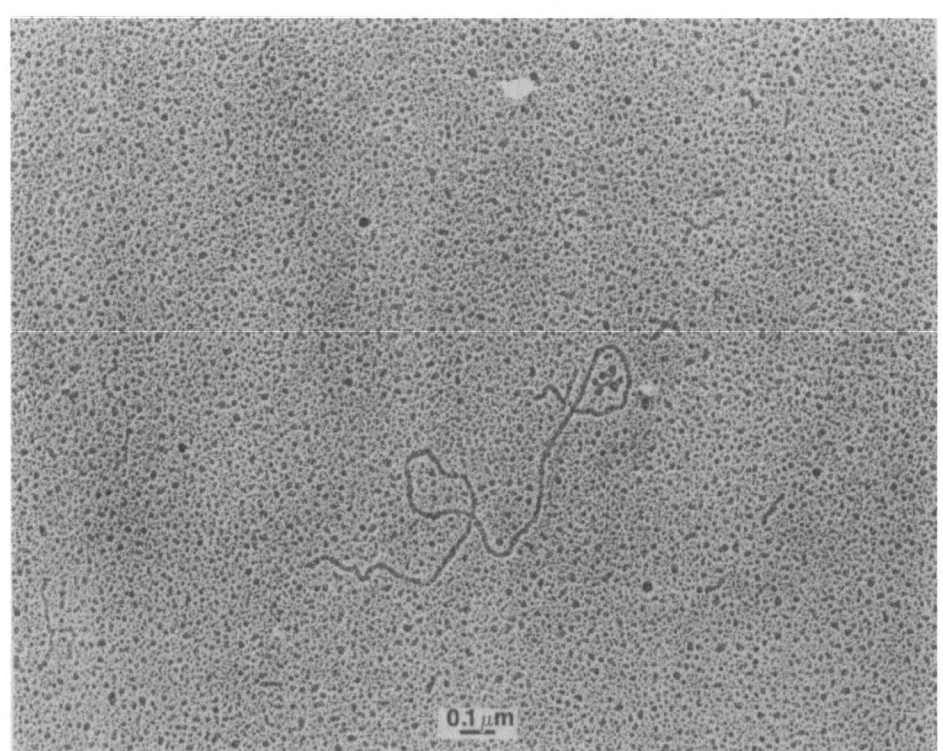

Fig. 6.   Electron micrograph (90,000X) by Pt/C shadowing of 7.9KB DNA
            strand made during the endogenous reverse transcription
            reaction of EIAV capsids.   The capsids (130 µg/ml total
            protein) were incubated for 5 hrs at 37°C in: 0.05M NaCl,
            0.1M Tris.HCl, 6mM MgCl$_2$, 10mM DTT, 0.05% Triton X-100,
            200uM each of dATP, dCTP and dGTP, 50 µM dTTP, 0.05 µCi/µl
            3H-dTTP, 0.5mM EDTA.

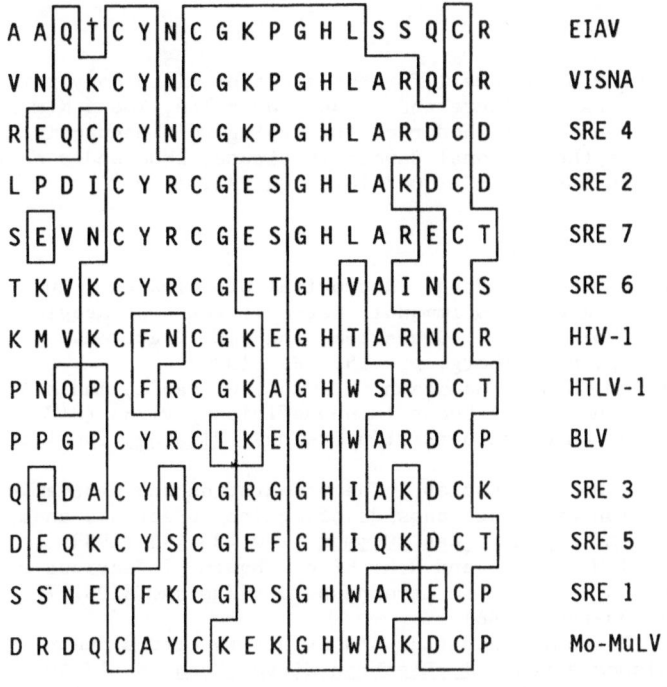

A A Q T C Y N C G K P G H L S S Q C R   EIAV

V N Q K C Y N C G K P G H L A R Q C R   VISNA

R E Q C C Y N C G K P G H L A R D C D   SRE 4

L P D I C Y R C G E S G H L A K D C D   SRE 2

S E V N C Y R C G E S G H L A R E C T   SRE 7

T K V K C Y R C G E T G H V A I N C S   SRE 6

K M V K C F N C G K E G H T A R N C R   HIV-1

P N Q P C F R C G K A G H W S R D C T   HTLV-1

P P G P C Y R C L K E G H W A R D C P   BLV

Q E D A C Y N C G R G G H I A K D C K   SRE 3

D E Q K C Y S C G E F G H I Q K D C T   SRE 5

S S N E C F K C G R S G H W A R E C P   SRE 1

D R D Q C A Y C K E K G H W A K D C P   Mo-MuLV

P E Q K C Y R C G K P G H L A R D C T   Consensus sequence

Fig. 7.   Alignment of the zinc finger domains of the sterol regulatory element binding protein with the first zinc finger domains of the NC proteins from a selection of retroviruses. The sequences are clustered according to the pairwise alignment procedure of GENALIGN (W.R. Taylor, Birkbeck College, London) using the "region" method (16). All amino acids matching the consensus sequence are boxed.

during the reverse transcription, and may be required *in vivo* for the production of the DNA provirus. Now that an infectious clone of EIAV has become available (13), it will be possible to construct mutants at the NC protein cleavage sites to make them noninfectious. If this is so, then the action of the proteinase during the early phase of the viral life cycle presents a more attractive target site for inhibitors than at the maturation process in the late phase. At the late phase the provirus is integrated, resulting in an irreversible generation of viral progeny that has to be continuously kept in check by high levels of inhibitor. Lower, non-toxic levels of inhibitor administered in one dose at the early phase would not allow the cell to be infected.

Proteins containing the zinc finger domain occur extensively in eukaryotic systems. One such example is the protein that binds to the sterol regulatory element, which shows a striking degree of sequence homology with the retroviral zinc finger domains (Ref. 14 and Fig. 7), and are therefore potential cleavage sites for retroviral aspartic protease-like enzymes. There are reports of chromatin-associated proteases (15), and if a role for the protease can be established during reverse transcription, then this provides a rationale to investigate proteolysis in the regulation of gene expression.

ACKNOWLEDGEMENTS

We wish to thank Diane Krell for the electron microscopy work, Stacey Bricker for tissue culture and virus production, Young Kim for protein sequencing, and Cheri Rhoderick for typing this manuscript. Research sponsored by the National Cancer Institute, DHHS under contract No. NO1-CO-74101 with ABL and NO1-CO-74102 with PRI.

REFERENCES

1. S. Oroszlan, and R.B. Luftig. Retroviral proteinases. in "Current Topics Microb. & Immunol. Retroviruses - Strategies of Replication", Vol. 157, R. Swanstrom, and P.K. Vogt, eds, Springer-Verlag, Heidelberg, pp. 153-185 (1990).

2. H.R. Gelderblom, E.H.S. Hausmann, M. Ozel, G. Pauli, and M.A. Koch. Fine structure of human immunodeficiency virus (HIV) and immunolocalization of structural proteins. Virology 156:171-176 (1987).

3. M.M. Roberts, and S. Oroszlan. The preparation and biochemical characterization of intact capsids of equine infectious anemia virus. Biochem. Biophys. Res. Commun. 160:486-494 (1989).

4. R. Stephens, J.W. Casey, and N.R. Rice. Equine infectious anemia virus gag and pol genes: Relatedness to visna and AIDS virus. Science 231:589-594 (1986).

5. M.A. Gonda. Molecular genetics and structure of the human immunodeficiency virus. J. Electron Micro. Tech. 8, 17-40 (1988).

6. M.M. Roberts, and S. Oroszlan. The action of retroviral protease in various phases of virus replication. in "Retroviral Proteinases: Maturation and Morphogenesis", L.H. Pearl, ed., MacMillan Press, London, pp.131-139 (1990).

7. M.M. Roberts, T.D. Copeland, and S. Oroszlan. In situ processing of the retroviral nucleocapsid protein by the viral aspartic proteinase. Protein Engineering, to be submitted (1990).

8. R.L. Karpel, L.E. Henderson, and S. Oroszlan, Interactions of retroviral structural proteins with single-stranded nucleic acids. J. Biol. Chem. 262:4961-4967 (1987).

9. M.R. Summers, T.L. South, B. Kim, and D.R. Hare. High-resolution structure of an HIV zinc fingerlike domain via a new NMR-based distance geometry approach. Biochemistry 29:329-340 (1990).

10. A. Wlodawer, M. Miller, M. Jaskolski, B.K. Sathyanarayana, E. Baldwin, I.T. Weber, L.M. Selk, L. Clawson, J. Schneider, and S.B.H. Kent. Conserved folding in retroviral protease: Crystal structure of a synthetic HIV-1 protease. Science 245:616-621 (1989).

11. H. Varmus, and P.O. Brown. Retroviruses. in "Mobile DNA", M. Howe and D. Berg, eds, American Society Microbiology, Washington, DC, pp 53-108 (1989).

12. J.M. Coffin. Retroviridae and Their Replication, in "Virology" B.N. Fields, D.M. Knipe et al., eds, Raven Press, Ltd., New York, pp 1437-1500 (1990).

13. L. Whetter, D. Archambault, S. Perry, A. Gazit, L. Coggins, A. Yaniv, D. Clabough, J. Dahlberg, F. Fuller, and S. Tronick. Equine infectious anemia virus derived from a molecular clone persistently infects horses. J. Virol. 64:5750-5756 (1990).

14. T.B. Rajavashisth, A.K. Taylor, A. Andalibi, K.L. Svenson, and A.J. Lusis. Identification of a zinc finger protein that binds to the sterol regulatory element. Science 245:640-643 (1989).

15. H. Holzer, and P.C. Heinrich. Control of proteolysis. Ann. Rev. Biochem. 49:63-91 (1980).

16. E. Sobel, and H.M. Martinez. A multiple sequence alignment program. Nucleic Acids Res. 14:363-374 (1985).

RED BLOOD CELLS EXPOSING CD4, AS COMPETITIVE INHIBITORS OF

HIV-1 INFECTION

Claude Nicolau[1+], David J. Volsky[#], Pierre-François Tosi[+], Youssef Mouneimne[+], Michael Zeira[#], Jaime Lazarte[+], Loyd Sneed[+]

[+] Section of Cell Biology, Institute of Biosciences and Technology, Texas A&M University, College Station, TX 77843-2401
[#] Molecular Virology Laboratory, St. Luke's Roosevelt Hospital Center and College of Physicians and Surgeons, Columbia University, New York, NY 10019

Infection by human immunodeficiency virus (HIV) [1-3] is mediated by binding to CD4, a glycoprotein expressed on the surface of HIV-susceptible cells.[4] The interaction between CD4 and the HIV envelope glycoprotein, gp120, was demonstrated by McDougal et al.,[5] who coimmunoprecipitated the two proteins from infected cells. The affinity constant of the gp120-CD4 complex is about 4×10^{-9} M which is comparable to other virus-receptor interactions.[6] Maddon et al.[7] provided further evidence for the essential role of CD4 in HIV infection by introducing a functional CD4 gene into CD4⁻ human cells, thereby conferring HIV susceptibility to previously resistant cells.

Because of the role of CD4 as the HIV receptor, several strategies have been proposed for using the CD4 protein to block infection. A soluble fragment of the CD4 molecule (sCD4) which contains the extracellular domain but not the hydrophobic transmembrane domain, has been produced by various expression systems. sCD4 binds gp120 and neutralizes HIV, as measured by in vitro assays, which include syncytium induction, virus replication, and target cell cytotoxicity.[8-13] sCD4 also has been used as a targeting agent for delivering cytotoxins to HIV-infected cells that express gp120.[14]

Soluble CD4 injected intramuscularly[15] or intravenously[16] has a short life span in circulation. Even hybrid molecules of the immunoadhesive types have half life times of up to 48 hrs only.[16] This was one of the reasons why we thought[17,18] that the realization of a long-lived CD4 carrier in circulation, exposing the receptor with its full activity might be a useful approach on two levels:

---

[1] Corresponding author

Advances in Molecular Biology and Targeted Treatment for AIDS
Edited by A. Kumar, Plenum Press, New York, 1991

1. It might attach circulating free virus, competing with the circulating T4 cells or HIV.

2. It might attach gp120-expressing, HIV-infected cells and form aggregates which might be subsequently phagocyted by the reticulo-endothelial system cells.

The red blood cell (RBC) is comparatively long-lived in circulation with half life times of ~12 days in mice, ~20 days in rabbits and ~60 days in man. It therefore seemed that an ideal candidate for a long lived CD4 carrier would be the red blood cell provided that:

a. CD4 could be inserted in its membrane in a nondestructive way, not altering the cell's life span;
b. the insertion would be stable _in vivo_
c. the inserted CD4 would retain its high affinity for gp120 (and for HIV-1) thus being a competitor of the T4 cells for the virus;
d. _in vitro_ studies should demonstrate that RBC-CD4 would specifically aggregate with gp120-expressing, HIV-infected cells and prevent infection of the T4 cells.

EXPRESSION AND PURIFICATION OF FULL LENGTH CD4

Preliminary studies in our laboratory had indicated that membrane proteins like glycophorin or the complete CD4 molecule extracted from CEM cells (a line of human T-lymphoblasts) could be associated with RBC by low pH exposure,[19,20] could be inserted in the red blood cell membrane using a novel technique which we developed, called electroinsertion.[21] Soluble proteins like bovine or human serum albumin, β2 microglobulin or soluble CD4 could not be electroinserted in the plasma membrane of the RBC, apparently because they lacked the hydrophobic, membrane spanning sequence.[22] We therefore expressed the full length CD4 molecule in _Spodoptera frugiperda_ 9 cells using a baculovirus vector.[23]

Expression, Extraction and Purification of Full Length CD4[26]

A cDNA encoding full-length CD4 (gift from R. Axel, Columbia University, New York, NY) was inserted into the genome of _Autographa californica_ nuclear polyhedrosis virus under transcriptional regulation of the viral polyhedrin gene promoter. The recombinant virus was used to infect insect cells which resulted in the abundant expression of CD4 as evaluated by flow cytometry and immunoblot analysis (Figure 1). Recombinant CD4 expressed on the surface of infected insect cells was immunologically indistinguishable from human CD4 when using eleven different anti-CD4 monoclonal antibodies. The extraction of infected cells by phase-transition separation with Triton X-114 followed by immunoaffinity chromatography yielded a single protein detected by NaDodSO$_4$/PAGE using silver staining (Figure 2a,b). N-terminal sequence analysis of the purified recombinant protein showed that CD4 produced in sf9 cells is efficiently cleaved from the precursor protein. Immunoblot analysis under nondenaturing conditions showed that the purified protein reacted with the anti-CD4 monoclonal antibody Leu-3a.

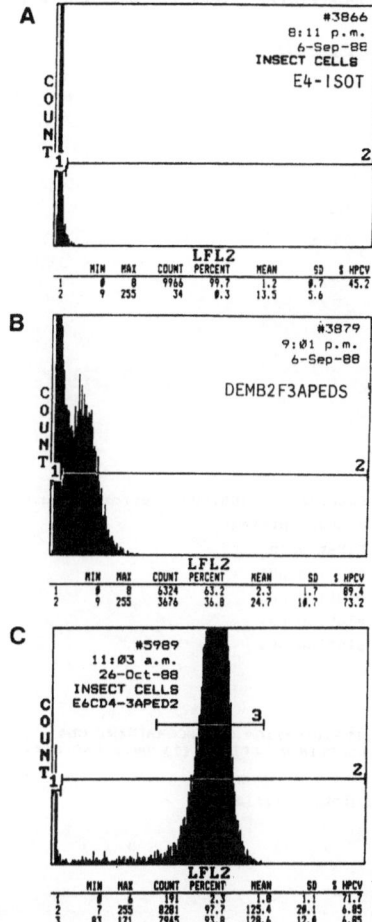

Fig. 1. Flow cytometry analysis of membrane-bound CD4.
Fluorescence histograms of Ac-CD4-infected insect
(sf9) cells incubated with secondary antibody goat
anti-mouse antibody alone (A). Human acute
lymphoblastic leukemia CEM-CM3 cells incubated with
Leu 3a, a CD4-specific monoclonal antibody that
inhibits HIV-induced syncytium formation[24] (B), and
Ac-CD4-infected sf9 cells 48 hrs. after infection
incubated with Leu 3a (C). Histograms were
generated from 10^4 cells. Similar histograms were
obtained with the following monoclonal antibodies,
each of which bind to distinct epitopes on the CD4
molecule; OKT4A, -B, -C, -D, -E and -F; 13B8.2;
BL4/10T4 and MT151. No measurable immunofluorescence was detected on AcMNPV or on mock-
infected sf9 cells.[23] On ordinate cell number
and on abscissae log red fluorescence.

Electrophoresis Analysis

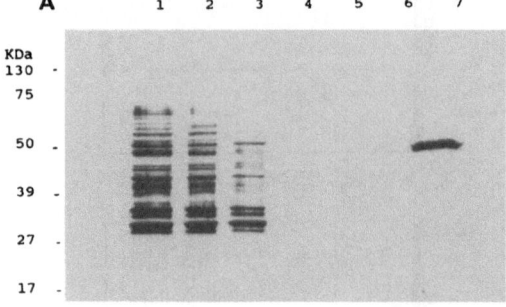

1 — Supernatant 100,000 g before column
2 — Unbound protein
3 — First wash
4 — elution pH6
5 — elution pH5
6 — elution pH4
7 — elution pH3

PURIFICATION OF RECOMBINANT CD4
ON IMMUNOAFFINITY COLUMN (13 B8.2 mAB antiCD4)

Westernblot Analysis

1 — Supernatant 100,000 g before column
2 — Unbound protein
3 — First wash
4 — elution pH6
5 — elution pH5
6 — elution pH4
7 — elution pH3

Analysis of the amino acid sequence predicted by the CD4 cDNA reveals two potential N-linked glycosylation sites.[25] A study by König et al.[26] shows that tunicamycin treatment of CEM cells decreases the apparent molecular mass of CD4 from 52 kDa to 46 kDa, which is consistent with glycosylation at both putative N-linked sites. Since N-terminal sequence analysis of purified recombinant CD4 showed that the signal peptide was efficiently cleaved from the CD4 precursor expressed in insect cells, we assume that the nonglycosylated CD4 observed in tunicamycin-treated insect cells is similar, if not identical, to the nonglycosylated CD4 reported for tunicamycin-treated human cells.[23] Direct chemical analysis of the oligosaccharide moieties present on authentic and recombinant CD4 has not been done.

The level of expression of the recombinant protein was compared to the level expressed in human lymphocytes by two methods. Flow cytometry analysis with eleven different CD4-specific monoclonal antibodies indicated that the cell-surface expression of recombinant CD4 is ~20-fold higher than that measured on human CEM-CM3 cells (Figure 1). Immunoblot analysis of total cell extracts suggested that the expression of CD4 in infected sf9 cells is at least 100-fold higher than in CEM cells. The discrepancy between the two assays may be due to the inherent sensitivity of each quantitation method used or the method by which each was standardized. Alternatively it is possible that Ac-CD4-infected sf9 cells contain a significant amount of CD4 not expressed on the cell membrane.

The extraction and purification methods described here yielded a highly purified membrane-derived CD4 with a molecular mass of 55 kDa and immunoreactivity to Leu-3a monoclonal CD4 antibody.

Electroinsertion of Full Length CD4 in Red Blood Cell Membranes[22]

The electroinsertion process consists of the application of pulsed electrical field, of microseconds' duration, on a suspension of cells, in the presence of a selected membrane protein having a membrane spanning sequence.[21] It results in the implantation of the protein in the cells' plasma membrane.

In previous studies we reported the electroinsertion of human glycophorin into mouse RBC membranes. Insertion of 10^5 molecules per cell was observed, but only 30% of the cells had inserted glycophorin.[21] On the basis of these results, we further developed the electroinsertion technique with the aim

---

Fig. 2A,B. Gel Electrophoresis (A) and immunoblot analysis (B) of purified recombinant CD4. CD4 was extracted and purified as described. An aliquot from the 100.000xg supernatant of the infected cell extract (lane 1) or the pH 3.0 eluant from an immunoaffinity column (lane 2) was detected by silver staining on NaDodSO$_4$/10% PAGE (A) and by immunoreactivity with Leu 3a monoclonal antibody under nondenaturing conditions (B).[23]

to insert membrane proteins in 100% of the RBC subjected to the electrical pulses. The rationale behind these experiments is that insertion of CD4 into RBC membrane by a non damaging method may provide a long lived CD4 carrier in circulation. Other methods of association of CD4 (or glycophorin) with RBC have been explored.

Studies had shown that after brief exposure at low pH (4.7), CD4 associates with human RBC [19,20] and such RBC-CD4 were shown to bind specifically to gp120-coated wells and to aggregate with gp120-expressing cells (Arvinte et al., J. AIDS, 1990, in the press). Nevertheless, the low pH treatment of the RBC, under conditions of maximum CD4 association, leads to a significant reduction in the RBC life-span in mouse (results not shown).

RBC subjected to pulsed electric fields have been shown to maintain a normal life span,[31] and the results previously obtained with glycophorin[21] indicated that significant levels of insertion may be attained.

In our electroinsertion experiments,[22] the pulse generator was a 606 Cober device, and we used a cylindrical teflon chamber.

Human erythrocytes were separated from fresh whole blood of healthy donor with citrate buffer as anticoagulant. Mouse erythrocytes were separated from heparinized fresh whole blood obtained from BALB/C strain mouse by retro orbital sinus puncture. The blood was washed with the electroinsertion medium.

Highly purified, lyophilized, full length recombinant CD4 obtained in our laboratory was dissolved into the electroinsertion medium and the RBC were suspended in this solution. The CD4 solution, at 8 mg/ml concentration, was added to the erythrocyte suspension. After 20 minutes incubation on ice, the temperature was raised to 37°C and four square electrical pulses, 1 ms long, were applied at 15 minutes interval.

A control sample where RBC were subjected to all steps of the described procedure but CD4 was replaced by BSA, was used as reference. Another control for the possibility of insertion of protein without membrane spanning sequence was performed replacing CD4 by β_2 microglobulin and by sCD4 (courtesy of SmithKline Beecham Laboratory). The experimental details of the electroinsertion procedures have been published.[22]

To assay different epitopes of the CD4 glycoprotein at the surface of RBC membranes, cells were incubated, at room temperature during 30 minutes, with different monoclonal antibodies, OKT4A, OKT4C, OKT4D, BL4/10T4 or Leu 3a, washed twice and incubated with secondary antibody Gam-IgG-PE. After washing, the cells were examined by fluorescence microscopy and flow cytometry.

In order to compare the inserted CD4 into the red blood cells with the native CD4 exposed by human cells (T-lymphoblasts, line CEM-CM3), we studied the possibility to accumulate successive immune complex as described.[22] Briefly,

RBC-CD4 or CEM cells already stained with OKT4D and Gam-IgG-PE were incubated with a mouse monoclonal antibody anti-PE (clone PE-85), washed and then incubated with Gam-IgG-PE. The enhancement of the fluorescent signal was quantified by flow cytometry after application of every new cycle of PEAPE[2] complexes.

The CD4 molecules inserted into the red blood cell membrane were able to react with the following different anti CD4 monoclonal antibodies: OKT4A, OKT4C, OKT4D, Leu 3a, and BL4/10T4. It appears that the electroinserted CD4 exposes the different active epitopes in the proper orientation including the Leu 3a and OKT4A epitopes characteristic for the binding of the viral envelope protein gp120.[22,27] Knowing that these epitopes are distributed over all the external sequence of CD4 [24,28,29] and that the fluorescence intensity, measured by flow cytometry, upon reaction of each monoclonal antibody separately with the same RBC-CD4 sample was identical, it appears that electroinserted CD4 exposes different active epitopes in the proper orientation and ratio. This further supports the view that adsorption of CD4 is unlikely to be the result of subjecting RBC and CD4 to electrical fields. Should the CD4 molecules have been adsorbed, some epitopes would have been less detectable than others. Finally, the presence of the different epitopes indicates that the CD4 molecule was not denatured during the electroinsertion procedure.

Examination of the RBC-CD4 under the fluorescent microscope, with blue light excitation (488nm), yielded the images shown in Figure 3a-c.[22] The reference sample, where CD4 was replaced by BSA, did not show any fluorescence (Figure not shown). The punctate fluorescent pattern observed on the RBC-CD4[+] after immunofluorescence reaction is due to the patching of the inserted CD4 receptors upon reaction with monoclonal antibodies (Leu 3a and Gam-IgG-PE). The same phenomenon is observed with CEM cells (a human T-lymphoblast cell line) upon reaction with mAb (Leu 3a and a Gam-IgG-PE) (Figure 1e). In order to check that patching occurred only after reaction with mAb, FITC-labeled human glycophorin was electroinserted in mouse RBC membrane. It showed uniformly fluorescent cells under the fluorescent microscope. Upon reaction with 10F7 anti-human glycophorin monoclonal antibody, patching was observed (data not shown). This indicates that the patchy distribution does not occur upon insertion of the antigen into the RBC membrane, it only happens due to antibody cross linking and is possible because of the lateral mobility of the inserted antigen. Patching of CD4 may be considered as a clear indication of insertion of CD4 molecules in the RBC lipid bilayer.

After reaction of RBC-CD4 with mAB Leu 3a and Gam-IgG-PE, the flow cytometry analysis shows single peak of fluorescence with a complete shift of the cell population towards the higher fluorescence intensity (Figure 4). This means that the red blood cells in the population subjected to the electroinsertion procedure expose CD4 molecules. Figure 4 indicates a mean

---

[2] PEAPE = Phycoerythrin-antiphyco-erythrin complex.

concentration of 4800 epitopes per mouse RBC. The reference sample showed the normal background level of the flow cytometry measurements.[22]

Figure 5 shows the variation of fluorescence of RBC-CD4 when successive immune complex were accumulated. The similarity between this curve and that of CEM cells treated with the same procedure, shows that the inserted CD4 molecules into the RBC membrane behaves like the naturally expressed CD4 on CEM cells, where three cycles PEAPE can be formed before saturation. This may be additional proof that CD4 molecules are inserted into the RBC membrane and not adsorbed. Adsorbed CD4 on the surface of RBC would not withstand the mechanical stress caused by the four successive immune complex accumulated and would very probably be desorbed from the RBC surface.

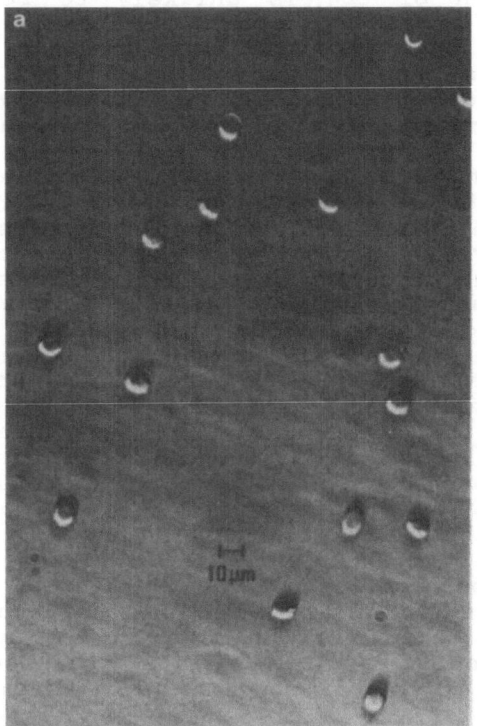

Fig. 3.   Immunofluorescence imaging of RBC-CD4 (a,c) and CEM cells (e). RBC-CD4 and CEM cells were stained with monoclonal antibody Leu 3a then Phycoerythrin conjugated goat anti-mouse.
a) Human RBC-CD4. Excitation wavelength 488 nm.
b) The same human RBC-CD4. Direct light.
c) CEM cells. Excitation wavelength 488 nm. Notice the capping of CD4 upon reaction with monoclonal antibodies in both cases (RBC-CD4 and CEM cells). Fluorescence microscope VANOX0T, Olympus Inc. (San Antonio, TX, USA).[18]

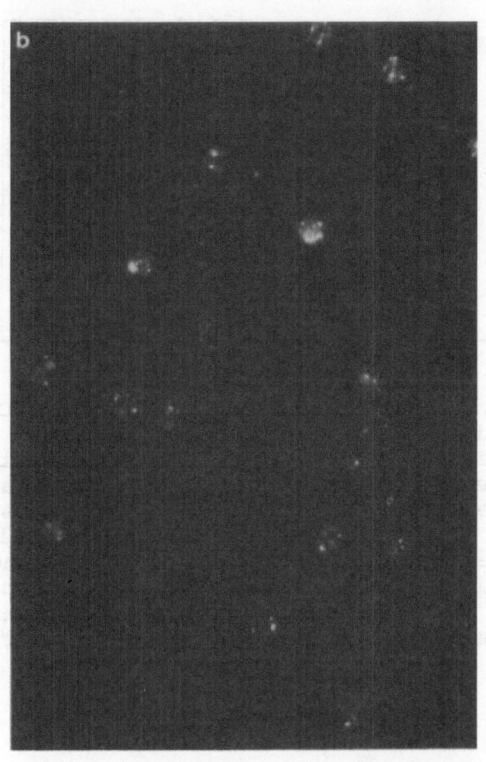

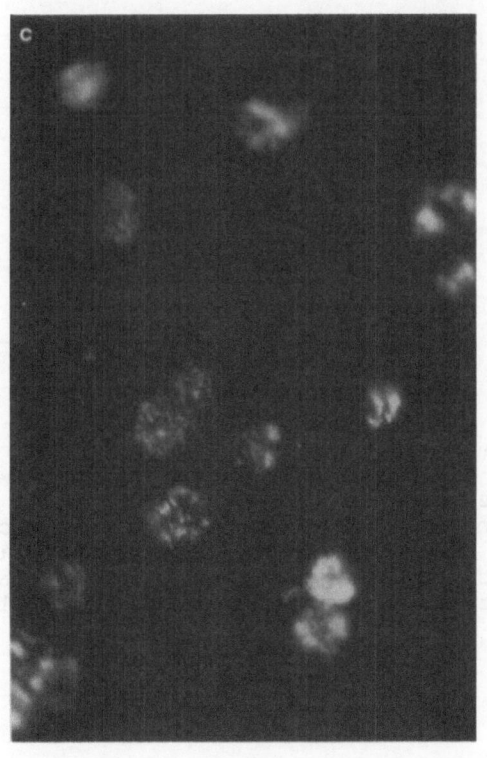

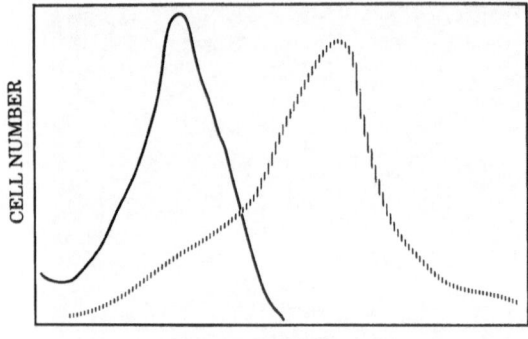

LOG FLUORESCENCE

Fig. 4. Flow cytometry histograms. Mouse RBC subjected to electroinsertion with BSA (solid line) or CD4 (broken line) were stained with monoclonal antibody Leu 3a and subsequently with Phycoerythrin conjugated goat anti-mouse secondary antibodies. Histograms were analyzed for red fluorescence intensity and cell number. The mean peak channel of the control RBC and RBC-CD4 were respectively 5.3 and 48.1. This corresponds to a mean peak of 4800 epitopes per cell. Epics Profile flow cytometer; Coulter Epics flow cytometer (Hialeah, FL, USA).[22]

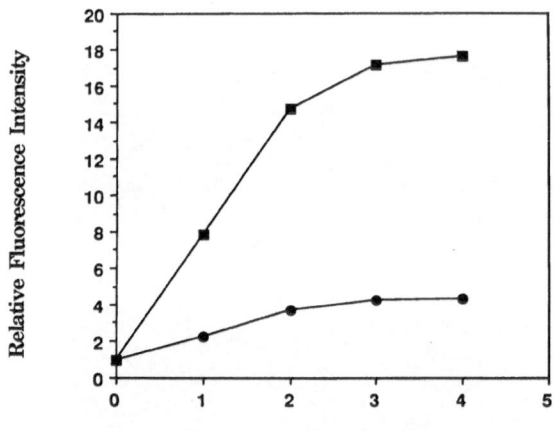

Number of Anti-Pe labeling

Fig. 5. Variation of fluorescence intensity with the number of Anti-PE successive labeling of RBC-CD4 (O) and CEM cells (■). Mouse RBC-CD4 and CEM cells, labeled with OKT4D and Gam-Ig-PE, were labeled successively with Anti-Pe and Gam-IgG-PE till saturation. The fluorescence intensity was plotted relatively to the intensity of the first Gam-IgG-PE antibody.[22]

In our experience, only membrane proteins, for example full length rCD4 or glycophorin, could be electroinserted. Proteins without membrane spanning sequence such as human $\beta2$ microglobulin could not be electroinserted into RBC membrane. $\beta2$ microglobulin ($\beta2$-m) is a soluble protein (MW 11,800) that was found associated with cell surface antigens of the major histo compatibility complex. Attempts to electroinsert $\beta2$-m as a control or sCD4 failed. No $\beta2$-m or sCD4 epitopes following interaction with antibodies were ever detected by flow cytometry analysis or by fluorescent microscope observations (data not shown) on the RBC subjected to this procedure in the presence of these proteins. Our failure to associate CD4 at the surface of RBC which allowed insertion of both full length CD4 and glycophorin with RBC suggests that this technique does not favor nonspecific adsorption of proteins on the plasma membrane of RBC.[22]

Electroinsertion of CD4 in RBC is not expected to result in all CD4 molecules being exposed in the "correct" way, i.e. displaying all epitopes presented by CD4 on human T4 cells. In order to determine the number of "correctly" oriented CD4 molecules on the surface of the RBC, fluorescein isothiocyanate (FITC) was covalently attached to CD4 prior to electro-insertion. FITC-CD4 was electroinserted in RBC membrane and after cytofluorimetric assay of the number of FITC-CD4 molecules per cell, the RBC-CD4-FITC were reacted with anti FITC antibodies.[27] Cytofluorimetric assay of the fluorescence emission indicated that 27% of the fluorophores could not be quenched by the anti FITC antibodies. This data indicates that ~70% of the CD4 molecules electroinserted are oriented in the proper direction in the RBC membrane, displaying the epitopes detected on naturally occurring CD4 molecules on human T cells.[27]

Interaction of HIV-1 with RBC-CD4[27]

To determine whether CD4 molecules inserted into the RBC membrane are functionally equivalent (with regard to their interaction with HIV-1) to native CD4 expressed on the T cell surface, we evaluated three functions normally associated with membrane-bound viral receptors: i) their ability to mediate HIV-1 entry; ii) their ability to inhibit infection of CD4$^+$-T cells with cell-free HIV-1; iii) their ability to mediate the aggregation of RBC-CD4 and HIV-1 infected T cells.

The potential of CD4 exposed on the RBC surface to mediate HIV-1 binding and "entry" was determined using the membrane fluorescence dequenching (DQ) technique[30-32]. HIV-1 enters CD4-positive T cells and monocytes by fusion between viral envelopes and target cell membranes,[31] and this process can be analyzed quantitatively by measuring an increase in fluorescence resulting from the intermixing between quenched-labelled virions and unlabelled target membranes.[31] As measured by the DQ method, HIV-1 fused in a cell number-dependent manner with human RBC-CD4 but not with control human RBC (Figure 6a,b).

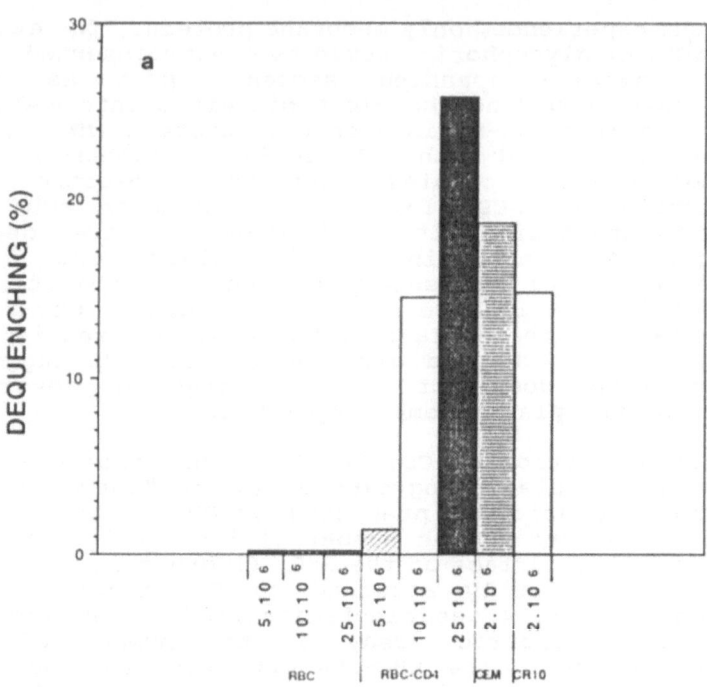

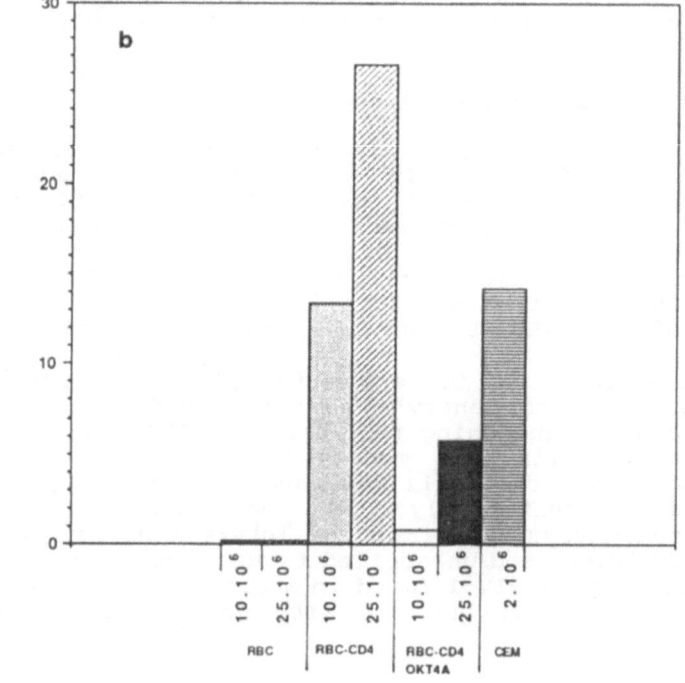

Results of two separate experiments using different preparations of RBC-CD4 and labelled virions are shown. The DQ value obtained with 2×10^6 CD4-positive CEM or CR10 cells was equivalent to that obtained with 25×10^6 RBC-CD4 (Figure 6). This is consistent with the 2,000 CD4 epitopes/cell in RBC-CD4 versus about 50,000 molecules/CEM cell.[23] Fusion of HIV-1 with RBC-CD4, like that with CD4 cells, occurred at 37°C but not 4°C and had rapid kinetics with half-DQ time of 5-8 min. at 37°C. It is noteworthy that pretreatment of RBC-CD4 with OKT4A antibodies significantly reduced the DQ (Figure 6b). This suggests that the binding and fusion of HIV-1 with RBC-CD4 is mediated by the same functional region(s) of the CD4 molecule that serve as HIV receptor on T cells.

Inhibition by RBC-CD4 of HIV-Infection of T4 Cells[27]

Next, the capacity to prevent infection in vitro with cell-free HIV-1 of RBC-CD4 was tested. This test was crucial for the strategy to use RBC-CD4 as competitors with circulating T4 cells in order to reduce the in vivo titers of HIV in circulation and to possibly eliminate gp120-expressing, HIV-infected cells from circulation. RBC-CD4, RBC-CD4/OKT4A, RBC, or CEM cells were preincubated with HIV-1/III$_B$ for 30 min. at 37°C; the cells were removed by low-speed centrifugation and supernatants were filtered through 0.45 Millipore filters. Filtered supernatants were then used for infection of target CD4-positive T cells by standard protocol.[34] Figure 3 shows the results of experiments of inhibition of HIV-1 infection by RBC-CD4. Clearly, pretreatment with either RBC-CD4 or CEM cells drastically reduced the infectious titer of HIV-1 present in the virus preparation. RBC-CD4 in which HIV-1 binding CD4 epitopes were blocked with OKT4A antibodies were significantly less effective in "clearing" infectious HIV (Figure 8). Similar results were obtained when C-8166 cells which are exquisitely susceptible to HIV infection served as targets in these experiments (Figure 7).

Fig. 6.   Measurement of fusion between HIV-1 and RBC-CD4 by the membrane fluorescence dequenching method (DQ). Purified and R-18 labelled HIV-1 was incubated with the designated amounts of RBC-CD4, RBC-CD4 saturated with OKT4A mAB, control RBC, and CD4 positive cell lines, CEM and CR10 as control. When fusion between the labelled virus and the target cells occurred, an increase in fluorescence signal (i.g. fluorescence dequenching) was measured by the fluorometer. This increase in signal is directly proportional to the extent of viral-cell fusion (30-32).[27,30-32]

It thus appears that RBC in whose membrane full-length recombinant CD4 molecules have been inserted can serve as a long term carrier of functional HIV-1 receptors. In *in vitro* experiments, CD4 exposed on RBC is indistinguishable from native CD4 on T cells in its capacity to mediate HIV-1-target membrane fusion, to clear infectious HIV from cell-free virus preparations

Interaction of RBC-CD4 with HIV-infected CEM Cells[27]

One of the possible mechanisms of action of RBC-CD4 was envisioned to be the formation of aggregates with gp120-expressing, HIV-infected cells.

Studies of the interaction of RBC with a line of genetically engineered CHO cells constitutively expressing gp120, showed strong aggregation of RBC-CD4 with those cells (Arvinte et al. *J. AIDS*, in the press).

The major potential advantages of this system are, besides the relatively long $t_{\frac{1}{2}}$ of CD4 in circulation, the much smaller amounts of CD4 needed to achieve the "therapeutical concentration" than those used with sCD4, and the aggregation and probable clearance of gp120-expressing, HIV-infected cells from circulation as well as the lack of immune response to the RBC-CD4.

To evaluate whether RBC-CD4 can interact with HIV-1 infected T lymphocytes, RBC-CD4 were mixed with the chronically infected CEM/N1T-E [24] cells at the numerical ratio of 100:1, respectively, and examined under a light microscope after 5-10 min. of incubation at 37° C. Numerous aggregates between CEM/N1T-E cells and RBC-CD4, but not control RBC, were seen within a short time of incubation were observed between RBC-CD4 and T cells and occasionally among T cells themselves (Figure 8). Thus, RBC-CD4 appears to recognize the native gp120 molecules on the surface of HIV-1 infected cells, and thus aggregate with these cells.

It appears that the RBC-CD4 are capable of attaching and aggregating with gp120-expressing, HIV-infected cells *in vitro*. If such aggregates were to be formed *in vivo*, they would be phagocyted from circulation by the recticuloendothelial system cells. Lysosomal digestion of the phagocyted material should be able to prevent largely survival of HIV in those cells. Moreover, should autologous RBC-CD4 be administered to patients, this system would act concomitantly with AZT (or other nucleoside analogs) which might further prevent additional HIV infection of the phagocytic cells.

Fig. 7. Reduction in the infectivity titre of HIV-1 by preincubation with the RBC-CD4. (a) Kinetics of intracellular HIV-1 antigen expression after exposure of CEM cells to HIV-1 preincubated with RBC-CD4 (o), RBC-CD4 blocked with OKT4A (△), and untreated RBC (□); (b) Supernatant p24 antigen levels in parallel experiment using C-8166 cells as targets; assay time: 6 days post-infection. (c) Supernatant RT levels on day 12 post infection of CEM cells.[27]

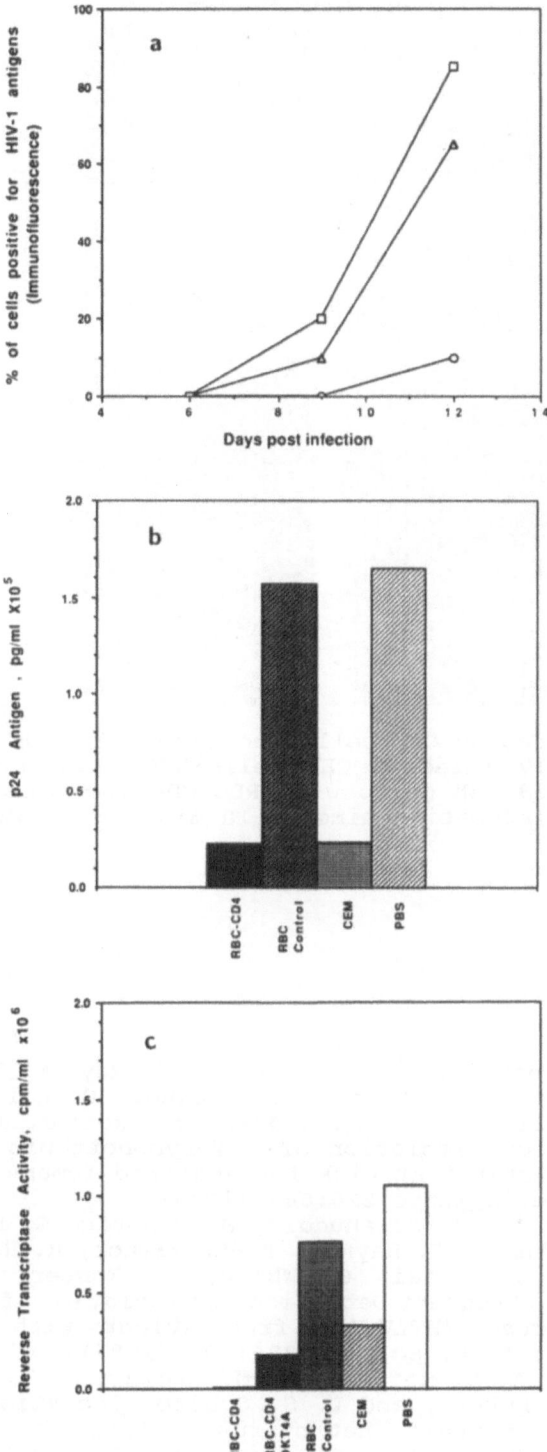

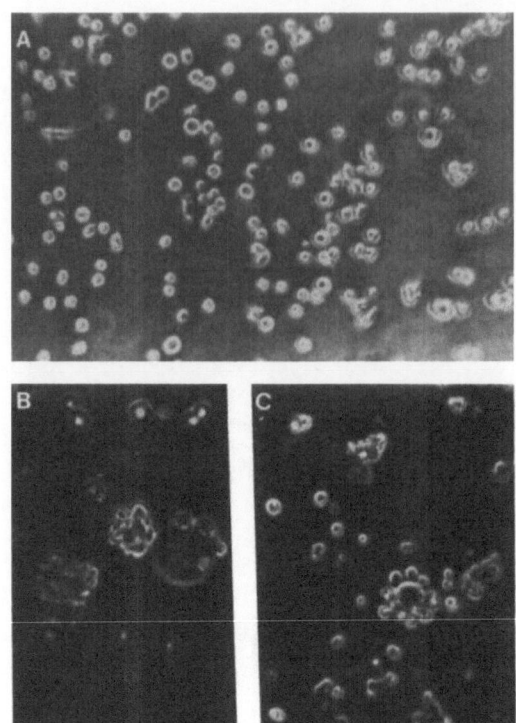

Fig. 8. HIV-infected CEM cells incubated with RBC-CD4.
A) HIV infected CEM cells-RBC. B) and C) HIV-infected CEM cells with RBC-CD4 incubated with RBC-CD4. Incubation time: 5-10 min. at 37° C.

References

1. F. Barré-Sinoussi, J.-C. Chermann, F. Ray, M. T. Nugeyre, S. Chamaret, J. Gruest, C. Dauguet, C. Aaler-Blin, F. Brun-Vézinet, C. Rouzioux, W. Rozenbaum, and L. Montagnier, Isolation of a T-Lymphotropic Retrovirus from a Patient at Risk for Acquired Immune Deficiency Syndrome, Science 220:868 (1983).
2. R. C. Gallo, S. Z. Salahuddin, M. Popovic, G. M. Shearer, M. Kaplan, B. F. Haynes, T.-L. Palker, R. Redfield, J. Oleske, B. Safai, G. White, P. Foster, and P. D. Markham, Frequent Detection and Isolation of Cytopathic Retroviruses (HTLV-III) from Patients with AIDS and at Risk for AIDS, Science 224:500 (1984).
3. J. A. Levy, A. D. Hoffman, S. M. Kramer, J. A. Landis, J. M. Shimabukuro, and L. S. Oshiro, Isolation of Lymphocytopathic Retroviruses from San Francisco Patients with AIDS, Science 225:840 (1984).
4. Q. J. Sattentau and R. A. Weiss, The CD4 antigen: physiological ligand and HIV receptor, Cell 52:631 (1988).

5. J. S. McDougal, M. S. Kennedy, J. M. Sligh, S. P. Cort, A. Mawle, and J. K. A. Nicholson, Binding of HTLV-III/LAV to T4 + T cells by a complex of the 110K viral protein and the T4 molecule, <u>Science</u> 231:382 (1986).

6. L. A. Lasky, G. Nakamura, D. H. Smith, C. Fennie, C. Shimasaki, E. Patzer, P. Berman, T. Gregory, and D.J. Capon, Delineation of a region of the human immunodeficiency virus Type 1 pg120 glycoprotein critical for interaction with the CD4 receptor, <u>Cell</u> 50:975 (1987).

7. P. J. Maddon, A.G. Dalgleish, J. S. McDougal, P.R. Clapham, R.A. Weiss, and R. Axel, The T4 gene encodes the AIDS virus receptor and is expressed in the immune system and the brain, <u>Cell</u> 47:333 (1986).

8. D. H. Smith, R.A. Byrn, S.A. Marsters, T. Gregory, J. E. Groopman, D. J. Capon. Blocking of HIV-1 infectivity by a soluble, secreted form of the CD4 antigen, <u>Science</u> 238:1704 (1987).

9. E. A. Berger, T. R. Fuerst, and B. Moss, A soluble recombinant polypeptide comprising the amino-terminal half of the extracellular region of the CD4 molecule contains an active binding site for human immunodeficiency virus, <u>Proc. Natl. Acad. Sci. USA</u> 85:2357 (1988).

10. K. C. Deen, J. S. McDougal, R. Inacker, G. Folena-Wasserman, J. Arthos, J. Rosenberg, P. J. Maddon, R. Axel, and R. W. Sweet, A soluble form of CD4 (T4) protein inhibits AIDS virus infection, <u>Nature</u> 331:82 (1988).

11. R. A. Fisher, J. M. Bertonis, W. Meier, V. A. Johnson, D. S. Costopoulos, T. Liu, R. Tizard, B. D. Walker, M. S. Hirsch, R. T. Schooley, R. A. Flavell, HIV infection is blocked <u>in vitro</u> by recombinant soluble CD4, <u>Nature</u> 331:76 (1988).

12. R. E. Hussey, N. E. Richardson, M. Kowalski, N. R. Brown, H. C. Chang, R. F. Siciliano, T. Dorfman, B. Walker, J. Sodroski, and E. L. Reinherz, A soluble CD4 protein selectively inhibits HIV replication and syncytium formation, <u>Nature</u> 331:78 (1988).

13. A. Traunecker. W. Lüke, and K. Karjalainen, Soluble CD4 molecules neutralize human immunodeficiency virus type 1, <u>Nature</u> 331:84 (1988).

14. M. A. Till, V. Ghetie, T. Gregory, E. J. Patzer, J. P. Porter, J. W. Uhr, D. J. Capon, and E. S. Vitetta, HIV-infected cells are killed by rCD4-ricin A Chain, <u>Science</u> 242:267 (1988).

15. M. Watanabe, K. A. Reimann, P. A. DeLong, T. Liu, R. A. Fisher, and N. L. Letvin, Effect of recombinant soluble CD4 in rhesus monkeys infected with simian immunodeficiency virus of macaques, <u>Nature</u> 337:267 (1989).

16. D. J. Capon, S. M. Chamow, J. Mordenti, S. A. Marsters, T. Gregory, H. Mitsuya, R. A. Byrn, C. Lucas, F. M. Wurm, J. E. Groopman, S. Broder, and D. H. Smith, Designing CD4 immunoadhesins for AIDS therapy, <u>Nature</u> 337:525 (1989).

17. C. Nicolau, G. M. Ihler, J. L. Melnick, C. A. Noonan, S. K. George, F. Tosi, T. Arvinte, and A. Cudd, Targeted drug delivery via protein mediated membrane fusion, Specific killing of HIV-infected human lymphocytes, <u>J. Cell Biochem</u> 12B:254 (1988).

18. C. Nicolau, P.-F. Tosi, T. Arvinte, Y. Mouneimne, A. Cudd, L. Sneed, C. Madoulet, B. Schulz, and R. Barhoumi, CD4 inserted in Red Blood Cell membranes or reconstituted in liposome bilayers as a potential therapeutic agent against AIDS, in "Horizons in Membrane Biotechnology," C. Nicolau and D. Chapman eds., Wiley-Liss, London, New York (1990).

19. T. Arvinte, B. Schulz, A. Cudd, and C. Nicolau, Low-pH association of proteins with the membranes of intact red blood cells. I. Exogenous glycophorin and the CD4 molecule, Biochim. Biophys. Acta 981:51 (1989).

20. T. Arvinte, A. Cudd, B. Schulz, and C. Nicolau, Low-pH association of proteins with the membranes of intact red blood cells. II. Studies of the mechanism, Biochim. Biophys. Acta 981:61 (1989).

21. Y. Mouneimne, P. F. Tosi, Y. Gazitt, and C. Nicolau, Electro-Insertion of Xeno-glycophorin into the red blood cell membrane. Biochem. Biophys. Res. Comm. 159:34 (1989).

22. Y. Mouneimne, P. F. Tosi, R. Barhoumi, and C. Nicolau, Electroinsertion of full length recombinant CD4 into red blood cell membrane, Biochim. Biophys. Acta 1027:51 (1990).

23. N. R. Webb, C. R. Madoulet, P. F. Tosi, D. R. Broussard, L. Sneed, C. Nicolau, and M. D. Summers, Cell-surface expression and purification of human CD4 produced in baculovirus-infected insect cells, Proc. Natl. Sci. USA 86:7731 (1989).

24. Q. J. Sattentau, A. G. Dalgleish, R. A. Weiss, and P. C. L. Beverley, Epitopes of the CD4 antigen and HIV infection, Science 234:1120 (1986).

25. P. J. Maddon, D. R. Littman, M. Godfrey, D. E. Maddon, L. Chess, R. Axel, The isolation and nucleotide sequence of a cDNA encoding the T cell surface protein: A new member of the immunoglobulin gene family, Cell 42:93 (1985).

26. R. König, G. Ashwell, and J. A. Hanover, Glycosylation of CD4, Tunicamycin inhibits surface expression, J. Biol. Chem. 263:9502 (1988).

27. M. Zeira, P.-F. Tosi, Y. Mouneimne, L. Sneed, J. Lazarte, D. J. Volsky, and C. Nicolau, submitted for publication (1990).

28. B. A. Jameson, P. E. Rao, L. I. Kong, B. H. Hahn, G. M. Shaw, L. E. Hood, and S. B. H. Kent, Location and chemical synthesis of a binding site for HIV-1 on the CD4 protein, Science 240:1335 (1988).

29. N. R. Landau, M. Warton, and D. R. Littman, The envelope glycoprotein of the human immunodeficiency virus binds to the immunoglobulin-like domain of CD4, Nature 334:159 (1988).

30. A. Loyter, V. Citkovsky, and R. Blumenthal, The use of fluorescence dequenching measurements to follow viral membrane fusion events, Meth. of Biochem. Anal. 33:129 (1988).

31. F. Sinangil, A. Loyter, and D. J. Volsky, Quantitative measurement of fusion between human immunodeficiency virus and cultured cells using membrane fluorescence dequenching, FEBS Lett. 239:88 (1988).

32. D. Hoekstra and J. W. Kok, Entry mechanisms of enveloped virus. Implications for fusion of intracellular membranes, Biosci. Reports 9:273 (1989).

33. J. D. Lifson, M. B. Feinberg, G. R. Reyes, L. Rabin, B. Banapour, S. Chakrabarti, B. Moss, F. Wong-Staal, K. S. Steimer, E. G. Engelman, Induction of CD4-dependent cell fusion by the HTLV 14/LAV envelope glycoprotein, Nature 323:725 (1986).
34. X. Ma, K. Sakai, E. Golub, and J. D. Volsky, Interaction of a non cytopathic human immunodeficiency virus type 1 with target cells: efficient virus entry followed by delayed expression of its RNA and protein, Virology 176:184 (1990).
35. K. Kinoshita and T. Y. Tsong, Survival of sucrose-loaded erythrocytes in the circulation, Nature 272:258 (1978).

GENE THERAPY FOR AIDS

Richard A. Morgan and W. French Anderson

Molecular Hematology Branch, National Heart, Lung
and Blood Institute, National Institutes of Health
Bethesda, MD   USA

With the number of HIV infected individuals world-wide now estimated in the millions and with no curative treatment or protective vaccine available in the immediate future, all efforts should be pursued to develop novel treatment modalities for AIDS.   One potential anti-HIV technology which is rapidly moving from the laboratory to the clinic is commonly refereed to as gene therapy. The nature of this technology, the potential methods by which it could be applied to combat HIV infection, as well as issues related to the safety and the practicality of administrating this technology on a large scale will be discussed.

Gene Transfer Technology

Gene therapy can be defined as: the transfer of new genetic material to the cells of an individual with resulting therapeutic benefit to the individual. The broadness of this definition should be contrasted to the more commonly conceived notion that gene therapy is strictly aimed at curing human genetic diseases.   There are currently three categories of approaches for the transfer of new genetic material (usually in the form of cloned DNA) into cells; physical transfer techniques such as microinjection, chemical mediated transfer procedures such as calcium phosphate coprecipitation, and biological transfer via viruses.   At present, only biological transfer via viral vectors is efficient enough to be contemplated for use in clinical gene transfer procedures.   The viral gene transfer system that is currently the most advanced is based on disabled murine retroviruses (1).

The first approved clinical trial to use retroviral-mediated gene transfer was designed not as a gene therapy experiment, but as a cell marking experiment.   In this protocol, human tumor infiltrating lymphocytes (TIL) were genetically marked with a retroviral vector and subsequently given to patients undergoing TIL cell immunotherapy for metastatic cancer (2).   The purpose of gene marking the TIL cells with a retroviral vector was to permit the study of the in vivo distribution and longevity of TIL.   While this particular protocol did not involve any therapeutic benefit to the patient, the gene transfer technology used in the TIL cell marking experiment is directly applicable to the development of therapeutic gene transfer.

Advances in Molecular Biology and Targeted Treatment for AIDS
Edited by A. Kumar, Plenum Press, New York, 1991

Retroviral mediated gene transfer

Retroviral mediated gene transfer (3) is dependent on a two component system, the packaging cell and the viral vector (figure 1). A packaging cell is a cell which contains a retroviral genome from which the signals that are responsible for the encapsidation of the viral genome into virus particles have been removed. Thus, the viral genome in the packaging cell line produces all the proteins necessary for virus replication and assembly, but can not create an infectious particle due to the lack of a transferable viral genome. Into the packaging cell line can be introduced a genetically engineered retroviral genome known as a retroviral vector. The retroviral vector is the complement to the packaging cell genome in that, all the normal viral protein coding sequences have been removed, but the encapsidation and replication signals have been retained. Any gene or promoter/gene combination that the investigator wishes to transfer can be inserted into the retroviral vector by molecular cloning (the absolute amount of genetic material that can be inserted into the retroviral vector is less than 7.0 kilobase pairs of DNA).

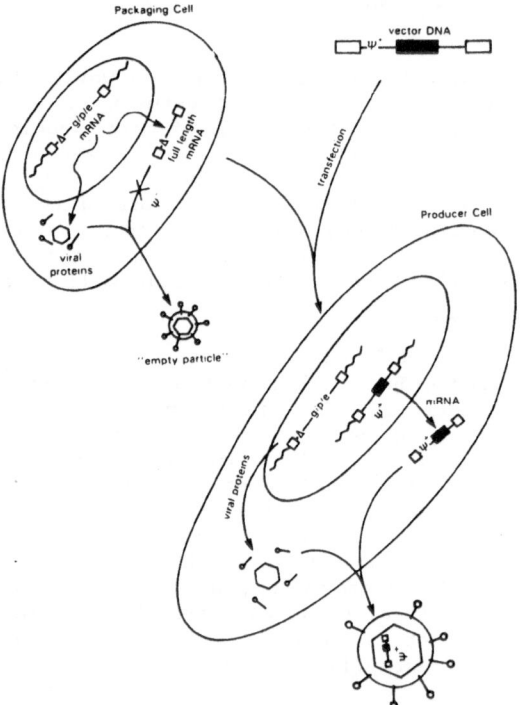

Figure 1.   Production of Retroviral Vector Particles. This figure diagrams the generation of retroviral vector particles by the introduction of the retroviral vector DNA into a packaging cell line. The packaging cell line (top left) makes the proteins necessary for viral assembly but can not package its own RNA. The vector DNA (top right) is introduced into the packaging cell line to make a producer cell line (bottom right), which will assemble a functional viral particle around the vector RNA as shown.

The virus particle assembled by the producer cell line (a producer cell line is defined as a packaging cell line containing a retroviral vector) can then be used to transfer the retroviral vector sequence to another cell. Upon addition of the retroviral vector to a culture of cells, the vector sequence is integrated into the host genome. Because of the defective nature of the retroviral vector, this transfer procedure is termed transduction to contrast it with infection, which is reserved for replication competent viruses. Once integrated, the internal sequences of the retroviral vector can be expressed leading to the production of a new protein in the transduced cell.

Retroviral vectors as anti-HIV agents

The most promising feature of retroviral vectors as anti-HIV agents, is their ability to deliver a gene or gene product precisely where you want it, at or in the cell. This can have enormous advantages for biological compounds which may only function in the intracellular environment or for compounds which need to be present in a high local concentration. Because retroviral vectors are stably integrated into the host cell chromosomes, it can be expected that they will express a given protein on a continuous basis (this has obvious benefits for any molecule with a limited biological half-life). In retroviral-mediated gene transfer the use of ex vivo transduction protocols can permit the engineering of specific cell populations. Thus, it may be possible to protect CD4+ T-cells from HIV infection by first isolating the cells from the bulk lymphocyte population, engineering them, and then returning them to the body. It may also be possible to engineer cells in the macrophage lineage, and then use the natural properties of theses cells to transfer anti-HIV agents to the brain (much as they are suspected of delivering the HIV virus itself). All of these factors combine to make retroviral vectors a potentially promising anti-HIV system.

Intracellular Intervention

A direct attack on the HIV virus is possible at any stage in its intracellular life cycle. The first set of strategies to be described necessitate that the retroviral vector and the HIV virus genome be present in the same cell for anti-HIV activity to be mediated (figure 2). Each of these intracellular intervention strategies takes advantage of HIV's own regulatory or structural proteins which can be modified and expressed by the retroviral vector to specifically inhibit HIV replication (4). The general principle of these strategies is that of using competitive HIV gene products (nucleic acid or protein). Dominant negative mutants of the HIV genes products could compete with the normal HIV viral gene products. The results of this competition would lead to the inhibition of HIV replication (5-7).

The first stage in the HIV life cycle (following entry into the CD4+ target cell) is reverse transcription and subsequent integration of the viral genome into the host chromosomes. It is possible that proteins produced by a retroviral vector could be designed to inhibit HIV at this point. It is generally believed that at this early stage of viral propagation, the replication mechanism occurs as a complex (within the viral core) of preassociated nucleic acid/protein complexes. This replication complex may thus not be amenable to intradiction by diffusible protein products. Following integration HIV proteins and nucleic acids become accessible to diffusible factors. Transcription and post-transcriptional events could be inhibited by expression of altered HIV regulatory proteins such as nef, tat or rev. This would lead to a decrease in the levels of critical HIV proteins, or to alteration in the balance of HIV regulatory to structural proteins. Further, nucleic acid decoys for cis-acting HIV regulatory motifs (eg. TAR or RRE) and anti-sense RNAs or ribozymes may also be envisioned as candidate inhibitory factors. At later stages of infection, it is possible that expression of altered core and/or envelope proteins could inhibit virus assembly and or lead to the production of defective virus particles. Finally, it may be possible to design a retroviral vector such that it could produce a suicide gene product if, and only if, HIV regulatory proteins (eg. tat and rev) are present in the cell. This last strategy could afford the possibility of ridding an individual of HIV infected cells, assuming that these cells could be appropriately targeted.

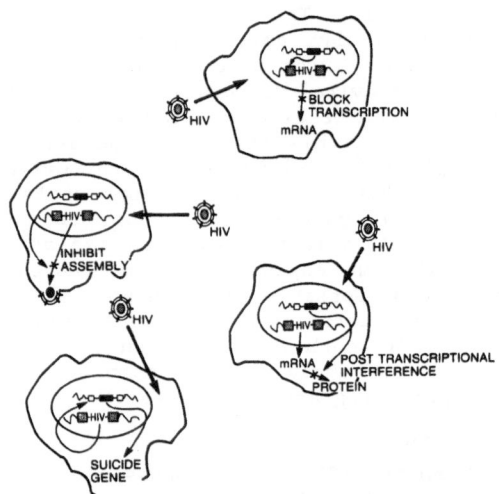

Figure 2.   Intracellular Intervention.
Shown are four different methods by which an integrated retroviral vector (black box) could inhibit HIV replication. The vector could produce proteins which would block HIV transcription (top right), inter-fere at the post-transcriptional level (center right), inhibit viral assembly (center left), or destroy the cell by expressing a suicide gene (bottom left).

Extracellular Intervention

Target cell independent protection involves the transduction of cells which are not necessarily targets (CD4- cells) for HIV infection (figure 3).   Strategies can be envisioned which are either directly targeted at HIV infected cells, methods which would result in an indirect anti-HIV action, or strategies which may ameliorate the secondary effects of HIV infection (CD4+ cell depletion). Two approaches which are specific anti-HIV strategies are retroviral expression of HIV antigens which stimulate anti-HIV immune responses and retroviral vector mediated expression of a free floating form of the HIV cell surface receptor, CD4 (8-10). A more indirect attack at inhibiting HIV infection would be through retroviral vector directed expression of a general anti-viral agent such as alpha-interferon. Lastly, it is a theoretical possibility that vector mediated synthesis of the appropriate cytokine (or combination of cytokines) could counter-act the HIV mediated depletion of CD4+ T-cells.

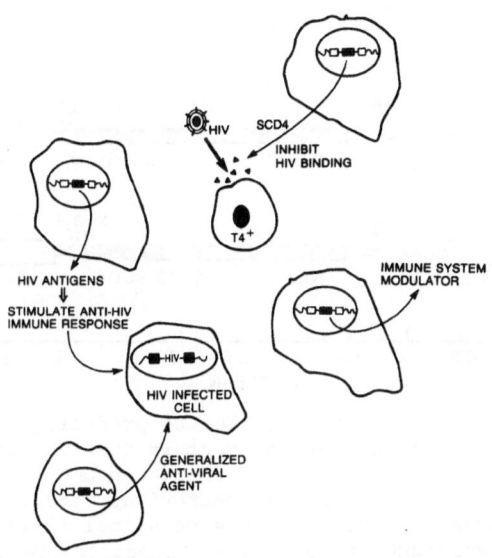

Figure 3. Target cell independent protection. Shown are four different strategies by which retroviral vectors (black box) could be used to engineer cells and have these cell secrete gene products which inhibit HIV replication. These schemes include inhibition of HIV binding by production of sCD4 (top right), secretion of cytokines to modulate immune system function (bottom right), production of HIV antigens with the subsequent stimulation of anti-HIV immune response (center left), and the secretion of a generalized anti-viral agent such as alpha interferon (bottom left).

Choice of cell delivery system

In evaluating the choice of cells for potential anti-HIV engineering, the most desirable candidate would be the hematopoietic stem cell of the bone marrow. Because the stem cell is self renewing and gives rise to all hematopoietic lineages, its engineering could afford life time protection. Hematopoietic stem cells have been purified from the mouse (11) but not yet from primates or humans. A more likely source of cells for anti-HIV engineering would be lymphocytes (12). The experience gained by our gene marking of human TIL cells has demonstrated that lymphocytes can be removed from an individual, transduced with retroviral vectors, and returned to the individual successfully. Alternatively, any of the target cell independent protection schemes are amenable to cell implant technology. Cell implants can be successfully maintained in vivo by inducing neovessel formation with appropriate angiogenesis factors (13).

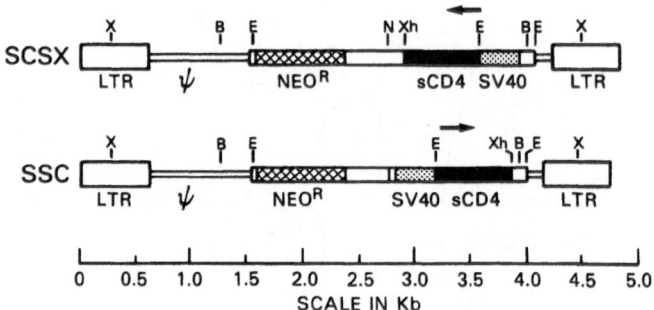

Figure 4. Diagram of soluble CD4 producing retroviral vectors. Retroviral vectors SCSX and SSC were produced by inserting an SV40 promoted sCD4 cassette into the N2 (14) retroviral vector. The viral promoter (LTR), packaging signal (Psi), neomycin resistance gene (NEO^R), truncated CD4 gene (sCD4), and SV40 virus promoter (SV40) are as indicated. The arrow indicates the direction of transcription of the SV40 promoter. Restriction enzyme sites are defined as follows: X = Xba I, B = Bst NI, E = Eco RI, N = Nru I, and Xh = Xho I.

sCD4 expressing retroviral vectors

We have demonstrated the feasibility of using retroviral mediated gene transfer as an anti-HIV system by designing retroviral vectors which express the soluble CD4 protein (sCD4, 14). In these experiments, the portion of the human CD4 antigen coding for the HIV binding domain was linked to a promoter element from the SV40 virus and then inserted into a retroviral vector (figure 4). These retroviral vectors can used to transduce primary cell cultures (such as fibroblasts or endothelial cells) and these cells express the secreted CD4 (sCD4) gene product (figure 5). The cell types used in this experiment are potential candidate cell types for in vivo gene transfer via cell implants. Further, we have shown that transduced cells producing the sCD4 gene product can protect HIV susceptible cells from infection by HIV in vitro (14). In these experiments, a human T-cell line was cocultured with sCD4 transduced cells and then the coculture challenged by the addition of HIV-1 virus. The results of this experiment demonstrated that virus replication was significantly decreased, thus proving that sCD4 engineered cells can protect (in trans) human T-cells from infection by HIV in vitro.

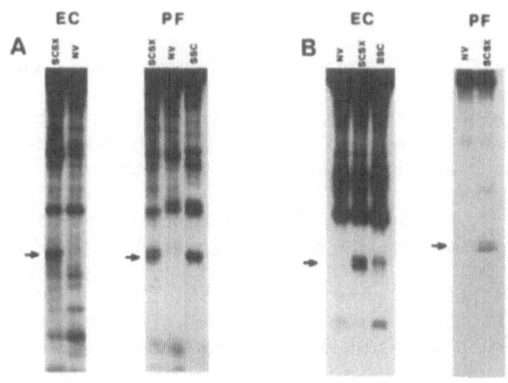

Figure 5. Soluble CD4 production by vector transduced cells. Shown are the resulting autoradiograms
of immunoprecipitated sCD4 from retroviral vector
transduced primary cells (resolved on 12% acrylamide
SDS-PAGE gels). Samples were conditioned medium
from approximately one half of a near confluent
100mm tissue culture dish containing primary culture
endothelial cells or, primary fibroblasts (PR), from
either rat (part A), or rabbit (part B). Cells were
transduced with the SSC vector (SSC), the SCSX vector
(SCSX), or no vector (NV). Arrows point to the
soluble CD4 gene product.

Conclusions

Retroviral vector-mediated gene transfer has now seen its first
therapeutic applications. As with any new clinical protocol, the
potential risks of the procedure should be assessed before consideration
of treatment. One of the potential risks which can not be avoided in the
use of retroviral vectors is the potential for the disruption of the
regulation (both activation and inactivation are possible) of a normal
cellular gene by integration of the vector into the host genome. The
insertional mutation risk posed by the vector is likely much smaller than
the similar risk posed by the HIV retrovirus (which being replication
competent, is thus capable of multiple integration events).

The ability to perform the gene transfer is not in itself a complex
or technically elaborate procedure, and it has been performed on
approximately a dozen patients with no unforseen side effects. The major
obstacle for efficacious gene therapy may involve various aspects of the
cell biology necessary to maintain and engineer cells ex vivo and still
permit efficient in vivo reconstitution of long-lived engineered cells.
If gene therapy is to become a reasonable treatment modality for AIDS, it
will have to be based on a simple cell transfer system (such as involves
the transduction of lymphocytes) and not be dependent on heroic procedures
such as bone marrow transplantation or the surgical implantation of neo-
organs. Efficacy is and always will be the major determining factor in
the decision to attempt any given therapy. The initial in vitro
experiments to test anti-HIV gene therapy strategies from this laboratory
and others have been encouraging and tests in AIDS animal models are in
progress. Given the speed to which this field has progressed in the last
few years, it is conceivable that several safe and efficacious AIDS gene
therapy protocols will be proposed in the near future.

References

1. Anderson, W.F. Prospects for human gene therapy. _Science_ 1984;226:401-409.
2. Rosenberg, S.A., Aebersold, P., Cornetta, K., et al. Gene transfer into humans: immunotherapy of patients with advanced melanoma, using tumor-infiltrating lymphocytes modified by retroviral gene transduction. _N. Engl. J. Med._ 1990;323:570-578.
3. Eglitis, M.A., and Anderson, W.F. Retroviral vectors for introduction of genes into mammalian cells. _BioTechniques_ 1988;6:608-614.
4. Baltimore, D. Intracellular immunization. _Nature_ 1988;335:395-396.
5. Green, M., Ishino, M., and Loewenstein, P.M. Mutational analysis of HIV-1 Tat minimal domain peptides: identification of trans-dominant mutants that suppress HIV-LTR driven gene expression. _Cell_ 1989;58:215-223.
6. Malim, M. H., Bohnlein, S., Hauber, J., and Cullen, B.R. Functional dissection of the HIV-1 Rev trans-activator: derivation of a trans-dominant repressor of Rev function. _Cell_ 1989;58:205-214.
7. Trono, D., Feinberg, M.B., and Baltimore, D. HIV-1 Gag mutants can dominantly interfere with the replication of the wild-type virus. _Cell_ 1989;59:113-120.
8. Smith, D.H., Byrn, R.A., Marsters, S.A., et al.: Blocking of HIV-1 infectivity by a soluble, secreted form of the CD4 antigen. _Science_ 1987;238:1704-1707.
9. Deen, C.D., McDougal, J.S., Inacker, R., et al.: A soluble form of CD4 (T4) protein inhibits AIDS virus infection. _Nature_ 1988;331:82-84.
10. Berger, E.A., Fuerst, T.R., and Moss, B. A soluble recombinant polypeptide comprising the amino-terminal half of the extracellular region of the CD4 molecule contains an active binding site for human immunodeficiency virus. _Proc. Natl. Acad. Sci. U.S.A._ 1988;85:2357-2361.
11. Spangrude, G.J., Heimfeld, S., and Weissman, I.L. Purification and characterization of mouse hematopoietic stem cells. _Science_ 1988;241:58-62.
12. Culver, K.W., Morgan, R.A., Osborne, W.R.A., Lee, R.T., Lenschow, D., Able, C., Cornetta, K., Anderson, W.F., and Blaese, R.M. In vivo expression and survival of gene-modified T lymphocytes in rhesus monkeys. _Human gene therapy_, in press.
13. Thompson, J.A., Haudenschild, C.C., Anderson, K.D., DiPietro, J.M., Anderson, W.F., and Maciag, T. Heparin-binding growth factor 1 induces the formation of organoid neovascular structures in vivo. _Proc. Natl. Acad. Sci. U.S.A._ 1989;86:7928-7932.
14. Morgan, R.A., Looney, D., Muenchau, D., Wong-Staal, R., Gallo, R., and Anderson, W.F. Retroviral vectors expressing soluble CD4: a potential gene therapy for AIDS. _AIDS Res. and human retroviruses._ 1990;6:183-191.

DISCOVERY OF A NOVEL POTENT INHIBITOR OF HIV INFECTION

J.P. Bader*, J.B. McMahon*, R.J. Schultz*, V.L. Narayanan*,
J.B. Pierce†, W.A. Harrison§, O.S. Weislow‡, C.F. Midelfort*,
and M.R. Boyd*

*Developmental Therapeutics Program, National Cancer
Institute, Bethesda, Maryland; †Uniroyal Chemical Company,
Inc., Middlebury, Connecticut; §Uniroyal Chemical Ltd.,
Guelph, Ontario, Canada; ‡Program Resources, Inc., Frederick
Cancer Research and Development Center, Frederick, Maryland

A number of agents active against the human immunodeficiency virus
(HIV) currently are under consideration for clinical use in AIDS.
Nonetheless, the only drugs exhibiting unequivocal clinical efficacy are those
which inhibit the activity of viral reverse transcriptase, specifically
certain dideoxynucleosides including azidothymidine (AZT)(1), dideoxycytidine
(DDC)(2), and dideoxyinosine (3). The failure to effect a complete cure of
the disease with dideoxynucleosides, the toxicities associated with long-
term treatments with these compounds (2,4), and the occurrence of AZT-
resistant mutants (5), are considerations which necessitate the discovery and
development of new drugs effective against AIDS.

The National Cancer Institute operates an AIDS antiviral screening
program (6) which tests compounds submitted from diverse sources for anti-
HIV activity (7). Among a variety of compounds submitted by the Uniroyal
Chemical Company for testing, oxathiin carboxanilide (OC) (NSC 615985)
exhibited potent activity (8,9). OC was designed and synthesized as a
possible fungicide and its anti-HIV activity could not be anticipated from its
structure (Fig 1).

Oxathiin carboxanilide (OC) is a derived name for benzoic acid, 2-
chloro-5-[[5,6-dihydro-2-methyl-1,4-oxathiin-3-yl)carbonyl]amino]-, 1-
methylethyl ester (NSC 615985). The synthesis of OC is outlined in Fig. 2.

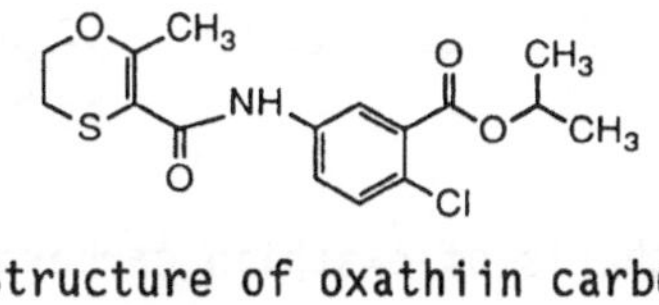

Fig 1.   Structure of oxathiin carboxanilide.

Advances in Molecular Biology and Targeted Treatment for AIDS
Edited by A. Kumar, Plenum Press, New York, 1991

The essential steps involved are (a) the preparation of the aminobenzoate 1, (b) the conversion of the oxathiin carboxylic acid 2 (10) to the acid chloride 3, and (c) the coupling of 3 with 1 to yield NSC 615985. Methanesulfonic acid (99%, 318 g, 3.3 mole) was added slowly to a stirred mixture of 5-amino-2-chlorobenzoic acid (85%, 215 g, 1.1 mole) in 2-propanol (about 1100 ml). The mixture was heated under reflux with stirring for 6 hours, then the excess 2-propanol was evaporated under reduced pressure. Water (about 1000 ml) was added to the residue and the mixture was neutralized with solid sodium bicarbonate and extracted with methylene chloride (about 1200 ml). The extract was washed twice with water, dried over magnesium sulfate and evaporated to give a purple oil, which crystallized on seeding. The product was reprecipitated from dilute hydrochloric acid solution by slowly basifying with concentrated ammonium hydroxide and seeding. The resulting crystals were dried to give 170 g (75%) of 1: m.p. 50.5-52 degrees C. 1,4-Oxathiin-3-carboxylic acid, 5,6-dihydro-2-methyl- 2 (10) (8.0 g, 0.050 mole) and methylene chloride (50 ml) were placed in a reaction flask, which was set up in a water bath, equipped for magnetic stirring, and attached via a reflux condenser to a gas absorption trap. Thionyl chloride (4.0 ml, 6.5 g, 0.055 mole) was added to the stirring slurry, and the water bath was warmed to 35-45 degrees C for 4 hours, during which the solid completely dissolved. The reaction flask was refitted for distillation and the green solution evaporated under reduced pressure at about 35 degrees C, using a water aspirator and gas absorption trap to remove hydrogen chloride, sulfur dioxide and unreacted thionyl chloride. The solid residue, crude acid chloride 3, was dissolved in methylene chloride (50 ml). The solution was chilled in ice and treated dropwise with a solution of benzoic acid, 5-amino-2-chloro-1-methylethyl ester 1 (10.7 g, 0.050 mole) and triethylamine (5.5 g, 0.055 mole) in methylene chloride (50 ml). The addition was carried out over about 2 hours, after which the reaction mixture was left stirring overnight at room temperature. The reaction mixture was worked up by washing the methylene chloride solution with water (50 ml), dilute hydrochloric acid (3.5%, 50 ml), water (50 ml), dilute sodium hydroxide (2%, 25 ml) and water (50 ml). The methylene chloride solution was then filtered through anhydrous sodium sulfate (about 5 g) and evaporated. The residue solidified. The crude product (17.0 g, m.p. 123-128 degrees C) was recrystallized from 95% ethanol (175 ml) to give 12.9 g (72.5%) of OC as light tan crystals: m.p. 130-132 degrees C; IR (KBr) 3260, 1710, 1655 cm^{-1}; ^{1}H NMR (CDCl$_3$) δ1.37 (d, 6H), 2.23 (s, 3H), 2.93 (dd, 2H), 4.37 (dd,2H), 5.23 (heptet, 1H), 7.27-7.93 (m, 3H), 8.22 (br s, 1H). Anal. Calcd for C$_{16}$H$_{18}$ClNO$_4$S: C, 54.01; H, 5.10; N, 3.94. Found: C, 53.96; H, 4.98, N, 3.94.

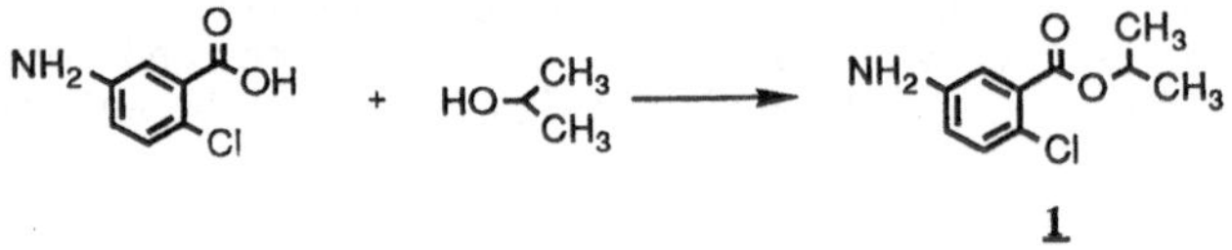

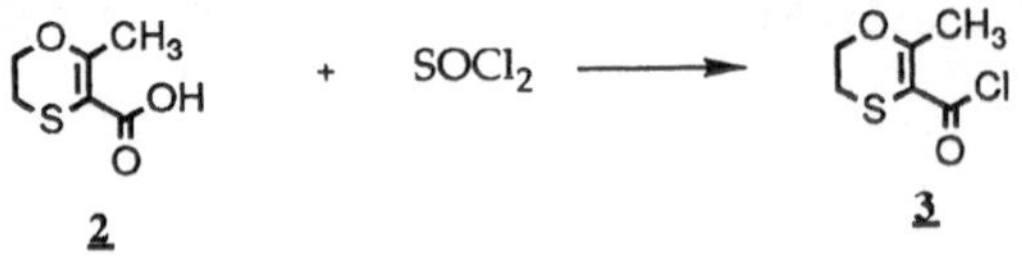

Fig 2. Synthesis of oxathiin carboxanilide.

The screening procedure used to detect anti-HIV activity basically involves the killing of T4 lymphoid cells (CEM-SS line) by HIV-1 (RF-strain), and inhibition of this cell-killing by active compounds (7). Cell viability is assessed six days after infection by the metabolic reduction of a colorless tetrazolium salt, XTT, to the colored formazan form. The assay distinguishes between cytotoxic and antiviral effects of compounds under test. Confirmatory tests for inhibition of virus production are performed on the same cell cultures.

OC was found to be potent inhibitor of HIV-induced cytopathology, completely preventing cell killing at concentrations ranging from 1 to 100 μM. In various experiments the effective antiviral concentration (EC_{50}) of OC ranged between 0.1 and 1.0 μM, and the cytotoxic concentration (IC_{50}), when seen, was over 100 μM; the effective, noncytotoxic range (in vitro therapeutic index) therefore was over 100 (8, 9, 11).

The inhibition of cytopathology correlated well with inhibition of virus production. Production of infectious HIV, virion-associated reverse transcriptase, and HIV p24 antigen, all were substantially decreased in cell cultures treated with OC (11).

Several other T cell lines of different origin were tested for efficacy of OC against HIV using the XTT viability assay. All of these lines (MT-2, H9, LDB-7, C-344, and C-8166) were protected from HIV-induced cell killing by OC. Also, susceptible CEM cells were protected from infection by the addition of virus-producing cells in a cocultivation assay (7).

In addition to the RF strain of HIV-1, used in most of these studies, another strain, HIV-1 IIIB, also was inhibited by OC. However, OC failed to inhibit the cell killing effect of HIV-2 (NIH-D_2 strain), and reproduction of another retrovirus, Rauscher murine leukemia virus, was unaffected by OC (11).

A series of experiments focusing on the mechanism of action of OC were performed, including possible direct effects on virus or cellular receptors for virus (11). Addition of OC to a stock HIV-1 preparation, and subsequently diluting the OC to less than an effective inhibitory concentration, failed to affect the infectivity of the virus. Also, addition of OC to cells for a brief period (up to four hours) with removal before infection failed to protect cells against HIV-induced cell killing. Likewise, exposure of cells to HIV in the presence of OC, with separation of the cells from the OC-containing medium within a few hours, failed to inhibit virus-induced cell killing. These studies demonstrate that OC has no direct effect on virions, and fails to prevent the binding of HIV to susceptible cells. In addition, these experiments showed that OC failed to accumulate in cells in concentrations sufficient to prevent infection with HIV-1.

A decrease in cellular expression of CD4 receptors could affect susceptibility of cells to HIV, or eventual recruitment of cells into cytopathic syncitia. Examination of levels of CD4 receptors revealed no changes after brief or extended exposures of cells to OC.

A possible effect of OC on the enzymatic activity of HIV reverse transcriptase was examined; no direct effect of OC on reverse transcriptase was seen using concentrations of OC substantially in excess of that required for inhibition of HIV infections.

Limited exposures of newly infected cells to DNA precursor analogs, or to inhibitors of reverse transcriptase, may interfere with reproduction of retroviruses and prevent cellular transformation by transforming retroviruses (12). Treatment of cells with dideoxycytidine for as little as one hour immediately after initial infection with HIV inhibited cell-killing.

In contrast, exposure to OC for up to eight hours after infection failed to inhibit HIV-induced cell killing, if OC was removed. Exposure of infected cells for twenty-four hours or longer reduced the cytopathic effects of HIV. These data suggested that OC either acted at some late stage of virus reproduction, or that OC had a reversible effect at an earlier stage.

When treatment with OC was delayed for twenty four hours after initial infection with HIV, no effect was seen on the production of HIV p24 antigen one and two days later. Also, when chronically infected H9 cells (13) which persistently produced HIV were exposed to OC, production of p24 antigen and virion-associated reverse transcriptase was unimpeded (11). Therefore, OC could not be shown to inhibit a late stage of virus reproduction, and a direct effect on synthesis or activity of a variety viral gene products, such as tat protein, rev protein, env glycoprotein, protease, or vif protein, is untenable. In fact, OC had no effect on the activity of HIV protease when tested directly against the enzyme. We conclude from these results, that OC has a reversible effect on some early stage of HIV reproduction after adsorption of virus but preceding transcription of viral DNA into RNA. Subsequent experiments to be described later have shown OC to interfere with the fusion of virion and cellular membranes.

The unique structural features of OC, its potency and what appears to be a novel mechanism of action, have prompted the National Cancer Institute to select OC as a lead compound for high priority developmental studies. A large series of related compounds are currently under test, and structure-activity relationships are being defined. Also, pharmacokinetic and toxicological studies are in progress which will be determinants in the potential clinical evaluation of OC.

REFERENCES

1. Yarchoan R, Kelecker RW, Weinhold KJ, Markham PD, Lyerly HK, Durack DT, Gelmann E, Lehrman SN, Blum RM, Barry DW, Shearer GM, Fischl MA, Mitsuya H, Gallo RC, Collins JM, Bolognesi DP, Myers CE, Broder, S. Administration of 3′-azido-3′-deoxythymidine, an inhibitor of HTLV-III/LAV replication, to patients with AIDS or AIDS-related complex. Lancet 1986; 1:575-80.
2. Yarchoan R, Perno CF, Thomas RV, Klecker RW, Allain JP, Wills RJ, McAtee N, Fischl MA, Dubinsky R, McNeely MC, Mitsuya H, Pluda JM, Lawley TJ, Leuther M, Safai B, Collins JM, Myers CE, Broder S. Phase I studies of 2′,3′-dideoxycytidine in severe human immunodeficiency virus infection as a single agent and alternating with zidovudine (AZT). Lancet 1988; 1:76-81.
3. Ahluwalia G, Conney DA, Mitsuya H, Fridland A, Flora KP, Hao Z, Dalal M, Broder S, Johns DG. Initial studies on the cellular pharmacology of 2′,3′-dideoxyinosine, an inhibitor of HIV infectivity. Biochem Pharmacol 1987; 36:3797-800.
4. Richman DD, Fischl MA, Grieco MH, Gottlieb MS, Jackson GG, Durack DT, Lehrman-Nusinoff S. The toxicity of azidothymidine (AZT) in the treatment of patients with AIDS and AIDS-related complex. A double-blind, placebo-controlled trial. N Engl J Med 1987; 317:192-7.
5. Larder BA, Darby G, Richman DD. HIV with reduced sensitivity to zidovudine isolated during prolonged therapy. Science 1989; 243:1731-4.
6. Boyd MR. Strategies for the identification of new agents for the treatment of AIDS: A national program to facilitate the discovery and preclinical development of new drug candidates for clinical evaluation. In Devita VT Jr, Hellman S, Rosenberg SA, eds. AIDS Etiology, Diagnosis, Treatment and Prevention. Philadelphia: Lipencott 1988, 305-317.

7. Weislow OS, Kiser R, Fine DL, Bader J, Shoemaker RH, Boyd MR. New soluble-formazan assay for HIV-1 cytopathic effects: application to high-flux screening of synthetic and natural products for AIDS-antiviral activity. J Natl Cancer Inst 1989; 81:577-86.
8. Bader J, McMahon J, Schultz R, Narayanan V, Boyd M. Oxathiin carboxanilide: A novel potent inhibitor of HIV reproduction. Third Int Conf on Antiviral Research, April 22-27, 1990.
9. Midelfort F, Schultz R, Narayanan V, Weislow O, McMahon J, Bader J, Boyd M. AIDS antiviral activity of oxathiin benzoic acid ester (NSC 615985). Sixth Int Conf on AIDS, San Francisco, June 20-24, 1990.
10. Harrison WA, Kulka M, Thiara DS, von Schemeling B. US Patent 3, 249, 499, 1966.
11. Bader JP, McMahon JB, Schultz RJ, Narayanan VL, Pierce JB, Harrison WA, Weislow OS, Midelfort CF, Stinson SF, Boyd MR. Oxathiin carboxanilide, A potent inhibitor of human immunodeficiency virus reproduction. Submitted for publication.
12. Bader JP. Transformation by Rous sarcoma virus: a requirement for DNA synthesis. Science 1965; 149:757-758.
13. Matsukura M, Zon G, Shinozuka K, Robert-Guroff M, Shimada T, Stein CA, Mitsuya H, Wong-Staal F, Cohen JS, Broder S. Regulation of viral expression of human immunodeficiency virus in vitro by an antisense phosphorothioate oligodeoxynucleotide against rev (art/trs) in chronically infected cells. Proc Natl Acad Sci USA 1989; 86:4244-8.

ANTIRETROVIRAL ACTIVITY OF CARBOCYCLIC NUCLEOSIDES AND POTENTIAL USE OF

INOSINATE DEHYDROGENASE INHIBITORS

Larry L. Bondoc Jr., and Arnold Fridland

Department of Biochemical and Clinical Pharmacology, St.
Jude Children's Research Hospital, 332 N. Lauderdale
Memphis, TN 38101

1. INTRODUCTION

Analogs of nucleosides and nucleotides are currently receiving increased attention for the treatment of human immunodeficiency virus-induced diseases. Zidovudine (AZT), an inhibitor of the activity of reverse transcriptase of human immunodeficiency virus (HIV), is presently the only approved drug available for the treatment of AIDS. Significant dose-related toxicity has been associated with the administration of AZT to patients with AIDS. Anemia and neutropenia are the most common toxicities observed in such patients and may result in dose reductions or discontinuation of treatment (1,2). Several new compounds are being evaluated both clinically and experimentally as potential therapeutic agents against HIV infectivity (3). The purpose of this paper is to provide some background information on the activity and metabolism of a new carbocyclic guanosine nucleoside analog carbovir (carbocyclic 2',3'-dideoxy-2',3'-didehydroguanosine, NSC-614846) leading up to the interaction of such compound with inhibitors of the rate-limiting enzyme IMP dehydrogenase of guanine nucleotide metabolism.

2. ANTIVIRAL ACTIVITY OF CARBOCYCLIC NUCLEOSIDES

The carbocyclic nucleoside analog carbovir (Fig 1) was synthesized by Dr R. Vince and his associates at the University of Minnesota . Among a large number of carbocyclic nucleosides, carbovir was found to have good activity against HIV cytopathic effects in ATH-8, MT-2 and CEM cells at noncytotoxic concentrations (4). Thus, the 50% inhibitory concentration (EC50) of carbovir against HIV in ATH-8 cells is 0.19 μg/ml, compared to a cytotoxic concentration of 35 μg/ml : the in vitro therapeutic index (TI) is 184. This compares with TI of >300 and 75 for AZT and ddC, respectively, in the same assay system. At least 40 other related carbocyclic purine nucleoside analogs have been tested against HIV and only a handful have exhibited antiviral activity and none are as active as carbovir. Interestingly, the true nucleoside (2',3'-dideoxy-2',3'-didehydroguanosine; D4G) is inactive as an HIV antagonist. This suggest that the cyclopentane ring and the double bond is critical for activity.

Advances in Molecular Biology and Targeted Treatment for AIDS
Edited by A. Kumar, Plenum Press, New York, 1991

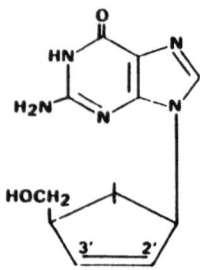

Fig. 1   Chemical structure of carbovir (NSC-614846).

3.   MODE OF METABOLISM OF CARBOVIR TO ITS PUTATIVE ACTIVE TRIPHOSPHATE

Because no animal model is presently available that is predictive for clinical efficacy , biochemical and pharmacokinetic studies are important in order to provide a rational basis for the clinical development of a novel compound. To exert an antiviral effect nucleoside analogs are first enzymatically phosphorylated in cells to their corresponding triphosphate which then blocks virus replication by inhibition of viral specified reverse transcriptase (3). Hao et al. (5) have shown that an important factor limiting the antiretroviral activity of nucleoside analogs is their ability to generate the 5'-triphosphate rather than relative inhibitory activity of the triphosphate for viral reverse transcriptase. Thus, predictions of efficacy could be derived by investigating the levels of carbovir metabolites in cells. We initially determined whether at pharmacological relevant concentrations, the putative active anabolites formed from carbovir could be determined in extracts of human T cell CCRF-CEM. A typical HPLC profile of the anabolites of carbovir formed in CEM cells incubated with 10 μM [3H]carbovir and analyzed by ion exchange HPLC is shown in Fig 2 (lower panel). It can be seen that carbovir mono-, di- and triphosphate elutes after guanine nucleotides under the conditions used. It is also apparent that the concentrations of these anabolites is rather low; i.e., in the experiment shown in Fig 2, after 6 h incubation the levels of carbovir di- and triphosphate detected are about 0.2 pmol/10^6 cells, (which is equivalent to 0.08 μM). This is still about 2-3-fold higher than the ID50 of 0.03 μM for inhibition of HIV reverse transcriptase. The remainder (about 15 to 20%) of the radioactivity found in the nucleotide fractions is attributed to guanine nucleotides.

In human lymphoid cells, nucleosides are converted to their phosphorylated products via several enzymes of varying specificities. When it became apparent that carbovir was converted to nucleotides in CEM cells, we sought to determine whether carbovir was activated by the same enzymes which convert 2'-deoxyguanosine (dGuo) to the 5'-phosphates. Cytosolic deoxycytidine kinase has been shown to be the primary enzyme in CEM cells that phosphorylate dGuo to the monophosphate (6-8). However, as shown in Table 1, this enzyme is not apparently responsible for the activation of carbovir since a mutant of CEM cells severely deficient in dCyd kinase (dCK⁻) was not altered in any way in the ability to phosphorylate carbovir to the corresponding nucleotides. The only enzyme that we have found that catalyzes the phosphorylation of carbovir is a cytosolic 5'-nucleotidase which apparently acts in the reverse direction as a phosphotransferase utilizing the monophosphate IMP or GMP as a phosphate donor (9).

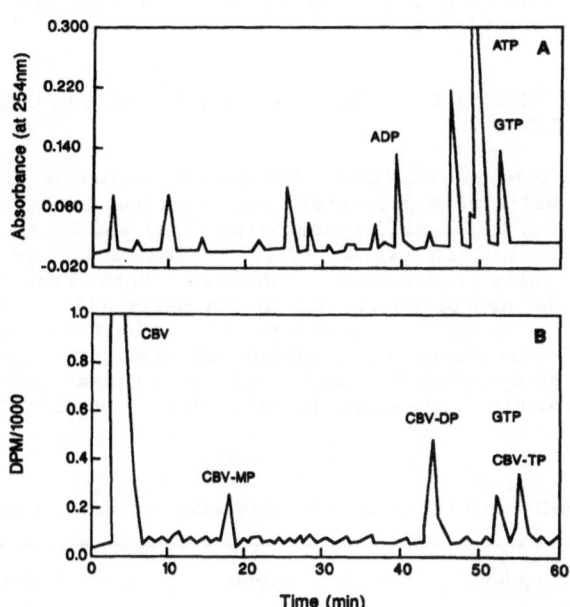

Fig. 2.   Separation of labeled metabolites arising from [³H]-carbovir. CEM cells (10⁶ cells/ml) were incubated with 10 μM [³H-8]carbovir for 6 h. After the incubation, cells were extracted with ice-cold 60% methanol and subjected to ion exchange HPLC (Partisil 10-SAX) using an elution program previously described (5).

Table 1. Phosphorylation of carbovir in kinase deficient CEM cells

cell line	CBVMP	CBVDP	CBVTP
	pmol/106 cells		
wild type	0.075	0.091	0.039
dCK⁻/AK⁻	0.094	0.93	0.026

Wild type and kinase deficient cells were incubated with 10 μM [^{3}H]carbovir for 6 h. After the incubation, cell extracts were prepared and analyzed by HPLC as described under Fig. 1.

4. POTENTIAL USE OF INOSINATE DEHYDROGENASE INHIBITORS FOR COMBINATION THERAPY WITH CARBOVIR

The inosinate dehydrogenase (IMPD) inhibitors including ribavirin, tiazofurin and mycophenolic acid potentiate the antiviral activity of the purine dideoxynucleosides 2',3'-dideoxyadenosine (DDA) and 2',3'-dideoxy-guanosine (DDG) against HIV in vitro (10,11). The mechanism of this potentiation is not fully understood. However, both ribavirin and tiazofurin stimulate the anabolism of DDG in CEM cells (11). We asked, whether, a similar effect could be achieved for the anabolism of carbovir. As shown in Table 2, treatment with either of the three inhibitors ribavirin, tiazofurin or mycophenolic acid resulted in marked increases in the formation of carbovir anabolites (CBVMP, CBVDP and CBVTP) in CEM cells.

Table 2. Effect of IMPD inhibitors on carbovir nucleotides accumulation

Inhibitor	CBVMP	CBVDP	CBVTP
		pmol/109 cells	
None	44	54	27
ribavirin (10 μM)	411	1,131	1,001
mycophenolic (0.5 μM)	291	5,148	1,616
tiazofurin (40 μM)	711	2,513	833

Cells in logarithmic growth were incubated with 10 μM [^{3}H-8]carbovir for 6 h in presence or absence of the named compounds. After the incubation, cells were extracted with 60% cold methanol and extracts analyzed by HPLC.

An intriguing question is the mechanism for this potentiation of the antiviral activity and anabolism for the aforementioned ddNs and an interesting hypothesis is that this is mediated by an increase in the phosphorylating activity of 5'-nucleotidase. Inhibitors of IMPD, such as those studied here, increase markedly the levels of IMP by preventing the utilization of the latter nucleotide for formation of xanthylate and

guanylate.   Johnson   and   Fridland   (9)   have   demonstrated   that   the
5'-nucleotidase is a regulatory enzyme and its activity is controlled by
the level of IMP and ATP. Under normal unperturbed conditions in the cell
the nucleotidase is undersaturated with respect to the co-substrate IMP ,
but susceptible to increased activation as IMP level are increased. To test
this possibility , the effect of IMPD inhibition by ribavirin and
tiazofurin on IMP levels were determined. As shown in Fig 3, IMP levels
increased progressively and in parallel with the increase seen in CBVMP and
CBVDP. In addition, the levels of GTP were decreased over 80% by these
inhibitors.   Of interest was also the observation that the increased
formation of mono- and diphosphorylated products was significantly higher
than that of CBVTP , i.e., 20-fold for CBVDP versus 4-fold for CBVTP,
suggesting that under the experimental conditions, the final step of
phosphorylation CBVDP → CBVTP is rate-limiting.   These results of the
effects of IMPD dehydrogenase inhibitors on 5'-phosphorylation of CBV and
the other purine dideoxynucleosides is in marked contrast to the effect
seen with pyrimidine analog AZT. In the latter case, the increase in dTTP
pool which accumulate with ribavirin results in inhibition of the
phosphorylation of AZT and a resulting antagonism of the antiviral activity
of AZT against HIV in human T cells (12).

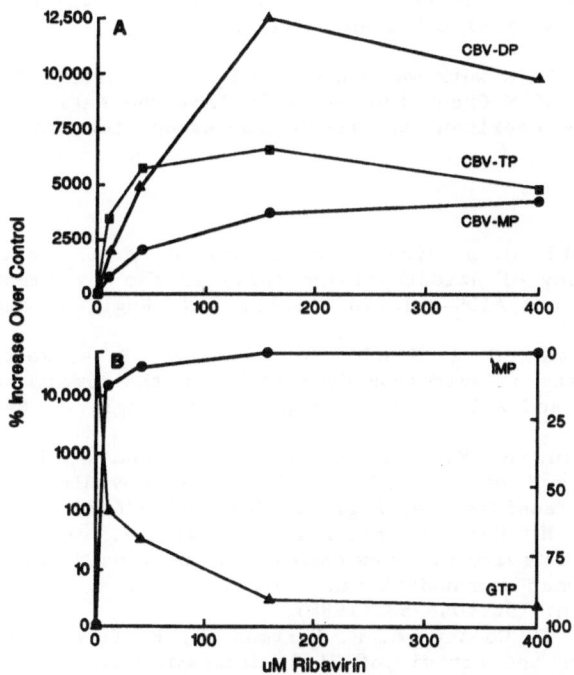

Fig. 3.   Effect of IMPD inhibitors on IMP and
               GTP pool sizes.   CEM cells incubated
               with ribavirin for 6 h were extracted
               and analyzed by ion-exchange HPLC
               (Partisil-10 SAX) as described in
               Fig. 1.

5. DISCUSSION

Carbovir is a new carbocyclic nucleoside which is regarded as a promising clinical candidate for AIDS treatment because of its selectivity in vitro against HIV. Biochemically this compound has some interesting attributes. Its 5'-triphosphate CBVTP is as active as the 5'-triphosphate of AZT or DDI against HIV reverse transcriptase but much less inhibitory than the latter two against the cellular DNA polymerases β and γ. Thus, if the toxicities of AZT (i.e, myelosuppression) or peripheral neuropathy for DDC is due to inhibition of either of the said polymerases, then carbovir could offer a therapeutic alternative for the treatment of AIDS which should not exhibit the same toxicity. The important cellular parameters involved in the metabolism of carbovir have also been defined in human CEM lymphoid cells, a cell line that can be infected with HIV. Thus, these biochemical and genetic studies demonstrate that a 5'-nucleotidase rather than usual nucleoside kinase may initiate the metabolism and subsequent effects of carbovir in human T cells. Moreover, these experiments have provided a possible explanation for the previously empirical observation that ribavirin potentiates the anti-HIV activity of dideoxypurine nucleosides. Thus, by demonstrating that two other inhibitors of IMPD share with ribavirin the properties of augmenting the anabolism of carbovir and augmentation of IMP levels they suggest that the latter of these effects is causative of the former. The finding of the 5'-nucleotidase as a route for activation for carbovir, one that is amenable to pharmacological manipulation by ribavirin or other inhibitors of IMPD, has important implications for combination regimens for AIDS. Thus, although ribavirin may have limited value as an antiHIV agent, it may play a significant role as a biochemical tool and an agent which can be used to modulate the activity of other antiviral agents.

Acknowledgments- This work was supported by USPHS grants 1 RO1 AI 27652, 1 RO1 CA 43296, CORE Grant P30 CA 21765 from the National Institute of Health and by the American Lebanese Syrian Associated Charities.

REFERENCES

1. M. A. Fischl, D. D. Richman, M. H. Grieco, P. A., Gottlieb, et al., The efficacy of azidothymidine (AZT) in the treatment of patients with AIDS and AIDS-related complex. N. Engl. J. Med. 317:185-91 (1987).
2. D. D. Richman, M. A. Fischl, M. H. Grieco, P. A. Gottlieb, et al., The toxicity of azidothymidine (AZT) in the treatment of patients with AIDS and AIDS-related complex. N. Engl. J. Med. 317:192-97 (1987).
3. R. Yarchoan, H. Mitsuya, C. E. Myers, and S. Broder, Clinical pharmacology of 3'-azidothymidine (zidovudine) and related dideoxynucleosides. N. Engl. J. Med. 321:726-38 (1989).
4. R. Vince, M. Hua, J. Brownell, S. Daluge, et al., Potent and selective activity of a new carbocyclic nucleoside analog (Carbovir) against human immunodeficiency virus in vitro. Biochem. Biophys. Res. Commun. 156:1046-53 (1988).
5. Z. Hao, D. A. Cooney, N. R. Hartman, C. F. Perno, et al., Factors determining the activity of 2',3'-dideoxynucleosides in suppressing human immunodeficiency virus in vitro. Mol. Pharmacol. 34:431-35 (1988).
6. W. R. A. Osborne, Kinase activities of T and B lymphoblasts distinguished by autoradiography. Proc. Natl. Acad. Sci. USA 83:4030-34 (1986).

7.   M. C. Hurley, T. D. Patella, and I. H. Fox, Human placental deoxyadenosine and deoxyguanosine phosphorylating activity. J. Biol. Chem. 258:15021-27 (1983).

8.   J. C. Sarup and A. Fridland, Identification of purine deoxyribonucleoside kinases for human leukemia cells. Biochemistry 26:590-97 (1987).

9.   M. A. Johnson and A. Fridland, Phosphorylation of 2',3'-dideoxyinosine by cytosolic 5'-nucleotidase. Mol. Pharmacol. 36:291-95 (1989).

10.   M. Baba, R. Pauwels, J. Balzarini, P. Herdewijn, et al., Ribavirin antagonozes the inhibitory effects of pyrimidine dideoxynucleosides but enhances the inhibitory effects of purine dideoxynucleosides on replication of human immunodeficiency virus. Antimicrob. Agents Chemother. 31:1613-17 (1987).

11.   G. Ahluwalia, D. A. Cooney, L. L. Bondoc, M. J. Currens, et al., Inhibitors of IMP dehydrogenase stimulate the phosphorylation of the antiviral nucleoside 2',3'-dideoxyguanosine. Biochem. Biophys. Res. Commun. (1990) (in press).

12.   M. W. Vogt, K. L. Hartshorn, P. A. Furman, T. C. Chou, J. A. Fyfe, L. A. Coleman, C. Crumpacker, R. T. Schooley, and M. S. Hirsch, Ribavirin antagonizes the effect of Azidothymidine on HIV replication. Science 235:1376-80 (1987).

LIPOPHILIC 6-HALO-2',3'-DIDEOXYPURINE NUCLEOSIDES: POTENTIAL ANTIRETROVIRAL AGENTS TARGETING HIV-ASSOCIATED NEUROLOGIC DISORDERS

Takuma Shirasaka, Kazuhiro Watanabe, Hidetoshi Yoshioka,
Eiji Kojima, Shizuko Aoki, Kunichika Murakami,
and Hiroaki Mitsuya

*The Clinical Oncology Program, National Cancer Institute
NIH, Bethesda, Maryland 20892*

*Research Laboratory of Bioresources
Sanyo-Kokusaku Pulp Co., Iwakuni 740*

Introduction

In the past seven years, a number of potentially useful approaches for the therapy of human immunodeficiency virus (HIV) infection have emerged.[1-3]   One such approach is the use of the broad family of 2',3'-dideoxynucleosides (ddN) as therapeutic agents.[4,5]   3'-Azido-2',3'-dideoxythymidine (AZT or zidovudine), the first member of the dideoxynucleoside family administered to humans, has now been formally proven to reduce the morbidity and mortality of patients with acquired immunodeficiency syndrome (AIDS) and its related disorders.[6-8] AZT has also been proven to benefit certain asymptomatic individuals infected with HIV-1 when administered in the early stages of HIV infection.[9]   It is worth noting that AZT has most recently been shown to be active at lower doses than that initially administered.[10,11]   Two other analogs of the dideoxynucleoside family, 2',3'-dideoxycytidine (ddC)[12,13] and 2',3'-didehydro-2',3'-dideoxythymidine (D4T),[14] have also demonstrated activity against HIV-1 in some patients in Phase I clinical trials.   A purine member, 2',3'-dideoxyinosine (ddI or didanosine), has also recently been shown to be active against HIV-1 in patients with AIDS and advanced AIDS related complex (ARC) in short-term Phase I clinical trials, although notable adverse effects including painful peripheral neuropathy and acute pancreatitis have been observed.[15-17]   In this regard, certain doses of ddI which can confer antiviral effects on patients have been tolerated for more than 2 years without toxicity.[18]   ddI is now under phase II clinical trials in several countries.

HIV not only causes severe forms of immunodeficiency but also often causes a variety of neurologic disorders.[19,20]   The reported incidence of neurological diseases complicating AIDS ranges from 30 to 40 % although such figures are thought to be underestimated since they are based on obvious evidence of neurologic abnormalities in patients.   HIV in the central nervous system (CNS) may replicate more actively than in other tissues, and indeed, the CNS may serve as a principal reservoir of the virus in the whole body.[21-23] Thus, the capability of antiretroviral agents to penetrate into the CNS may constitute an important feature of therapeutics against HIV infection. However, the lipophilicity of purine 2',3'-dideoxynucleosides, especially ddI, is generally low and perhaps, in part, this limits the penetration of 2',3'-dideoxypurine nucleoside analogs into the CNS.

Advances in Molecular Biology and Targeted Treatment for AIDS
Edited by A. Kumar, Plenum Press, New York, 1991

In this report, we describe the synthesis and *in vitro* anti-HIV activity of 2- and 6-substituted 2',3'-dideoxypurine ribofuranoside (ddP) analogs. This novel class of compounds was synthesized through an enzymatic method employing *Escherichia coli*, and identified to have an enhanced lipophilicity and potent antiretroviral activity against various strains of HIV *in vitro*.[24,25]

E. Coli Mediated Biosynthesis of 6-Halo-ddP Analogs

A number of 6-halo-substituted ddP analogs were synthesized using a bacterial transglycosylation reaction mediated by nucleoside phosphorylases derived from intact *E. coli* JA-300 cells[24,25] (Figure 1). Pioneering studies using bacterial enzyme systems for the synthesis of nucleoside analogs through transribosylation,[26,27] transdeoxyribosylation,[28,29] transarabino-sylation,[30,31] and transdideoxyribosylation[32] have been reported. However, transdideoxyribosylation with transfer of the dideoxyribosyl group to modified bases, in particular 6-halo-purines, has not been reported. Eight 6-halo-substituted ddPs (compounds **1-8**) discussed in this report are illustrated in Table 1 and Figure 2. Compounds **1-5**, **7**, and **8** were newly synthesized,[24,25] while compound **6** in Table 1 has been described by Chu and Schinazi.[33]

The enzymatic method we have exploited for the synthesis of 6-halo-substituted ddPs has several advantages. (i) Its reaction steps are simple and less labor-intensive than conventional chemical methods. (ii) Product purification is straightforward. (iii) Reactive functional groups (e.g. iodo) can be handled easily. (iv) Minimal byproducts are formed such as 2'- or 3'-deoxynucleosides or the α-anomer of 2',3'-dideoxynucleosides.

Antiretroviral Activity against Various HIV Strains *In Vitro*

Four 2-amino-6-halo-ddPs, (compounds 1-4 in Table 1 and Figure 2) and three 6-halo-ddPs (compounds 5-7 in Table 1) exerted a potent anti-HIV-1 activity *in vitro*. In the HIV cytopathic effect inhibition assay, almost all the target CD4+ ATH8 cells were destroyed by the virus by day 7 after exposure to HIV-1 in the absence of drugs (Figure 2). However, when the cells were cultured in the presence of 20 and 50 µM of each compound, the ATH8 cells were virtually completely protected against the infectivity and cytopathic

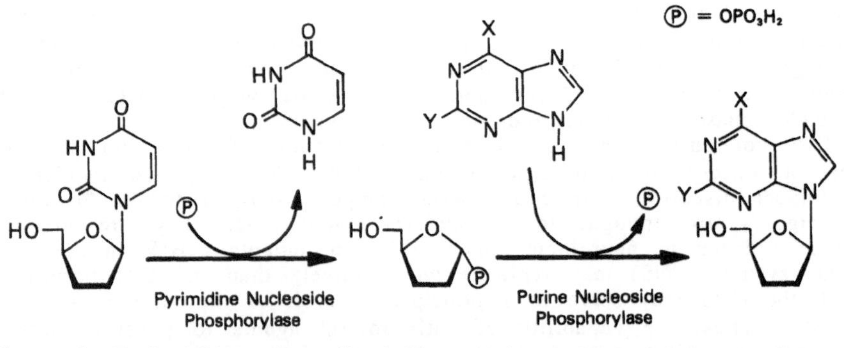

Figure 1. *Escherichia coli* mediated biosynthesis of 6-halo-substituted ddPs. 6-Halo-substituted ddPs were synthesized through a bacterial transglycosylation reaction mediated by nucleoside phosphorylases derived from intact *E. coli* JA-300 cells.

Table 1. *In Vitro* Antiretroviral Activity of 2',3'-Dideoxynucleosides Tested[a]

Compound		Concentration (μM)	Protective Effect (%)[b]	Cytotoxicity (%)[c]
1	2-NH$_2$-6-F-ddP	2, 5, 20, 50, 100, 200	48, 62, 100, 100, 77, 70	0, 0, 0, 0, 26, 32
2	2-NH$_2$-6-Cl-ddP	2, 5, 20, 50, 100, 200	38, 48, 100, 100, 84, 78	0, 0, 0, 0, 19, 25
3	2-NH$_2$-6-Br-ddP	2, 5, 20, 50, 100, 200	35, 35, 100, 100, 78, 64	0, 0, 0, 0, 20, 33
4	2-NH$_2$-6-I-ddP	2, 5, 20, 50, 100, 200	21, 21, 100, 93, 73, 47	0, 0, 0, 9, 26, 47
5	6-F-ddP	2, 20, 50, 100, 200	64, 100, 87, 61, 43	0, 5, 10, 32, 52
6	6-Cl-ddP	2, 20, 50, 100, 200	28, 100, 69, 56, 49	8, 1, 20, 41, 55
7	6-Br-ddP	2, 20, 50, 100, 200	25, 80, 63, 48, 36	7, 17, 17, 52, 61
8	6-I-ddP	2, 20, 50, 100, 200	17, 41, 39, 30, 20	4, 39, 37, 55, 81
9	ddI	1, 10, 100, 200, 1000	7, 97, 100, 100, 46	9, 6, 0, 0, 40
10	ddG	5, 10, 100, 200, 500	34, 90, 100, 91, 66	0, 0, 0, 1, 41
11	AZT	0.5, 1, 5, 10, 50	56, 100, 93, 93, 48	0, 0, 4, 4, 51

[a] ATH8 cells (2 x 10^5) were exposed to 4.3 x 10^3 TCID$_{50}$ doses of HIV-1/III$_B$ per cell and cultured in the presence of each compound. On days 5 to 7, the total viable cells were counted. Orders of numbers in the column for concentrations correspond to the orders of numbers in other columns.

[b] The percentage of protective effect of each compound was determined by the following formula: 100 x [(number of viable cells exposed to HIV-1 and cultured in the presence of the compound - number of viable cells exposed to HIV-1 and cultured in the absence of the compound) / (number of viable cells cultured alone - number of viable cells exposed to HIV-1 and cultured in the absence of the compound)]. Calculated percentages equal to or less than zero are expressed as 0%.

[c] The percentage of cytotoxicity was determined by the following formula: 100 x [1 - (number of total viable cells cultured in the presence of the compound / number of total viable cells cultured alone)]. Calculated percentages equal to or less than zero are expressed as 0%.
(From Ref. 24, by permission of the Proceedings Office of the National Academy of Sciences, U.S.A.)

effect of HIV-1 and grew comparably to virus-uninfected cells (Figure 2). These compounds exhibited comparable antiviral activity when MT2 cells were employed as target cells.[24] At concentrations ≥ 100 μM, however, the drugs appeared to be somewhat more suppressive to cell growth as compared to the reference compounds ddI and ddG. Among 6-halo-substituted ddPs tested, compounds with a bromine or an iodine appeared to be more toxic for cell growth than compounds substituted with a fluorine or a chlorine.

2-amino-6-fluoro- and 2-amino-6-chloro-ddPs were further tested for their activity to suppress the p24 Gag protein production in acutely HIV-1-exposed H9 cells. When H9 cells were exposed to the virus, by day 5 in culture, approximately 40 % of the cells expressed p24 Gag protein as assessed by an indirect immunofluorescence technique. On day 7, 60 % of the cells became positive. However, 10 μM 2-amino-6-fluoro- and 2-amino-6-chloro-ddPs suppressed the Gag protein expression by 75-100 % and at ≥ 20 μM both compounds virtually completely blocked the expression of HIV-1 throughout the 7 day period of time in culture. Potent antiretroviral activity was also seen when fresh monocytes/macrophages were employed as target cells.[24]

To further characterize the antiviral activity of 2-amino-6-fluoro- and 2-amino-6-chloro-ddPs, their capacity to suppress the foramtion of proviral DNA in PHA-stimulated peripheral blood mononuclear cells (PBM) following HIV-1 exposure was assessed using the Southern blot hybridization technique. In the absence of the drugs, proviral DNA was readily detectable on day 2, however, in the presence of 80 μM 2-amino-6-fluoro- or 2-amino-6-chloro-ddP, the synthesis of proviral DNA was virtually completely suppressed.[24]

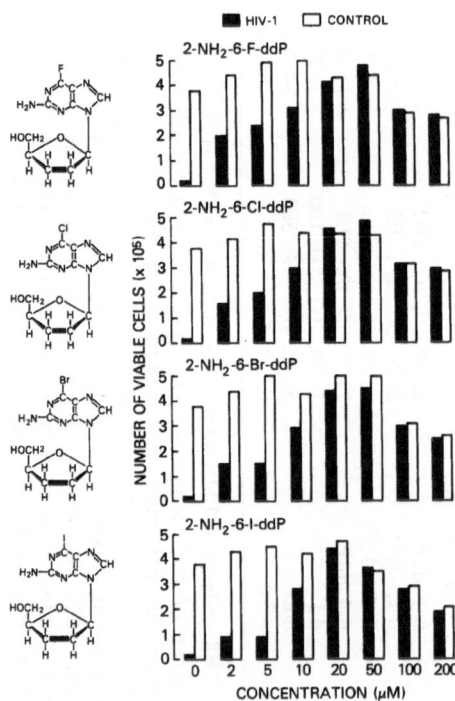

Figure 2. Inhibition of the cytopathic effect of HIV-1 by 2-amino-6-halo-ddPs. ATH8 cells (2 x 10^5) were exposed to HIV-1/III$_B$ in the presence or absence of various concentrations of 2-amino-6-halo-ddPs (solid bar). Control cells (open bar) were not exposed to the virus. On day 6, the total viable cells were counted. (From Ref. 24, by permission of the Proceedings Office of the National Academy of Sciences, U.S.A.)

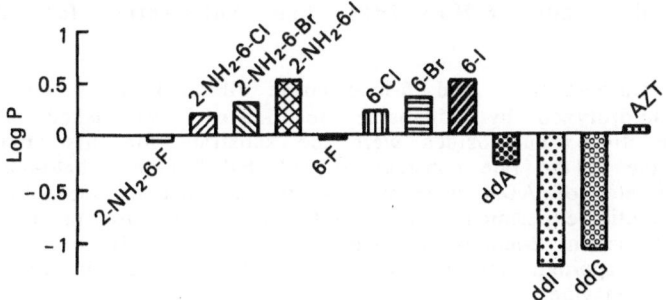

Figure 3. Octanol-water partition coefficients. An *n*-octanol/water partition coefficient (P) for each drug was determined by a micro shake-flask procedure.[43] All eight 6-halo-substituted ddPs (compounds **1-8**) had substantially higher *n*-octanol/water partition coefficients than ddI and ddG. Six of the 6-halo-substituted ddPs (**2-4, 6-8**) had higher log P values than AZT.

Lipophilicity of 2-Amino-6-halo- and 6-Halo-ddP

As we discussed above, one of the devastating features of HIV infection is in neurological abnormalities.[19,20] A recent controlled study of neurologic changes in men with asymptomatic HIV infection has revealed that abnormalities detected by electrophysiologic tests including electro-encephalography are far more common in asymptomatic carriers of HIV (67 % of the seropositive men: 18 of 27 tested) than in controls (10 % of the controls: 3 of 30 tested) (P <0.00005) and tend to progress over time.[20] While the pathogenesis of the HIV-caused neurological abnormalities is as yet incompletely understood, such a disorder may be directly linked to structural and functional changes due to infection and/or replication of HIV in the CNS, particularly in cells of a monocyte/macrophage lineage.[21-23] In this regard, HIV-associated neurological disorders in both adults and children with AIDS or ARC have been substantially improved during therapy with AZT.[34,35] These relatively prompt improvements of neurological abnormalities in some patients following therapy with AZT may represent the effect of antiretroviral activity of the drug in the CNS. Thus, lipophilicity of dideoxynucleosides could constitute an important feature of antiretroviral therapies against HIV infection, since antiretroviral drugs with high degree of lipophilicity may theoretically possess an enhanced ability to penetrate the blood-brain-barrier and may therefore inhibit the infectivity and replication of HIV-1 in the CNS.

It should be noted that replacement of the 6 position phenolic oxygen in a purine nucleoside with halogen atoms has been demonstrated to increase the lipophilicity under certain circumstances.[36] We, therefore, asked whether 2-amino-6-halo- and 6-halo-ddPs had a favorable lipophilicity (Figure 3). All eight 6-halo-substituted ddPs (compounds **1-8**) had substantially higher octanol/water partition coefficients than the reference purine dideoxynucleosides, ddI and ddG. Six of the 6-halo-substituted ddPs (compounds **2-4, 6-8**) had higher log P values than AZT, which has been proven to penetrate relatively well into the cerebrospinal fluid in patients with HIV-infection.[6,7] It should be stressed, however, that the lipophilicity of a given drug may not necessarily determine its CNS penetration potential, and the enhanced CNS penetration may not improve the therapeutic index of the drug.

2-Amino-6-halo- and 6-Halo-ddPs are Substrates for Adenosine Deaminase

Since 2-amino-6-halo- and 6-halo-purine ribonucleosides are known to be readily hydrolyzed by adenosine deaminase,[37] we asked whether the corresponding dideoxynucleosides were the substrate for this enzyme. As shown in Table 2, the rate constants of selected 2-amino-6-halo- and 6-halo-ddPs as substrates of ADA were very similar to that of adenosine or ddA. When the kinetic experiments were performed in culture media containing 15% fetal calf serum, 2-amino-6-halo-ddPs were still hydrolyzed to ddG, but at a rate that was approximately 60 times slower than the rate in the presence of excess isolated enzyme.

The antiviral activity of compounds **1**, **2**, and **5** against HIV-1 were tested in the HIV cytopathic effect inhibition assay using ATH8 cells as target cells in the presence of 2'-deoxycoformycin (2'-dCF), a potent inhibitor of adenosine deaminase. We found that the antiviral activity of all these compounds at a concentration of 50 µM was completely abrogated in the presence of 5 µM 2'-dCF, and essentially all the target ATH8 cells were destroyed by the virus.[25]

6-Halo-substituted ddP and Structure-Activity Relationships

The 6-halo versions of ddPs are of interest in view of structure-activity relationships, since the substitution with a halogen atom confers substantial lipophilicity on ddP analogs with a retention of antiretroviral activity. In this regard, 6-halo-substituted ddP analogs appear to exert antiviral activity only upon conversion to ddG and ddI (Figure 4). This is most plausible since a loss of the antiretroviral activity of 6-halo-substituted ddPs against HIV in the presence of deoxycoformycin was observed.[25]

These features of 6-halo-substituted ddPs were unexpected and intriguing in view that there is still no reliable algorithm for predicting which congeners will exert more antiretroviral activity against HIV or less toxicity to target human cells. For example, replacement of the aromatic oxygen of ddI by a hydrogen, generating 2',3'-dideoxypurine ribofuranoside, or 2',3'-dideoxynebularine, negates the potent antiretroviral activity of ddI even if this compound retains the intact 2',3'-dideoxyribose configuration.[24,25] The same replacement in ddG, generating 2-amino-2',3'-dideoxypurine ribofuranoside, also abolishes the antiretroviral activity of ddG.[24,25]

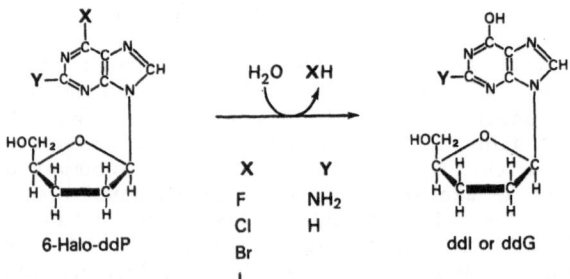

Figure 4. Enzymatic hydrolysis of 6-halo-substituted ddPs by adenosine deaminase.

Table 2. Rate Constants of 6-Halo-substituted ddPs as Substrates of Adenosine Deaminase

Substrate	Km (M)	Relative Vmax
Adenosine*	6.9×10^{-5}	100
2-NH2-6-F-ddP (1)	1.3×10^{-3}	127
2-NH$_2$-6-Cl-ddP (2)	6.2×10^{-3}	50
6-F-ddP (5)	1.0×10^{-3}	76
ddA	1.6×10^{-4}	65

*For comparison, the relative V_{max} of adenosine, the reference compound, is defined to be 100.

Enzymatic Synthesis of Dideoxynucleoside Analogs and Its Potential Usefulness

In the past twenty-five years, a number of chemical methods for synthesizing a variety of ddN analogs have been developed.[38-41] However, the requirement for a multiple-step synthesis using expensive reagents in such chemical methods has often prevented adaptation to a large-scale production of ddN analogs. In addition, many chemical methods can inherently have a serious disadvantage such as formation of unnecessary byproducts, i.e., 2'- or 3'-deoxynucleosides and the α-anomer of 2',3'-dideoxynucleosides.

As discussed at the outset, enzymatic synthesis of dideoxypurine nucleosides such as compounds containing halogen atoms described here is particularly useful, since it offers a great ease in handling such reactive functional groups.

In the enzymatic method we exploited in the present work, most ddP analogs are synthesized with yields in a range of 20-40 %. It is worth stressing, however, that the unused starting materials can be recovered simply by centrifugation and they can be used for the next reaction. Furthermore, the advantage of this enzymatic method can even be enhanced by using immobilized E. coli cells. We already adapted the present enzymatic method for large-scale production of the 6-halo-substituted ddP analogs by immobilizing JA-300 cells with strontium arginate gel in a bioreactor.[42] Briefly, 2.6 % (w/v) arginate sodium in 0.9 M NaCl was mixed with an equal volume of 17 % (v/v) E. coli emulsified in 0.9 M NaCl. The mixture was added dropwise to 2.5 volumes of 0.1 M strontium chloride solution. The resultant beads (with about 2 mm diameter) were soaked in 2 % polyethyleneimine solution for 2-3 min, treated in 0.5 % glutaraldehyde solution for 30 min, and extensively washed with 0.9 M NaCl.

Use of thus immobilized E. coli can offer several advantages: (i) enzymes can be used repeatedly; (ii) continual nucleoside synthesis is possible for at least one month; (iii) the purity of the products is higher with less labor in part because of ease in separating them from cells and cell-derived products; and (iv) enzymatic reactions can be carried out in buffers containing organic solvents such as dimethyl sulfoxide. It should be noted, however, that long term use tends to reduce the relative enzymatic activity, which often occurs in about a month.

6-Halo-substituted ddP Analogs as Potential Therapeutic Agents Targeting HIV-associated CNS Diseases

Some of 6-halo-substituted ddPs may be of interest in view of possible application of these agents for therapy targeting HIV-associated CNS diseases. We now have data that each of 6-halo-ddPs discussed in this report is as sensitive as ddI or ddA to solvolysis in acid reactions and decomposes to a purine base and a dideoxyribose, thus losing its antiretroviral activity.[42] However, it has been shown that ddI is orally bioavailable in humans when administered with antacids, and plasma concentrations higher than those that exert potent antiviral activity *in vitro* can be achieved.[15]

We then asked if 2-amino-6-chloro-ddP could be absorbed in mice when orally administered. The oral bioavailability of 2-amino-6-chloro-ddP (compound 2), as assessed as plasma levels of 2-amino-6-chloro-ddP plus its converted form ddG, was substantially higer than that of ddI (Figure 5). We have also found that when 2-amino-6-chloro-ddP was administered intravenously, a significant level of the intact compound was detected in homogenized brain as compared to ddI (Table 4). However, one must use caution in extrapolating data from animal experiments to humans. It should be stressed that only *in vivo* human studies could resolve these issues.

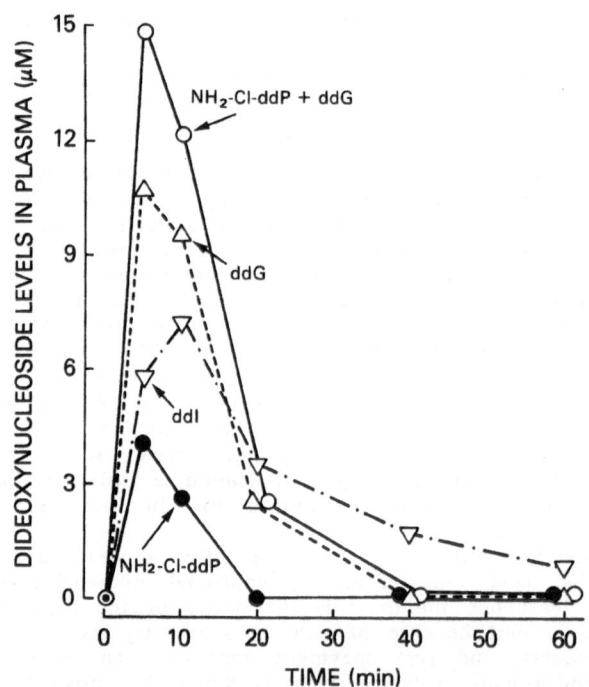

Figure 5. Levels of ddPs in plasma from mice receiving 2-amino-6-chloro-ddP (NH$_2$-Cl-ddP) or ddI orally. A twenty mg/kg dose of 2-amino-6-chloro-ddP or ddI was orally administered following administration of 10 mg/kg magnesium oxide to two fasting 6 wk-old ddY mice (weight 25-28 g). At various time points, venous blood was collected from retroorbital plexus, and levels of 2-amino-6-chloro-ddP, ddG, and ddI were determined by HPLC. Values shown represent the means (μM) of duplicate determinations.

Table 3. Levels of 2-Amino-6-chloro-ddP, ddG, and ddI in Brain Following Intravenous Administration of Various Dideoxypurine Nucleoside Analogs to Mice

Compound admnistered	Compound detected	3 min	10 min
2-NH$_2$-6-Cl-ddP	2-NH$_2$-6-Cl-ddP	1.34*	0.64
	ddG	0.42	0.45
	2-NH$_2$-6-Cl-ddP + ddG	1.76	1.09
ddG	ddG	0.37	0.16
ddI	ddI	0.32	0.16

Each compound was administered to ddY male mice (25-28 g) through tail vein at a dose of 20 mg/kg. At each time point, mice (one for each drug) were decapitated and the brain was excised out to be homogenized. Following methanol extraction, levels of 2-amino-6-chloro-ddP, ddG, and ddI in each extract sample were determined by HPLC. *μg/g brain.

Conclusions

Four 2-amino-6-halo- and four 6-halo-2',3'-dideoxypurine ribofuranosides (ddP) were synthesized using *E. coli*-mediated transdideoxyriosylation and tested for *in vitro* activity to suppress the infectivity and replication of HIV. The comparative order of *in vitro* anti-HIV activity of the eight 6-halo-ddPs was : 2-amino-6-fluoro, 2-amino-6-chloro, 6-fluoro > 2-amino-6-bromo > 2-amino-6-iodo, 6-chloro > 6-bromo > 6-iodo. 2-Amino-6-fluoro-, 2-amino-6-chloro- and 6-fluoro-ddPs showed a potent activity against HIV comparable to that of 2',3'-dideoxyinosine (ddI) or 2',3'-dideoxyguanosine (ddG), and completely blocked the infectivity of HIV without affecting the growth of target cells. The lipophilicity order was : 2-amino-6-iodo > 2-amino-6-bromo > 2-amino-6-chloro > 2-amino-6-fluoro >> ddG > ddI. All eight 6-halo-ddPs were substrates for adenosine deaminase (ADA). The relative rate of hydrolysis by ADA was : ddA, 2-amino-6-fluoro >> 2-amino-6-chloro, 2-amino-6-bromo > 2-amino-6-iodo. These 6-halo-substituted ddPs appear to exert their anti-retroviral activity against HIV *in vitro* upon conversion to ddI or ddG. Taken together, these compounds may represent a new class of lipophilic prodrugs for ddI and ddG, and may also provide a new strategy for endowing therapeutic purine nucleosides with desirable lipophilicity.

References

1. H. Mitsuya, R. Yarchoan, and S. Broder, Molecular targets for AIDS therapy. *Science*. 249:1533 (1990).
2. H. Mitsuya and S. Broder, Strategies of antiviral therapy in AIDS. *Nature*. 325:773 (1987).
3. R. Yarchoan, H. Mitsuya, C.E. Meyer, and S. Broder, Clinical pharmacology of 3'-azido-2',3'-dideoxythymidine (zidovudine) and related didexoynucleosides. *N. Engl. J. Med.* 321:726 (1989).
4. H. Mitsuya, K.J. Weinhold, F.A. Furman, M.H. St. Clair, S.N. Lehrman, R.C. Gallo, D. Bolognesi, D.W. Barry, and S. Broder. 3'-Azido-3'-deoxythymidine (BW A509U): an antiviral agent that inhibits the infectivity and cytopathic effect of human T-lymphotropic virus type III/lymphadenopathy-associated virus in vitro. *Proc. Natl. Acad. Sci. USA.* 82: 7096 (1985).
5. H. Mitsuya and S. Broder, Inhibition of the in vitro infectivity and cytopathic effect of human T-lymphotropic virus type III/lymphadenopathy associated virus (HTLV-III/LAV) by 2',3'-dideoxynucleosides, *Proc. Natl. Acad. Sci. USA.* 83:1911 (1986).
6. R. Yarchoan, R. Klecker, K.J. Weinhold, P.D. Markham, H.K. Lyerly, D.T. Durack, E. Gelmann, S.N. Lehrman, R.M. Blum, D.W. Barry, G.M. Shearer, M.A. Fischl, H. Mitsuya, R.C. Gallo, J.M. Collins, D.P. Bolognesi, C.E. Myers, and S. Broder, Administration of 3'-azido-3'-deoxythymidine, an inhibitor of HTLV-III/LAV replication, to patients with AIDS or AIDS-related complex, *Lancet*, i, 575 (1986).

7. M.A. Fischl, D.D. Richman, M.H. Grieco, M.S. Gottlieb, P.A. Volberding, O.L. Laskin, J.M. Leedom, J.E. Groopman, D. Mildvan, R.T. Schooley, G.G. Jackson, D.T. Durack, D. King, and the AZT Collaborative Group, The efficacy of azidothymidine (AZT) in the treatment of patients with AIDS and AIDS-related complex: a double-blind, placebo-controlled trial, *N. Engl. J. Med.* 317:185 (1987).

8   D.D. Richman, M.A. Fischl, M.H. Grieco, M.S. Gottlieb, P.A. Volberding, O.L. Laskin, J.M. Leedom, J.E. Groopman, D. Mildvan, M.S. Hirsch, G.G. Jackson, D.T. Durack, S. Nusinoff-Lehrman, and the AZT Collaborative Working Group. The toxicity of azidotymidine (AZT) in the treatment of patients with AIDS and AIDS-related complex. A Double-blind, placebo-controlled trial. *N. Engl. J. Med.*,317: 192 (1987).

9. P.A. Volberding, S.W. Lagakos, M.A. Koch, C. Pettinelli, M.W. Myers, D.K. Booth, H.H. Balfour, Jr., R.C. Reichman, J.A. Bartlett, M.S. Hirsch, R.L. Murphy, W.D. Hardy, R. Soeiro, M.A. Fischl, J.G. Bartlett, T.C. Merigan, N.E. Hyslop, D.D. Richman, F.T. Valentine, L. Corey, and the AIDS Clinical Tirals Group of the National Institute of Allergy and Infectius Diseases. Zidovudin in asymptomatic human immunodeficiency virus infection. A controlled trial in persons with fewer than 500 CD4-positive cells per cubic millimete. *N. Engl. J. Med.* 322: 941 (1990).

10. M.A. Fischl, C.B. Parker, C. Pettinelli, M. Wulfsohn, M.S. Hirsch, A.C. Collier, D. Antoniskis, M. Ho, D.D. Richman, E. Fuchs, T.C. Merigan, R.C. Reichman, J. Gold, N. Steigbigel, G.S. Leoung, S. Rasheed, A. Tsiatis, and the AIDS clinical trials group, A randomized controlled trial of a reduced daily dose of zidovudine in patients with the acquired immunodeficiency syndrome, *N. Engl. J. Med.* 323:1009 (1990).

11. A.C. Collier, S. Bozzette, R.W. Coombs, D.M. Causey, D.A. Schoenfeld, S.A. Spector, C.B. Pettinelli, G. Davies, D.D. Richman, J.M. Leedom, P. Kidd, and L. Corey, A pilot study of low-dose zidovudine in human immunodeficiency virus infection, *N. Engl. J. Med.* 323:1015 (1990).

12. R. Yarchoan, C.-F. Perno, R.V. Thomas, R.W. Klecker, J.P. Allain, J. Wills, N. McAtee, M.A. Fischl, R. Dubinski, M.C. McNeely, H. Mitsuya, J.M. Pluda, T.J. Lawley, M. Leuther, B. Safai, J.M. Collins, C.E. Myers, and S. Broder, Phase I studies of 2',3'-dideoxycytidine in severe human immunodeficiency virus infection as a single agent and alternating with zidovudine (AZT), *Lancet* i:76 (1988).

13. T.C. Merigan, G. Skowron, S.A. Bozzette, D.D. Richman, R. Uttamchandani, M. Fischl, R. Schooley, M. Hirsch, W. Soo, C. Pettinelli, H. Schaumburg, and the ddC Study Group of the AIDS Clinical Trials Group, Circulating p24 antigen levels and responses to dideoxycytidine in human immunodeficiency virus (HIV) infections. A phase I and II study, *Ann. Int. Med.* 110:189 (1989).

14. M. Browne, G. Curt, and Brown University Phase I working group; unpublished data.

15. R. Yarchoan, H. Mitsuya, R.V. Thomasm J.M. Pluda, N.R. Hartman, C.-F. Perno, K.S. Marczyk, J.-P. Allin, D.G. Johns, and S. Broder. In vivo activity against HIV and favorable toxicity profile of 2',3'-dideoxyinosine, *Science* 245:412 (1989).

16. J.S. Lambert, M. Seidlin, R.C. Reichman, C.S. Plank, M. Laverty, G.D. Morse, C. Knupp, C. McLaren, C. Pettinelli, F.T. Valentine, R. Dolin, 2',3'-Dideoxyinosine (ddI) in patients with the acquired immunodeficiency syndrome or AIDS-related complex, *N. Engl. J. Med.* 322:1333 (1989).

17. T.P. Cooley, L.M. Kunches, C.A. Saunders, J.K. Ritter, C.J. Perkins, C. McLaren, R.P. McCaffrey, H.A. Liebman, Once-daily administration of 2',3'-dideoxyinosine (ddI) in patients with the acquired immunodeficiency syndrome or AIDS-related complex, *N. Engl. J. Med.* 322:1340 (1990).

18. R. Yarchoan, J.M. Pluda, R.V. Thomas, H. Mitsuya, P. Browers, K.M. Wyvill, N. Hartman, D.G.Johns, and S. Broder. Long-term toxicity/activity profile of 2',3'-dideoxyinosine in AIDS or AIDS-related complex. *Lancet*, ii, 526 (1990).

19. W.D. Snider, D.M. Simpson, S. Nielsen, J.W.M. Gold, C.E. Metroka, J.B. Posner. Neurological complications of Acquired Immune Deficiency Syndrome: Analysis of 50 patients. *Ann. Neurol.* 14:403 (1983).

20. I.J. Koralnik, A. Beaumanoir, R. Häusler, A. Kohler, A.B. Safran, R. Delacoux, D. Vibert, E. Mayer, P. Burkhard, A. Nahory, M. Magistris, J. Sanches, P. Myers, F. Paccolat, F. Quoëx, V. Gabriel, L. Perrin, B. Mermillod, G. Gauthier, F. A. Waldvogel, B. Hirshel. A controlled study of early neurologic abnormalities in men with asymptomatic human immunodeficiency virus infection. *N. Engl. J. Med.* , 323:864 (1990).

21. S. Gartner, P. Markovits, D.M. Markovits, M. H. Kaplan, R.C. Gallo, and M. Popovic, The role of mononuclear phagocytes in HTLV-III/LAV infection, *Science*, 233: 215 (1986).

22. S. Koenig, H.E. Gendelman, J.M. Orenstein, M.C. Dal Canto, G.H. Pezeshpour, M. Yungbluth, F. Janotta, A. Aksamit, M.A. Martin, and A.S. Fauti, Detection of AIDS virus in macrophages in brain tissue from AIDS patients with encephalopathy, *Science*, 233: 1089 (1986).

23. B.A. Watkins, H.H. Dorn, W.B. Kelly, R.C. Armstrong, B.J. Potts, F. Michaels, C V. Kuft, M. Dubois-dalcq, Specific tropism of HIV-1 for microglial cells in primary human brain cultures. *Science*, 249:549 (1990).

24. T. Shirasaka, K. Murakami, H. Ford, Jr., J.A. Kelly, H. Yoshioka, E. Kojima, S. Aoki, S. Broder, and H. Mitsuya, Halogenated congeners of 2',3'-dideoypurine nucleosides active against HIV in vitro: A new class of lipophilic prodrugs. *Proc. Natl. Acad. Sci. USA.* (in press).

25. K. Murakami, T. Shirasaka, H. Yoshioka, E. Kojima, H. Ford, J.S. Driscoll, J.A. Kelly, and H. Mitsuya. *Escherichia coli*-mediated biosynthesis and in vitro antiviral activity of lipophilic 6-halo-2',3'-dideoxypurine nucleosides: A new class of prodrugs against HIV. (submitted).

26. H. M. Kalckar, The enzymatic synthesis of purine ribosides. *J. Biol. Chem.* 67:477 (1947).

27. J.L. Ott, C.H. Werkman, Formation of adenosine by cell-free extracts of *Escherichia coli, Arch. Biochem. Biophys.* 48:483 (1954).

28. J.L. Ott, C.H. Werkman, Coupled nucleoside phosphorylase reactions in *Escherichia coli, Arch. Biochem. Biophys.* 69:264 (1957).

29. A. Imada and S. Igarashi, Ribosyl and deoxyribosyl transfer by bacterial enzyme systems, *J. Bacteriol.* 94:1551 (1967).

30. K. Yokozeki, S. Yamanaka, T. Utagawa, K. Takinami, Y. Hirose, A Tanaka, K. Sonomoto, and S. Fukui, Production of adenine arabinoside by gel-entrapped cells of *enterobacter aerogenes* in water-organic cosolvent system, *European J. Appl. Microbiol. Biotechnol.* 14:225, (1982).

31. T. Utagawa, H. Morisawa, F. Yoshinaga, A. Yamazaki, K. Mitsugi, and Y. Hirose, Microbial synthesis of purine arabinosides and their biological activity, *Agric. Biol. Chem.* 49:2167 (1985).

32. H. Shirae, K. Kobayashi, H. Shiragami, Y. Irie, N. Yasuda, and K. Yokozeki, Production of 2',3'-dideoxyadenosine and 2',3'-dideoxyinosine from 2',3'-dideoxyuridine and the corresponding purine bases by resting cells of *Escherichia coli* AJ 2595, *Appl. Enviromental Micrbiol.* 55:419 (1989).

33. C.K. Chu, G.V. Ullas, L.S. Jeong, S.K. Ahn, B. Doboszewski, Z.K. Lin, J.W. Beach, R.F. Schinazi, Synthesis and structure-activity relationships of 6-substituted 2',3'-dideoxypurine nucleosides as potential anti-human immunodeficiency virus agents, *J. Med. Chem.* 33:1553 (1990).

34. R. Yarchoan, G. Berg, P. Brouwers, M.A. Fischl, A.R. Spitzer, A. Wichman, J. Grafman, R.V. Thomas, B. Safai, A. Brunetti, C.-F. Perno, P.J. Schmidt, S.M. Larson, C.E. Myers, and S. Broder, Response of human immunodeficiency virus-associated neurological disease to 3'-azido-3'-deoxythymidine, *Lancet.* i:132 (1987).

35. P.A. Pizzo, J. Eddy, J. Faloon, F. Balis, R.F. Murphy, H. Moss, P. Wolters, P. Browers, P. Jarosinski, M. Rubin, S. Broder, R. Yarchoan, A. Brunetti, M. Maha, S. Nusinoff-Lehrman, and D.G. Poplack, Effect of continuous intravenous infusion of zidovudine (AZT) in children with symptomatic HIV infection, *N. Engl. J. Med.* 319: 889 (1988).

36. P.N. Craig, Interdependence between physical parameters and selection of substituent groups for correlations studies, *J. Med. Chem,* 14:8 (1971).

37. B.M. Chassy and R.J. Suhadolnick. Adenosine aminohydrolase. *J. Biol. Chem.* 242:3655, (1967).

38. J.P. Horwitz, J. Chua, and M. Noel. Nucleosides, V. The monomesylates of 1-(2'-deoxy-b-D-lyxofuranosyl) thymine. *J. Org. Chem.* 29:2076 (1964).

39. M.J. Robins and R.K. Robins. The synthesis of 2',3'-dideoxyadenosine from 2'-deoxyadenosine. *J. Am. Chem. Soc.* 86:3585 (1964).

40. J.P. Horwitz, J. Chua, M. Noel, and J.T. Donatti, Nucleosides. XI. 2',3'-Dideoxycytidine, *J. Org. Chem.* 32:817 (1966).

41. T.S. Lin and W.H. Prusoff, Synthesis and biological activity of several amino analogues of thymidine, *J. Med. Chem.* 21:109 (1978).

42. K. Murakami, H. Yoshioka, E. Kojima, et al. *Unpublished data.*

43. H. Ford, Jr., C.L. Merski, and J.A. Kelley. *Abstracts of papers,* 41st Pittsburgh Conference, New York, NY; The Pittsburgh Conference: Pittsburgh, PA, Abstract, 1174 (1990).

ANTIVIRAL EVALUATION OF HIV-1 SPECIFIC RIBOZYME EXPRESSED

IN CD4+ HELA CELLS

John A. Zaia[1], Edouard M. Cantin[1], Pairoj S. Chang[2], Nava Sarver[3], and John J. Rossi[2]

[1]City of Hope National Medical Center, Duarte CA 91010
[2]Beckman Research Institute of the City of Hope, Duarte, CA
[3]National Institute of Allergy and Infectious Diseases, Bethesda, MD 20892

INTRODUCTION

Catalytic RNA, called ribozyme, has been shown to enzymatically cleave specific sites in RNA (for reviews, see Cech and Bass, 1986, and Uhlmann and Peymen, 1990). One group of ribozymes has an active site of consensus sequences forming a "hammerhead" structure. The active center of the hammerhead ribozyme consists of 11 essential bases juxtaposed near 3 target bases (Cech, 1988). The potential versatility of this system lies in the fact that simple ribozymes can be constructed which target sites in native RNA, completing the active site in a *trans* configuration, and inducing cleavage (Haseloff and Gerlach, 1988). Ribozymes, which have been developed to cleave human immunodeficiency virus type 1 (HIV-1) RNA, can be transcribed in both *in vitro* and *in vivo* systems (Chang et al., 1990; Sarver et al., 1990). The purpose of this report is to review the studies evaluating the effect of ribozyme on HIV-1 RNA cleavage and on the antiviral effect in mammalian cells. Using a hammerhead motif, ribozymes were constructed which targeted a *gag* cleavage site, were cloned into a mammalian expression vector, and used to transform HeLa-CD4+ cells. These cells expressed HIV-1-specific ribozyme and demonstrated inhibition of HIV-1 infection.

METHODS

Ribozyme Synthesis

Ribozyme synthesis and purification by gel electrophoresis was performed exactly as described (Chang et al., 1990). Briefly, cDNA complementary for the ribozyme sequence was cloned into the *Bam*HI site of the pBLUESCRIPT-KS(+) vector (Stratagene, LaJolla CA). *In vitro* transcription of RNA from the plasmid templates using the T7 promoter was carried out under conditions adapted from Lowary et al. (1986) and Zaug et al. (1986). The primer for this reaction consisted of a synthetic hemi-duplex formed from a synthetic DNA oligonucleotide containing both the sequence of the bacteriophage T7 RNA polymerase promoter and a sequence complementary to the ribozyme product (Chang et al., 1990). The two DNA oligonucleotides were designed to anneal under standard hybridization conditions forming the template for the T7 RNA polymerization reaction.

Advances in Molecular Biology and Targeted Treatment for AIDS
Edited by A. Kumar, Plenum Press, New York, 1991

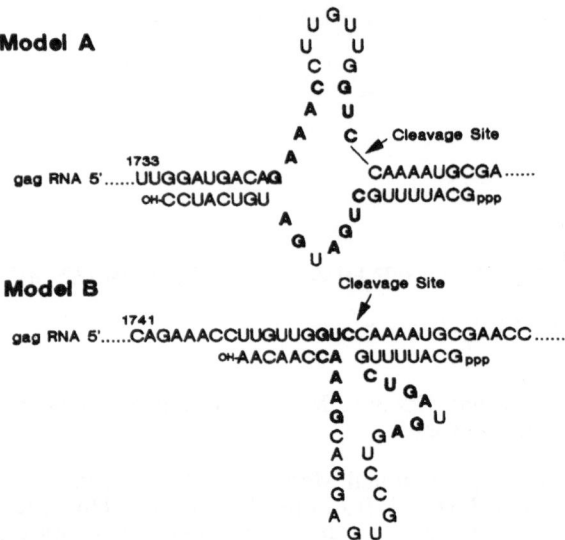

Model A

```
            U  G  U
          U       U
          C       G
          C       G
          A       U
       A           C  ← Cleavage Site
      1733         A
gag RNA 5'......UUGGAUGACAG        CAAAAUGCGA......
       OH-CCUACUGU         CGUUUUACG ppp
              A      U
              G      U
               G  A
               U
```

Model B

```
                          Cleavage Site
        1741                    ↓
gag RNA 5'......CAGAAACCUUGUUGGUCCAAAAUGCGAACC......
       OH-AACAACCA   GUUUUACG ppp
                   A   C  U
                   A     U  G A
                   A        G    U
                   G     G A  A
                   C      U
                   A      U
                   G      C
                   G      C
                   A        G
                      G  U
```

Fig. 1. Description of anti-HIV-1 *gag* ribozyme. The active center of hammerhead ribozymes, targeted to *gag* RNA of HIV-1, is shown. Certain of the bases of the active center (bold letters) can occur in a *cis* configuration upstream of the cleavage site (Model A) or can be in a *trans* position relative to the cleavage site (Model B). Note that the complementary flanking sequences determine the specificity of the structure.

Synthesis of target RNA substrate

A 9 kb *Sst*I and a 334bp *Hind*III-*Pst*I DNA fragment from HIV-1 (strain HIVHXB2; Ratner *et al.*, 1987), were cut from the plasmid pBH10-Re (Biotech Research Laboratories, Rockville, MD) and subcloned separately into the pBLUESCRIPT-KS(+) vector (Stratagene, LaJolla, CA). *In vitro* transcription from the T7 promoter, as described above, yielded HIV-1 sense RNAs which were purified by 8% polyacrylamide-7M urea gel electrophoresis (PAGE-U) for use in the ribozyme cleavage reactions.

Ribozyme Cleavage Reaction

The *in vitro* ribozyme cleavage reactions were carried out in 10 *u*l volume containing 1*p*mol ribozyme and 1*p*mol target RNA in 75mM Tris-Hcl pH 7.5 and 1.5mM Na-EDTA as described (Chang *et al.*, 1990). Prior to use, this mixture was heated to 95°C for 2 min followed by fast cooling on ice. To initiate the reaction, MgCl$_2$ was added and the final reaction contained 20mM MgCl$_2$, 50mM Tris-HCl pH 7.5, 1 mM Na-EDTA. The reaction was performed at 37°C for 14 hr, unless specified otherwise, and then stopped by the addition of 0.2 vol 200 mM Na-EDTA pH 7.9. The reaction was analyzed by PAGE-U and autoradiography as described (Chang *et al.*, 1990).

Expression of ribozyme in HeLa cells

HeLa-CD4+ cells (Maddon *et al.*, 1986) were obtained from the National Institute of Allergy and Infectious Diseases Research and Reference Reagent Program and grown in Dulbecco modified Eagle medium (DMEM) supplemented with 10% fetal calf serum. The mammalian vector, p-hu ß Apr-1-gpt (Gunning *et al.*, 1987), was used to clone the ribozyme cDNA between the human ß-actin promoter and an SV40 polyadenylation signal. HeLa cells were transfected with lipofectin reagent (BRL, Gaithersburg MD) using a 10*u*g:1*u*g ratio of lipid:DNA as described (Felgner *et al.*, 1987). Cells were cultured in medium containing xanthine, hypoxanthine, aminopterin, and mycophenolic acid to select for expression of xanthine-guanine phosphoribosyl-transferase (GPT) gene as described by Mulligan and Berg (1981). Selected cells were further screened for ribozyme expression by polymerase chain reaction (PCR) and for CD4+ expression by PCR and immunohistochemical staining.

Polymerase Chain Reaction Assays

PCR assays for HIV-1 sequences or for ribozyme were performed exactly as described by Murakawa *et al.* (1988) and by Holland *et al.* (1989).

Infection with HIV-1

HeLa-CD4+ cells or ribozyme-expressing transformants were grown in 24-well microtiter plates in DMEM supplemented with 10% FCS. Cells were washed, treated with 200 *u*l Polybrene (10*u*g/ml), and inoculated with HIV-1 strain III$_B$ in 20*u*l as described by Zaia *et al.* (1988). HIV-1 expression was evaluated with a p24 antigen assay using a commercially available kit (NEN-DuPont, Boston MA) or a histochemical stain with HIV-1 specific antiserum, and for viral RNA or DNA using PCR.

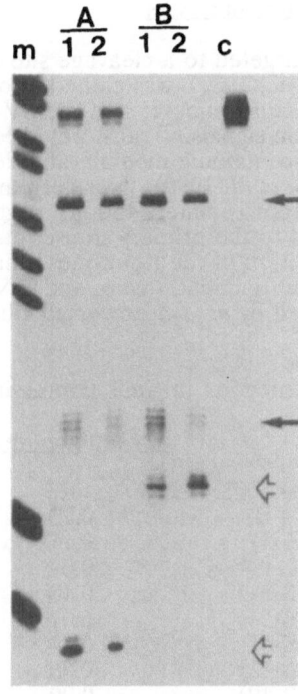

Fig. 2. Comparison of Model A and Model B ribozyme. Cleavage reactions were compared for Model A (A1, A2) or Model B (B1,B2) ribozymes with 8% polyacrylamide-7M urea and autoradiography. Lane M, *Hpa*II-digested pBR322 molecular markers, lane S, RNA substrate without ribozyme. S = substrate; solid arrow = 5' (upper) or 3' (lower) cleavage products; open arrow = ribozyme. Modified from Chang *et al.* (1990) with permission.

RESULTS

Efficiency of hammerhead ribozymes

As mentioned above, the active center of the hammerhead ribozyme consists of 11 essential bases juxtaposed near 3 target bases. The 11 bases CUGA[N]GAG [N]$_n$ GAAAC must lie near an XUX', where X is any ribonucleotide and X' is any U,A, or C. The cleavage occurs immediately after the X'. Based on the HIV-1 genome, there are 11 GAAAC sites which could be used for ribozyme targeting, and a single *gag* site was selected between coordinate 1743-1757 of HIVHXB2 (Ratner *et al.*, 1987) for the initial studies. As shown in Fig. 1, the bases in the active site of the ribozyme can come together not only in a *cis* configuration (Model A) but also in a *trans* configuration (Model B) with respect to the target sequence.

To determine the time course of cleavage, equimolar amounts of the ribozymes shown in Fig. 1 were reacted at 37°C for varying lengths of time. Substantial cleavage was observed after 30 min and progressed to complete cleavage after 14 hours of incubation (data not shown, see Chang *et al.*, 1990). These two ribozymes were compared for efficiency of HIV-1 RNA cleavage as shown in Fig. 2. Here, the substrate RNA consisted of nucleotides 1711-1789 and the synthetic ribozyme was transcribed with or without 5' GpppG-capping of the RNA. It is clear that the Model B ribozyme was more efficient then the Model A ribozyme and that the efficiency of the reaction was not improved with capping (compare Fig. 2, lanes A1, B1 and lanes A2, B2). Based on these results, subsequent mammalian expression studies were performed using the Model B ribozyme conformation.

Expression of ribozyme in mammalian cells

A Model B ribozyme, targeted to a cleavage site at 805-806 in the *gag* sequence of HIV-1 (Ratner *et al.*, 1987), was cloned into a mammalian expression vector containing a human ß-actin promoter and the SV-40 late transcriptional termination and polyadenylation signals. Transformed HeLa-CD4+ cells were selected in mycophenolic acid-containing medium and demonstrated to contain ribozyme-specific RNA (Chang *et al.* 1990). Northern analysis showed that these cells expressed an RNA with approximately 450 nts, suggesting that 200-400 polyadenosines were appended to the primary transcripts (Chang *et al.* 1990). Despite this extensive modification of the ribozyme, it has been shown that the molecule maintains its catalytic function *in vitro*, and RNA isolated from HIV-1-infected H9 cells can be cleaved by *in vivo* expressed ribozyme (Chang *et al.* 1990).

Table 1. HIV-1 Infection in Cells Expressing Ribozyme

Cell line	p24 antigen *ng*/ml
Parent HeLa CD4+	>10
Line 1 (A+B)	0.23
Line 2 (A3-3)	0.14
Line 3 (B2-11)	0.04
Line 4 (A3-31)	0.48
Line 5 (A6-54)	0.07
Line 6 (A6-6)	0.14
Line 7 (B6-42)	0.08
Line 8 (A2-19)	0.09

As shown in Table 1, eight of eight cell lines, selected from the transformed HeLa-CD4+ cells on the basis of ribozyme expression and continued CD4+ phenotype, demonstrated reduced p24 antigen compared to the parent cells. Culture media from infected transformants had between 0.07-0.48 *ng*/ml p24 compared to >10ng/ml in the infected parent culture.

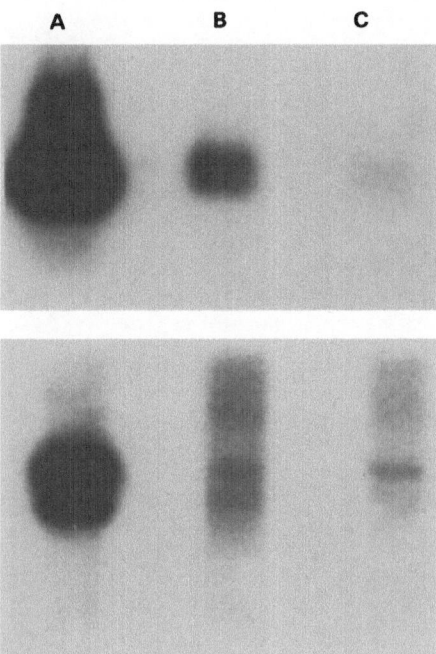

Fig. 3. Expression of HIV-1 DNA in ribozyme expressing HeLa-CD4+ cells. Parent cells (A) or transformants (B and C) were infected with HIV-1 and evaluated at 96 hours for HIV-1 DNA using a PCR assay specific for *nef* (upper panel) or *gag* (lower panel) sequences.

These cells were evaluated for HIV-1 specific DNA and RNA. As shown in Fig. 3, the presence of HIV-1 DNA in HeLa-CD4+ cells at 96 hr p.i. was variably affected in these anti-*gag* ribozyme expressing lines. Clones which were studied by PCR showed decreased HIV-1 DNA, and certain of these had markedly reduced viral DNA (see Fig. 3, lanes B and C). When the RNA was analyzed from these same cell lines, the amount of *gag*-specific RNA was decreased in the ribozyme transformants (Fig. 4, upper panel, lanes B, C, and D) compared with HIV-1 infected parent cells (Fig. 4, upper panel, lane E). The transformed lines continued to express CD4-specific RNA (Fig. 4, lower panel).

The state of the HIV-1 RNA in the infected cells was analyzed by PCR by comparing primer amplification across the putative cleavage site with that downstream from this site (Fig. 5). RNA from HIV-1 infected anti-*gag*-specific ribozyme-transformed cells did not support the amplification of a 480 nt sequence spanning the cleavage site but did permit amplification of a 200 nt sequence within this region but downstream from the putative cleavage site (Fig. 5, lower left and right, lanes B).

Toxicity of endogenous ribozyme

To demonstrate potential toxicity of expressed ribozyme, the growth rate for parent and transformed HeLa-CD4+ cells was compared by measurement of total cell number over time. There was no difference observed between the parent and the transformants (data not shown). As mentioned above, certain transformants no longer expressed CD4 antigen, and it is likely that this is explained by selection of revertant from the parent line rather than by direct or indirect ribozyme effect.

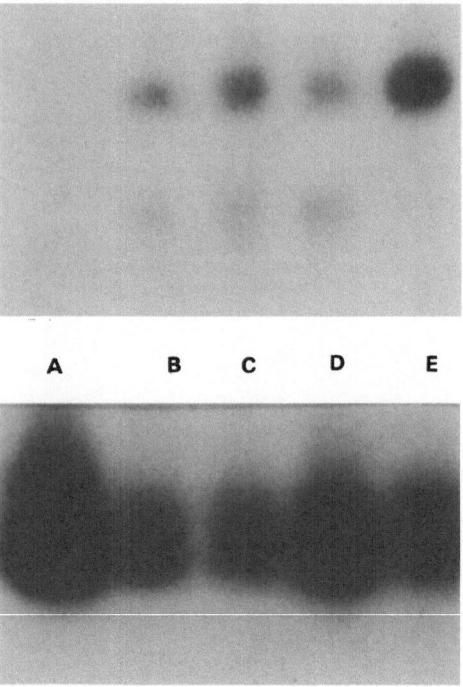

Fig. 4. Expression of HIV-1 RNA in ribozyme-expressing HeLa-CD4+ cells. Uninfected (A) or infected (E) parent cells or infected transformants (B, C and D) were evaluated at 96 hours p.i. for HIV-1 RNA using a PCR assay specific for *gag* (upper panel) or CD4 (lower panel) sequences.

DISCUSSION

Autocatalytic RNA is an important mechanism for processing of RNA in species bridging many forms of biologic life from plant viruses to vertebrates (Cech and Bass, 1986 and Rossi *et al.*, in press 1990). Because the hammerhead ribozymes utilize a similar active site consensus sequence, it might be possible to artificially produce ribozymes which could catalyze the cleavage of any RNA. With the demonstrationjh by Hazeloff and Gerlach (1988) that synthetic RNA can be used in *trans* to complete the functionally active site of ribozyme, the potential versatility of ribozyme has been extended. Virtually any XUX', a sequence present in all natural RNAs, can be targeted provided that secondary structure does not prevent access to the target sequence. Anti-HIV-1 ribozymes have been synthesized with activity against HIV-1 and have been expressed in mammalian cells with absence of toxicity. Cell which express an anti-*gag* ribozyme shown markedly diminished ability to support the replication of HIV-1 (Sarver *et al.* 1990).

These results raise several questions, the most important of which is whether the observed HIV-1 inhibition is ribozyme-specific. It will be important to

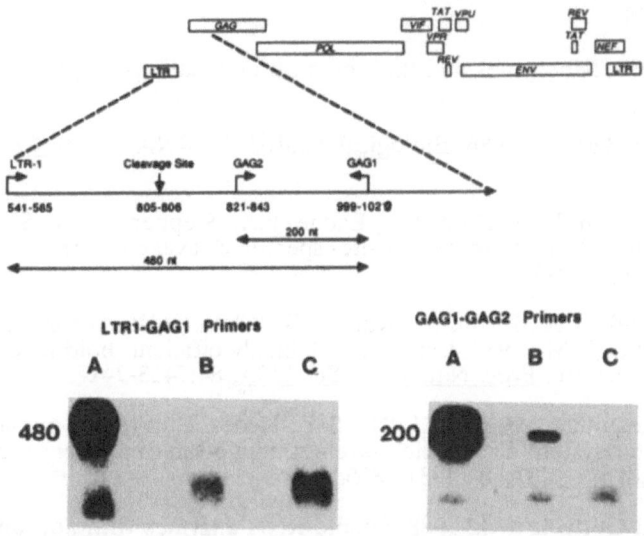

Fig. 5. PCR analysis of *gag* RNA target site. The HIV-1 genomic structure is shown schematically with regions of PCR amplification for a 480 nt-length sequence using LTR1 and GAG1 primers, which included the cleavage site at position 805-806, and for a 200 nt-length sequence using GAG1 and GAG2 primers downstream from the cleavage site. The RNA-based PCR of HIV-1-infected HeLa-CD4+ parent cells (A), pooled anti-*gag* ribozyme-transformants (B), or PCR control lacking template (C). Modified from Sarver *et al.* (1990), with permission.

determine if cells which express these anti-HIV-1 ribozymes can support the replication of another human retrovirus such as HIV-2. It is possible that the selection of cells transformed to express ribozymes will have alterations which produce non-specific anti-viral effects. In our experience, several of the clones produced after anti-*gag* ribozyme transformation no longer expressed CD4 antigen. Hence, in the anti-viral screening program, it is important to confirm the similarity of the selected clone to the parent line. Regarding the similarity of these ribozyme-expressing clones to the parent cells, it will be necessary to show that these clones can support the replication of other retroviruses. Since HeLa-CD4+ cells are not as optimal for growing human retroviruses as lymphoid cell lines, the specificity and mechanism of anti-HIV-1 action of ribozyme should be evaluated in these more permissive cells. Other questions such as the optimal method for delivery and uptake of synthetic ribozyme, the anti-viral activity of synthetic ribozyme, and the duration of auto-catalytic activity of exogenous ribozyme remain to be studied.

ACKNOWLEDGEMENTS

This work was supported by United States Public Health Service grants AI25959 and CA33572. The authors are grateful for the laboratory assistance of Delilah S. Stephens, Carolyn I. Lowery, Paula A. Ladne, and Jian Chen. The authors are particularly grateful to Joan Cass for the photo-ready preparation of this manuscript.

REFERENCES

Cech, T.R., 1988, Ribozymes and their medical implications. J. Am. Med. Assoc., 260:3030-3034.

Cech, T.R. and Bass, B., 1986, Biological catalysis by RNA, Annu. Rev. Biochem., 55:599-629.

Chang, P.S. , Cantin, E.M., Zaia, J.A., Ladne, P.A., Stephens, D.A., Sarver, N., and Rossi, J.J., 1990, Ribozyme-mediated site- specific cleavage of the HIV-1 genome, Clin. Biotechnol., 2:23-31.

Felgner,P., Gadek, T.R., Holm, M., Roman, R., Chan, H.W., Wenz, M., Northrop, J.P., and Ringold, G.M., 1987, Lipofectin: A highly efficient, lipid-mediated DNA-transfection procedure, Proc. Nat. Acad. Sci. USA, 84:7413-7417.

Gunning, P., Leavitt, J., Muscat, G., Ng, S-Y., Kedes, L., 1987, A human ß-actin expression vector system directs high-level accumulation of antisense transcripts, Proc. Nat. Acad. Sci. USA, 84:4831-4835.

Hazeloff,J. and Gerlach, W.l., 1988, Simple RNA enzymes with new and highly specific endoribonuclease activities, Nature, 334:585-591.

Holland, H.K., Saral, R., Rossi J.J., Donnenberg, A.D., Burns, W.H., Beschorner W.E., Farzadegan, H., Jones, R.J., Quinnan, G. V., Vogelsang, G.B., Vriesendorp, H.M., Wingard, J.R., Zaia, J.A., and Santos, G.W., 1989, Allogeneic bone marrow transplantation, zidovudine, and human immunodeficiency virus type 1 (HIV-1) infection, Ann. Intern. Med. 111:973-981.

Lowary, P., Sampson, J., Milligan, J., Groebe, D., and Uhlenbeck, O.C., 1986, A better way to make RNA for physical studies, In: van Knipperberg, P.H. and Hilbers, C.N. (eds), Structure and Dynamics of RNA, p. 69, Plenum Press, New York.

Maddon, P.J., Dalgleish, A.G., McDougal, J.S., Clapham, P.R., Weiss, R.A., and Axel, R., 1986, The T4 gene encodes the AIDS virus receptor and is expressed in the immune system and the brain, Cell, 47:333-348.

Mulligan, R.C. and Berg, P., 1981, Selection for animal cells that express the Escherichia coli gene coding for xanthine-guanine phosphoribosyltransferase, Proc. Nat. Acad. Sci. USA, 78:2072-2076.

Murakawa, G.J., Zaia, J.A.,Spallone, P.A., Stephens, D.A, Kaplan, B.E., Wallace, R.B., and Rossi, J.J., 1988, Direct detection of HIV-1 RNA from AIDS and ARC patient samples, DNA, 7:287-295.

Ratner, L., Fisher, A., Jagodzinski, L.L., Mitsuya, H., Liou, R.-S., Gallo, R.C., and Wong-Staal, F.,1987, Complete nucleotide sequence of functional clones of the AIDS virus, AIDS Res. Hum. Retroviruses, 3:57-69.

Rossi, J.J., Cantin, E.M., Zaia, J.A., Ladne, P.A., Chen, J., Stephens, D.A., Sarver, N., and Chang, P.S., 1990, Ribozymes as therapies for AIDS. Proc. N. Y. Acad. Sci., in press.

Sarver, N., Cantin, E.M., Chang, P.S., Zaia, J.A., Ladne, P.A., Stephens, D.A., and Rossi, J.J., 1990, Ribozymes as potential anti-HIV-1 therapeutic agents, Science 247:1222-1225.

Uhlmann, E. and Peyman, A., 1990, Antisense oligonucleotides: a new therapeutic principle, Chem. Rev., 90:5543-584.

Zaia, J.A., Rossi, J.J., Murakawa, G.J., Spallone, P.A., Stephens, D.A., Kaplan, B.E., Eritja, R., Wallace, R.B., and Cantin, E.M., 1988, Inhibition of human immunodeficiency virus by using an oligonucleoside methylphosphonate targeted to the *tat*-3 gene., <u>J. Virology</u>, 62:3914-3917.

Zaug, A.J., Been, M.D., and Cech, T.R., 1986, The *Tetrahymena* ribozyme acts like and RNA restriction endonuclease, <u>Nature</u>, 324:429-433.

RESISTANCE OF INFLUENZA A VIRUSES TO AMANTADINE AND

RIMANTADINE

Alan J. Hay, Setareh Grambas and
Michael S. Bennett

National Institute for Medical Research
The Ridgeway
Mill Hill, London, NW7 1AA

Numerous clinical studies carried out over the years have documented the efficacy of amantadine and rimantadine in the prophylaxis and treatment of influenza A infections [1-9]. Although initially licensed in 1966, the use of amantadine in both the United States and the United Kingdom has been limited. Rimantadine, more widely used in the Soviet Union in recent years[3] has recently been licensed in France and awaits approval for use in the United States. The reticence to use these agents has been due to several factors, not least the extreme specificity of the drugs for influenza A viruses and the lack of any benefit against influenza B infections which are responsible for a significant proportion of recurrent disease in the human population. This has not, however, precluded the successful prophylactic use of amantadine following identification of influenza A outbreaks to limit their spread within semi-closed communities such as boarding schools and nursing homes[8,9]. Initial concerns regarding adverse side-effects have now largely been allayed and rimantadine which causes a lower incidence of neurological reactions[1,10,11] is now generally perceived as the drug of choice.

Recently concern has focussed on the potential limitations to the use of these agents posed by the occurrence of drug-resistant virus strains. Amantadine-resistant mutants have been isolated from animals[12] and birds[13] administered the drug and in cell culture following growth of virus in the presence of drug[14]. More recently it has become evident that, as with the use of drugs against other viruses[15], drug-resistant strains readily arise in children and adults undergoing treatment with either amantadine or rimantadine[16-19]. Since these are currently the only drugs available for use against influenza, the impact of drug resistance on their usefulness is of particular concern. This article will briefly consider what we know regarding the basis of resistance to amantadine and rimantadine and the mechanism of their antiviral action in relation to the possible impact of drug resistance on the more widespread use of these agents.

Advances in Molecular Biology and Targeted Treatment for AIDS
Edited by A. Kumar, Plenum Press, New York, 1991

Table 1. Amino acid substitutions in the M2 proteins of amantadine and rimantadine resistant influenza A viruses

Virus	Amino acid residue				
	26	27	30	31	34
'In Vivo' Isolates					
A/New York/83 (H3N2)[16#]		Val>Ala(1)*	Ala>Val(2)	Ser>Asn(10)	
A/Virginia/88 (H3N2)[17]		Val>Ala(1)	Ala>Val(1) >Thr(1)	Ser>Asn(14)	
A/Shanghai/87 (H3N2)[18]		Val>Ala(3)	Ala>Val(1) >Thr(1)	Ser>Asn(1)	
A/90 (H3N2)[19]	Leu>Phe(2)				
A/Ck/Pennsylvania/83 (H5N2)[37]		Ile>Ser(1) >Thr(3)	Ala>Ser(1) >Thr(1)	Ser>Asn(1)	
'In Vitro' Isolates					
A/Singapore/57 (H2N2)[24]		Val>Ala(7)	Ala>Thr(6)	Ser>Asn(12)	
A/Ck/Weybridge/27 (H7N7)[24]		Val>Ala(3) >Gly(2) >Asp(1)	Val>Thr(7) >Pro(2)	Ser>Asn(4)	Gly>Glu(29)
A/Ck/Rostock/34 (H7N1)[24]		Ile>Ser(17) >Thr(8) >Ala(1)			
	Leu>Hist(1)+	Ile>Thr(6)+ >Ser(5) >Asn(3)	Ala>Thr(9)+ >Ser(1) >Glu(1)	Ser>Asn(5)+	Gly>Glu(4)+
A/Eq/Cambridge/73 (H7N1)				Ser>Asn(1)	

\# Reference
* Number of isolates
+ Viruses isolated by direct plaque selection without extensive passage in presence of drug.

346

Genetic basis of drug resistance

Drug-resistant viruses selected 'in vivo' or 'in vitro' in the presence of either amantadine or rimantadine exhibit a strict cross-resistance to either drug and a variety of related compounds indicating they share a common mechanism of action. Analyses of genetic reassortants between drug-resistant and drug-sensitive human virus strains have shown that for all subtypes (H1N1, H2N2 and H3N2) the M gene is the principle determinant of drug resistance[16,20,21]. Although apparent that viruses belonging to different subtypes, including those isolated from other species, do differ in their drug-susceptibility and that this may be influenced by other genetic parameters, in particular the HA gene[22,23], the total loss of susceptibility to either drug as a result of mutations in the M2 coding sequences show that these are the dominant resistance determinants[24]. Nucleotide sequence analyses of amantadine or rimantadine-resistant mutants of a variety of human, avian and equine viruses, whether isolated 'in vivo' or 'in vitro' have shown that drug resistance is determined by single amino acid changes in the membrane spanning domain of the M2 membrane protein, the second product, of a spliced mRNA, of the M gene[16-19,24,37]. To date, changes have been localized to only five closely apposed amino acids, residues 26, 27, 30, 31 and 34 (Table 1), defining this as the site of interaction of the drug with the protein. Since rimantadine-resistant variants with the same amino acid substitutions in M2 have been isolated from rimantadine-treated patients and selected following passage 'in vitro' in the presence of rimantadine[16], it is evident that the mechanism of action of the drugs in inhibiting virus replication in cell culture represents the pharmacological activity.

Mechanism of drug action

In cell culture two stages in the replicative cycle of influenza A viruses may be susceptible to inhibition by amantadine and rimantadine, although their relative importance depends on the particular virus strain[23,25]. The most common inhibitory activity and that principally effective against human strains is directed against a stage in virus uncoating and is due to inhibition of an as yet undefined function of M2[26,27] (see below). In the case of certain virus strains, in particular of the H7 subtype, the haemagglutinins (HA) of which are cleaved intracellularly in tissue culture, drug action causes an alteration in the maturation of the virus HA[28] which consequently blocks production of progeny virus[25,29]. This is due to an amantadine-induced change in HA structure from the native to low pH conformation, equivalent to that acquired following exposure to a pH of approximately 5[28]. Since the protein is directly implicated in drug action it is not surprising that differences in HA, especially those which affect the pH of the conformational transition, pH_{trans}, also influence drug susceptibility. The drugs, however, clearly do not interact directly with HA since for most drug-resistant mutants amino acid changes in their M2 proteins, in the absence of any changes in HA, abolish sensitivity to drug action. The change in HA is thus the indirect result of inhibition of M2 function. The very nature of the drug-induced alteration in HA and its reversibility by eg monensin, as well as the influence of the pH_{trans} on drug susceptibility indicates that

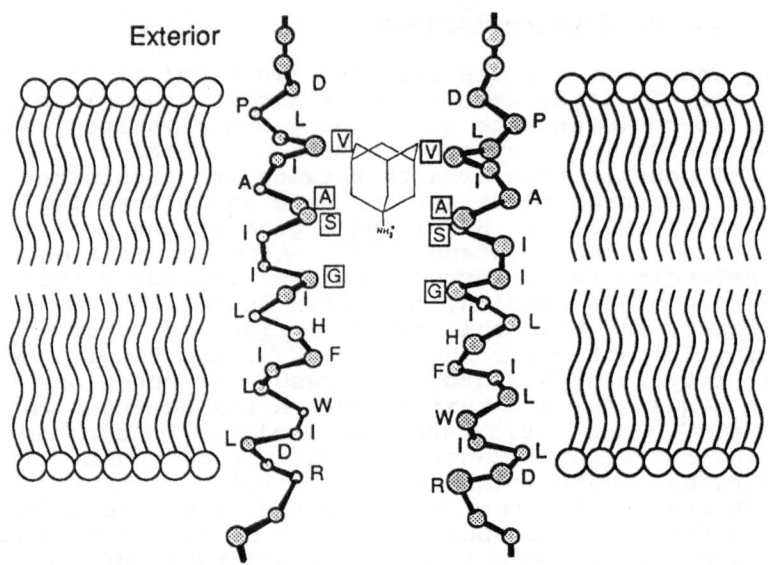

Fig. 1. Diagrammatic representation of the perceived interaction of amantadine with the α-helical transmembrane
domains (aspartic acid 24 to arginine 45) of two
diagonally opposed subunits of the M2 tetramer of
A/chicken/Germany/27 (H7N7). The N-termini are
exterior to the cell or virus. Single amino acids
altered in different amantadine-resistant mutants are
boxed.

the M2 protein plays a role in regulating the pH within
vesicles of the trans golgi network (TGN), specifically to
counteract increased acidity resulting from virus infection.
Inhibition of M2 function thus causes a drop in the vesicular
pH below that required to trigger the conformational change in
HA such that the low pH form of the molecule is transported to
the plasma membrane of infected cells. As a consequence of
this, although budding virus particles still appear on the cell
surface their release is blocked[29].

The M2 protein is a minor component of the virus envelope
although synthesized in relatively larger amounts in virus
infected cells[30]. It is a homotetramer composed of two
disulphide-linked dimers held together by non-covalent
interactions[31]. The amphiphilic nature of the membrane-
spanning α-helical domain and the locations of drug-resistance
determining amino acids on the hydrophilic face supports the
hypothesis that amantadine interacts directly with a channel
formed by the four subunits of the tetramer (Figure 1). Based
on the foregoing observations it has been suggested that the M2
protein may form a proton-selective channel which is 'blocked'
by the interaction of amantadine. The antiviral action of
amantadine thus appears analogous to its anticholinergic
activity in inhibiting neuromuscular transmission by
interacting with the ion channel of the nicotinic acetylcholine
receptor (nAchR)[32]. In this context, studies of the
interaction of the quaternary ammonium anaesthetic, QX-222,

with mutant forms of the nAchR have indicated that reciprocal changes of polar and apolar amino acids at two successive turns of the α-helical "M2" membrane-spanning domains of the subunits alter the affinity of drug binding[33]. By analogy with the suggested interaction of Qx-222 with the nAchR, the disposition of amantadine depicted in Figure 1 with the amino group interacting with the hydroxyl group of serine 31 and the hydrophobic ring moiety directed towards the previous turn of the helix is consistent with the drug resistance changes in the viral protein.

Since the same mutations in M2 abolish susceptibility to both early and late actions of amantadine and rimantadine, it is evident that the two actions have a common basis and are directed against a related function of M2. It has been postulated[27,34] therefore that during the infection process, in addition to triggering the conformational change in HA and consequent membrane fusion, the low pH within the endosome induces the M2 protein to transfer protons across the viral membrane causing the acidification of the virion interior and dissociation of the internal matrix structure. By inhibiting complete uncoating drug action would thus prevent the release of the free ribonucleoprotein and the initiation of virus infection.

Consequences of amino acid substitutions in M2

The amino acid changes in M2 which abolish its inhibition by amantadine or rimantadine may do so by altering the functional activity of the protein or the interaction of the drug with M2. In the absence of a direct assay of M2 function there is no clear evidence that these changes affect drug binding although the analogies mentioned above suggest that this is likely. Furthermore, most amantadine and rimantadine resistant variants isolated 'in vivo' or selected following passage with drug in cell culture provide no evidence of impaired M2 function. In contrast, however, some of these changes have been shown to influence the function of the M2 protein of the Rostock (A/chicken/Germany/34, H7N1) strain. Whereas passage of this virus in the presence of amantadine yielded resistant mutants exclusively altered at position 27 of M2, mutants isolated directly following plaque titration in the presence of drug possessed changes in one of 5 amino acids (Table 1). The adverse effects of changes at residues 26,30,31 and 34 were evident from the smaller plaque size and the elevated expression of low pHHA in cells infected by these mutants. Furthermore, some of the mutations were unstable and reversion to the wild, drug-sensitive phenotype occurred following passage in the absence of drug (Table 2). Not only are the substitutions of isoleucine 27 well tolerated and may even enhance certain features of M2 function, but the substitution by threonine also complements the adverse effects of changes in residues 30 and 31 (Table 2). Although these observations are not typical of most amantadine and rimantadine- resistant viruses including rimantadine-resistant isolates from children or adults, they do indicate that the mutations can adversely affect biological characteristics and serve to emphasize that the 'universal' susceptibility of natural isolates, apart from those associated with drug use, may not simply be coincidental.

Table 2.   Stability of amantadine-resistant mutations in M2 of Rostock.

Amantadine-resistant mutant	Changes following passage + Amantadine	No drug
R^R Ileu (27) → Ser	-*	
A26 Ileu (27) → Asn	-	
19-7 Ala (30) → Glu	-	
19-9 Ala (30) → Thr	additional Ileu(27)Thr→ (resistant)	reversion Thr→Ala (sensitive)#
19-3 Ser (31) → Asn	additional Ileu(27)→Thr (resistant)	additional Ileu(27)→Thr (resistant)
A17 Gly (34) → Glu	-	reversion Glu→Gly (Sensitive)

* No change
# Drug susceptibility

Implications of drug resistance

As already indicated, of concern is the prospect that
viruses causing influenza A epidemics within the human
population will acquire a drug-resistant phenotype.   A number
of recent studies have shown that rimantadine and amantadine-
resistant viruses are frequently isolated from patients treated
with these drugs.   In a study comparing the efficacies of oral
rimantadine and acetaminophen in the treatment of children
during an outbreak of influenza A (H3N2) rimantadine-resistant
viruses were isolated from 27% of children treated with the
drug and as many as 45% of those exhibiting prolonged (7 days)
shedding of virus[35].   More recently a multicentre family study
provided evidence for the transmission of drug resistant virus
from rimantadine-treated index patients to family members
receiving rimantadine prophylaxis[17].   In this study no
differences were noted in the symptoms of the illnesses caused
by the rimantadine-resistant as compared to those caused by the
drug-sensitive viruses.   Furthermore, no differences were
discernible in their pathogenicity for ferrets.   These
observations reflect the results of field studies of the
biological characteristics of amantadine-resistant avian
influenza A viruses which were shown to be similar in
virulence, were genetically stable, and were equivalent to the
drug-sensitive parent strains in their capacity to cause
disease in chickens in the absence of drug[36,37].   Amantadine-
resistant viruses have also been isolated from individuals in
nursing homes[18,19] and a boarding school (Joan R. Davies and John
S. Oxford, personal communication) when the drug was used
prophylactically.

In each instance, drug-resistance was shown to correlate with the presence of one of the amino acid changes in the M2 protein described above. To date, genetically confirmed resistance to these drugs has been observed only in association with drug use and there is no evidence for their escape into the general population[38]. Unlike the situation described above for certain Rostock mutants, there is no evidence to indicate that the human drug-resistant viruses are disabled in any respect and the question remains unanswered as to whether these viruses retain the potential to cause epidemics of human disease. It may prove that some, as yet unidentified, biological characteristic which correlates with the drug susceptible genotype together with the continual selection of antigenic variants will help to counteract the occurrence of drug-resistant epidemic strains.

REFERENCES

1. R. Dolin, R. C. Reichman, H. P. Madore, R. Maynard, P. N. Linton, and J. Webber-Jones, A controlled trial of amantadine and rimantadine in the prophylaxis of influenza A infection, N. Engl. J. Med. 307:580 (1982).
2. T. A. Bektimirov, R. G. Douglas Jr., R. Dolin G. J. Galasso, V.F. Krylov and J. Oxford, Current status of amantadine and rimantadine as anti-influenza A agents: Memorandum from a WHO meeting, Bull WHO 63:52 (1985).
3. D. M. Zlydnikov, O. I. Kubar, T. P. Kovaleva and L. E. Kamforin, Study of rimantadine in the USSR: a review of the literature, Rev. Infect. Dis. 3:408 (1981).
4. R. D. Clover, S. A. Crawford, T. D. Abell, C. N. Ramsey, W. P. Glezen and R. B. Couch, Effectiveness of rimantadine prophylaxis of children within families, Am. J. Dis. Child. 140:706 (1986).
5. R. L. Tominack and F. G. Hayden, Rimantadine hydrochloride and amantadine hydrochloride use in influenza A virus infections, Infect. Dis. Clin. North Am. 1:459 (1987).
6. W. S. Wingfield, D. Pollack and R. R. Grunert, Therapeutic efficacy of amantadine HCl and rimantadine HCl in naturally occurring influenza A2 respiratory illness in man, N. Engl. J. Med. 281:579 (1969).
7. L. P. Van Voris, R. F. Betts, F. G. Hayden, W. A. Christmas and R. G. Douglas, Successful treatment of naturally occurring influenza A/USSR/77 H1N1, J. Am. Med. Assoc. 245:1128 (1981).
8. D. K. Payler, P. A. Purdham, Influenza A prophylaxis with amantadine in a boarding school, Lancet 1:504 (1984).
9. W. L. Atkinson, N. H. Arden, P. A. Patraica et al., Amantadine prophylaxis during an institutional outbreak of type A (H1N1) influenza, Arch. Intern. Med. 146:1751 (1986).
10. F. G. Hayden, H. E. Hoffman and D. A. Spyker, Differences in side effects of amantadine hydrochloride and rimantadine hydrochloride relate to differences in pharmacokinetics. Antimicrob. Agents Chemother. 23:458 (1983).
11. P. A. Patriarca, N. A. Kater, A. P. Kendal, D. J. Bregman, J. D. Smith and R. K. Sikes, Safety of prolonged administration of rimantadine hydrochloride in the prophylaxis of influenza A virus infections in nursing homes, Antimicrob. Agents Chemother. 26:101 (1984).

12 J. S. Oxford, I. S. Logan and C. W. Potter, In vivo selection of an influenza A2 strain resistant to amantadine, Nature (London) 226:82 (1970).

13 R. G. Webster, Y. Kawaoka, W. J. Bean, C. W. Beard and M. Brugh, Chemotherapy and vaccination: A possible strategy for the control of highly virulent influenza virus, J. Virol. 55:173 (1985).

14 G. Appleyard, Amantadine-resistance as a genetic marker for influenza viruses, J. Gen. Virol. 36:249 (1977).

15 H. J. Field and S. E. Goldthorpe, Antiviral drug resistance, Trends Pharmacol. Sci. 10:333 (1989).

16. R. B. Belshe, M. Hall-Smith, C. B. Hall, R. Betts and A. J. Hay, Genetic basis of resistance to rimantadine emerging during treatment of influenza virus infection, J. Virol. 62:1508 (1988).

17. F. G. Hayden, R. B. Belshe, R. D. Clover, A. J. Hay, M. G. Oakes and W. Soo, Emergence and apparent transmission of rimantadine-resistant influenza A virus in families, N. Engl. J. Med. 321:1696 (1989).

18. E. E. Mast, J. P. Davis, M. W. Harmon, N. H. Arden, R. Circo and G. E. Tyszka, Emergence and possible transmission of amantadine-resistant viruses during nursing home outbreaks of influenza A (H3N2), Program and Abstracts of the 29th Interscience Conference on Antimicrobial Agents and Chemotherapy, Houston. Washington D.C.: American Society for Microbiology, 111, abstract.

19. F. Roumillat, E. Roacha, H. Regnery, D. Wells and N. Cox, Emergence of amantadine-resistant influenza A viruses in nursing homes during the 1989-1990 influenza season, Program and Abstracts of the Eighth International Congress of Virology, Berlin (1990).

20. M. D. Lubeck, J. L. Schulman and P. Palese, Susceptibility of influenza A viruses to amantadine is influenced by the gene coding for M protein, J. Virol. 28:710 (1978).

21. A. J. Hay, N. T. C. Kennedy, J. J. Skehel and G. Appleyard, The matrix protein gene determines amantadine-sensitivity of influenza viruses, J. Gen. Virol. 42:189 (1979).

22. C. Scholtissek, and G. P. Faulkner, Amantadine-resistant and sensitive influenza A strains and recombinants. J. Gen. Virol. 44: 807 (1979).

23. A. J. Hay and M. C. Zambon, Multiple actions of amantadine against influenza viruses, in "Antiviral drugs and interferon: The molecular basis of their activity", Y. Becker, ed. Martinus Nijhoff Publishing, Boston, MA, 301 (1984).

24. A. J. Hay, A. J. Wolstenholme, J. J. Skehel and M. H. Smith, The molecular basis of the specific anti-influenza action of amantadine, EMBO J. 4:3021 (1985).

25. A. J. Hay, M. C. Zambon, A. J. Wolstenholme, J. J. Skehel and M. H. Smith, Molecular basis of resistance of influenza A viruses to amantadine, J. Antimicrob. Chemother. 18 (Suppl.B): 19 (1986).

26. A. G. Bukrinskaya, N. E. Vorkunova and N. L. Pushkarskaya, Uncoating of a rimantadine-resistant variant of influenza virus in the presence of rimantadine, J. Gen. Virol. 60:61 (1982).

27. S. A. Wharton, A. J. Hay, R. Sugrue, J. J. Skehel, W. I. Weis and D. C. Wiley, Membrane fusion by influenza viruses and the mechanism of action of amantadine, in "Use of X-ray crystallography in the design of antiviral agents", W. G. Laver and G. M. Air, eds., Academic Press.

28. R. J. Sugrue, G. Bahadur, M. C. Zambon, M. Hall-Smith, A. R. Douglas and A. J. Hay, Specific structural alteration of the influenza haemagglutinin by amantadine, EMBO J. 9: 3469 (1990)

29. R. W. H. Ruigrok, E. M. A. Hirst and A. J. Hay, The specific inhibition of influenza A virus maturation by amantadine: an electron microscopic examination, J. Gen. Virol. 71 (1990).

30. S. L. Zebedee and R. A. Lamb, Influenza A virus M2 protein: monoclonal antibody restriction of virus growth and detection of M2 in virions, J. Virol. 62:2762 (1988).

31. R. J. Sugrue, and A. J. Hay, Structural characteristics of the M2 protein of influenza A viruses: evidence that it forms a tetrameric channel, Virology, in press.

32. J. E. Warnick, M. A. Maleque, N. Bakry, A. T. Eldefrawi and E. X. Albuquerque, Structure-activity relationships of amantadine: 1. interaction of the N-alkyl analogues with the ionic channels of the nicotinic acetylcholine receptor and electrically excitable membrane, Molec. Pharmacol. 22:82 (1982).

33. P. Charnet, C. Labarca, R. J. Leonard, N. J. Vogelaar, L. Czyzyk, A. Gouin, N. Davidson and H. A. Lester, An open-channel blocker interacts with adjacent turns of α-helices in the nicotinic acetylcholine receptor, Neuron 4: 87 (1990).

34. R. B. Belshe and A. J. Hay, Drug resistance and the mechanisms of action on influenza A virus, J. Resp. Dis. Suppl. S52 (1989).

35. C. B. Hall, R. Dolin, C. L. Gala, D. M. Markovitz, Y. Q. Zhang, P. H. Madore, R. A. Disney, W. B. Talpey, J. L. Green, A. B. Francis and M. E. Pickichero, Children with influenza A infection: treatment with rimantadine, Pediatrics 80:275 (1987).

36. C. W. Beard, M. Brugh and R. G. Webster, Emergence of amantadine-resistant H5N2 avian influenza virus during a simulated layer flock treatment program, Avian Dis. 31:533 (1987).

37. W. J. Bean, S. C. Threlkeld and R. G. Webster, Biologic potential of amantadine-resistant influenza A virus in an avian model, J. Infect. Dis. 159:1050 (1989).

38. R. B. Belshe, B. Burk, F. Newman, R. L. Cerruti and I. S. Sim, Resistance of influenza A virus to amantadine and rimantadine: results of one decade of surveillance, J. Infect. Dis. 159:430 (1989).

MVT-101, 247, 248
myc gene, 52
Mycophenolic acid, 227, 318, 337

NC, *see* Nucelocapsid proteins
nef gene, 107
  functional studies of, 189,
    195-198
  gene therapy and, 303
  structure and function of,
    203-212
nef-Responsive factor, *see* NRF
Negative regulator of splicing,
  52
Negative regulatory element,
  195-196, 208
neo gene, 71, 306
NF*k*B site, 127, 189
NH2T, *see* 3'-Amino-3'-
    deoxythymidine
NH2TTP, *see* 3'-Amino-dTTP
Nicotinic acehtylcholine
    receptor, 348-349
NRF, 210, 212
NTP, *see* Nucleotide triphosphate
Nucleic acid, RT binding by,
    35-48, *see also under*
    Reverse transcriptase
Nucleocapsid proteins (NC)
  cysteine arrays of, 257-269,
    273, 275
  proteolytic processing with,
    273-279
Nucleosides, carbocyclic,
    315-320
Nucleotides
  IN and, 21-22
  nef, 205, 206
  rev, 175, 177
  RT and, 2, 5, 8
  tat, 110, 123, 133
  terminal, 21-22
Nucleotide triphosphate (NTP),
    205, 212

OC, *see* Oxathiin carboxanilide
OKT4A antibodies, 286, 293
OKT4C antibodies, 286, 287
OKT4D antibodies, 286, 287, 290
Oligonucleotides
  IN and, 28, 29, 30, 31
  cysteine arrays and, 258, 265
  env regulation and, 185
  IN-mediated cleavage of, 32
  phosphorothioate, 3
  rev, 178
  ribozyme and, 335
  RNA splicing and, 51
  RT and, 63
  TAR and, 80, 81

Open reading frames (ORF), 189,
    196, 198, 217, 221
ORF, *see* Open reading frames
P-Orthophosphate, 261
Osteosarcoma cells, 69, 71, 73,
    227
Oxathiin carboxanilide (OC),
    309-312
Oxathiin carboxylic acid, 310

Papovaviruses, 225, *see also*
    specific types
PBMC, *see* Peripheral blood
    mononuclear cells
PDB program, 46
PE, *see* *Pseudomonas aeruginosa*
    exotoxin A
Peripheral blood mononuclear
    cells (PBMC), 240, 326
pH
  amantadine/rimantadine and,
    347-348, 349
  in RBC-CD4 preparation, 282,
    286
  rev processes and, 191
  tat processes and, 135, 136
PHA, *see* Phytohemaglutinin
Phenylalanine
  in protease inhibitors, 245,
    247, 248
  in RT, 56, 57, 65
Phorbol 12-myristate 13-acetate
    (PMA)
  CD4 exotoxin and, 238
  TAR function and, 87-89
  tat and, 98, 101
Phosphorothioate
    oligonucleotides, 3
Phycoerythrin, 288, 290
Phytohemaglutinin (PHA), 87, 88,
    240
PMA, *see* Phorbol 12-myristate
    13-acetate
pol gene, 93, 107, 145, 173, 183,
    203, *see also gag-pol*
    gene
  IN and, 21, 22, 27
  envelope pseudotyping and, 229
  NC protein and, 273, 274
  RNA splicing and, 53
  vif and, 217
Pol I, 2, 8, 16, 18
Polymerase, *see also* DNA
    polymerase; RNA
    polymerase
  RT and, 1, 63
  tat and, 124, 125, 129
Polynucleotide kinase, 41
Polypeptides, of RT, 14, 35, 40,
    43, 44, 57, 63

vpr gene, 107, 203
vpu gene, 107, 184, 203
  functional analysis of, 217,
       218, 221-223
vpx gene, 107, 203

Xanthine, 337
Xanthylate, 318

Zidovudine, *see* Azidothymidine
Zinc, 135, 258, 269, 275, 279